Introduction to Aircraft Flight Dynamics

Introduction to Aircraft Flight Dynamics

Louis V. Schmidt
Naval Postgraduate School
Monterey, California

EDUCATION SERIES
J. S. Przemieniecki
Series Editor-in-Chief
Air Force Institute of Technology
Wright–Patterson Air Force Base, Ohio

Published by
American Institute of Aeronautics and Astronautics, Inc.
1801 Alexander Bell Drive, Reston, VA 20191

American Institute of Aeronautics and Astronautics, Inc., Reston, Virginia

Library of Congress Cataloging-in-Publication Data

Schmidt, Louis V.
    Introduction to Aircraft Flight Dynamics / Louis V. Schmidt
      p. cm.—(AIAA education series)
    Includes bibliographical references and index.
    ISBN 1-56347-226-0 (alk. paper)
      1. Aerodynamics. 2. Stability of airplanes. I. Title. II. Series.
TL570.S33 1998
629.132'3—DC21   97-20180
                                                                                    CIP

Copyright © 1998 by the American Institute of Aeronautics and Astronautics, Inc. All rights reserved. Printed in the United States. No part of this publication may be reproduced, distributed, or transmitted, in any form or by any means, or stored in a database or retrieval system, without the prior written permission of the publisher.

Data and information appearing in this book are for informational purposes only. AIAA is not responsible for any injury or damage resulting from use or reliance, nor does AIAA warrant that use or reliance will be free from privately owned rights.

MATLAB™ is a registered trademark of The MathWorks, Inc.

Foreword

Introduction to Aircraft Flight Dynamics by Louis V. Schmidt is a comprehensive textbook that incorporates the latest theories and concepts used in the analysis and design of modern aircraft. The text evolved over a number of years, and it reflects the author's extensive knowledge of the subject and his practical experience initially as an engineer with the U.S. Navy and then as a professor while teaching the subject matter at the Naval Postgraduate School, Monterey, California.

The first four chapters provide an introduction to the basic definitions, stability derivatives, and static stability and control, followed by the formulation of the equations of motion for the airframe as a dynamic system (Chapters 1 through 4). Chapter 5 discusses the mathematical principles used to analyze the dynamics of the whole system, including its stability. Chapters 6 and 7 deal with the longitudinal and lateral-directional dynamics, while Chapter 8 deals with the nonlinear dynamics involved in inertial cross coupling, wing rock, and stall dynamics. Chapter 9 provides an introduction to atmospheric turbulence and to aircraft response due to random gusts. Several appendices provide useful supplementary information on atmospheric properties, aircraft stability derivatives, span load program, linear algebra principles, and the usage of MATLAB™ programs. A very useful feature of this text is the extensive use of MATLAB software in conjunction with personal desktop computers.

The Education Series of textbooks and monographs published by the American Institute of Aeronautics and Astronautics embraces a broad spectrum of theory and application of different disciplines in aeronautics and astronautics, including aerospace design practice. The series also includes texts on defense science, engineering, and management. The complete list of textbooks published in the series (more than 50 titles) can be found after the Index. The series serves as teaching texts as well as reference materials for practicing engineers, scientists, and managers.

J. S. Przemieniecki
Editor-in-Chief
AIAA Education Series

Table of Contents

Preface xi

Chapter 1. Introduction 1
  1.1 Background 1
  1.2 Definitions 1
  1.3 Coordinate Transformations 7
  1.4 Wing Properties 8
  1.5 Dimensions and Units 11
      References 12
      Problems 12

Chapter 2. Aerodynamic Principles 15
  2.1 Background 15
  2.2 Longitudinal Stability Derivatives 17
  2.3 Lateral-Directional Stability Derivatives 29
  2.4 Wing Theory 46
  2.5 Stability Derivatives Using Wing Theory 56
      References 62
      Problems 63

Chapter 3. Static Stability and Control 67
  3.1 Background 67
  3.2 Longitudinal Stability and Control 67
  3.3 Lateral-Directional Stability and Control 83
      References 90
      Problems 90

Chapter 4. Airframe Equations of Motion 93
  4.1 Background 93
  4.2 Euler Angle Transformations 93
  4.3 Rotation of a Rigid Body 100
  4.4 Airframe Equations of Motion 107
  4.5 Linearized Equations of Motion 107
  4.6 Matrix Formulation of the Equations of Motion 110
      References 118
      Problems 118

Chapter 5. Dynamic System Principles 119
  5.1 Background 119
  5.2 Laplace Transforms 119
  5.3 First-Order Linear System 128

5.4	Second-Order Linear System	132
5.5	Stability	142
5.6	State-Space Fundamentals	144
5.7	Stability in the Sense of Lyapunov	157
	References	161
	Problems	161

Chapter 6. Longitudinal Dynamics 165

6.1	Background	165
6.2	Aircraft Longitudinal Dynamics	166
6.3	Modal Approximations	177
6.4	Control Response	186
6.5	Maneuvering Flight	198
	References	204
	Problems	204

Chapter 7. Lateral-Directional Dynamics 207

7.1	Background	207
7.2	Pure Rolling Motion	208
7.3	Lateral-Directional Dynamics	214
7.4	Control Response	223
7.5	Design Guidelines	236
7.6	Yaw Damper for Stability Augmentation	237
7.7	Controllability and Pole Placement	248
	References	254
	Problems	254

Chapter 8. Nonlinear Dynamics 257

8.1	Background	257
8.2	Inertial Cross Coupling	258
8.3	Wing Rock	265
8.4	Stall Dynamics	275
	References	287
	Problems	288

Chapter 9. Atmospheric Turbulence 291

9.1	Background	291
9.2	Discrete Gust Response Solutions	293
9.3	Random Processes	302
9.4	Random Gust Response	313
9.5	Lyapunov Equation Usage for Gust Responses	319
	References	329
	Problems	329

Appendix A: Atmospheric Table 331

Appendix B: Aircraft Stability Derivatives 333

B.1	U.S. Navy A-4D Attack Aircraft	334
B.2	U.S. Navy A-7A Attack Aircraft	336

B.3	U.S. Air Force F-4C Fighter Aircraft	339
B.4	McDonnell–Douglas DC-8 Jet Transport	341
B.5	Convair CV-880M Jet Transport	345
B.6	Lockheed Jetstar Utility Jet Transport	349
B.7	North American Navion General Aviation Aircraft	352
	References	354

Appendix C: Span Load Program 355
   C.1 Aerodynamic Influence Coefficients 355
   C.2 Program Listing 358

Appendix D: Linear Algebra Principles 367
   D.1 Linear Algebra Usage 367
   D.2 Linear Algebra and Vector Concepts 369
   D.3 Eigenvalue Problem 373
   D.4 Orthogonality Principles 376
   D.5 Diagonalization of a Square Matrix 377
   D.6 Eigenvalue Problem for a Symmetric Matrix 379
   D.7 Eigenvalue Problem for an Unsymmetric Matrix 379
   D.8 Cayley–Hamilton Theorem 381
   D.9 Linear Independence and Rank 383
         References 384

Appendix E: Usage of MATLAB Programs 385
   E.1 Introductory Remarks 385
   E.2 Elementary Use of MATLAB 385
   E.3 Use of MATLAB Function Files 389
         References 394

Index 395

Preface

This book is intended as an introductory text relating to the flight mechanics of aerospace vehicles for use in either a senior undergraduate or first year graduate level aeronautical engineering program. It is assumed that the student has familiarity with fluid mechanics, aircraft performance, and static stability and control. Although desirable as background topics, these subjects are not considered as absolute prerequisites for understanding the content of this text.

Contemporary analysis methods are used while developing topical details including 1) a study of the aerodynamics involved in-flight vehicle motion in order to remove the mystery of what an influence coefficient means in a set of equations, 2) the use of linear algebra concepts with varying degrees of detail in order to develop a rationale for modeling and solving related problems in aircraft dynamics, 3) an introduction to modern control theory methods, and 4) problems and examples that illustrate the use of the powerful desktop computational tools that are currently available for solving problems previously considered tractable for solution only by analog or mainframe digital computers.

Stability derivative principles are briefly described in Chapter 2 because of their import when developing the governing relations for airframe dynamics. To bring recognition to the fact that stability and control derivatives involve other factors such as the aerodynamic dependence on damping or rate dependent terms and static aeroelasticity influences on rigid body aerodynamics, a section in Chapter 2 is devoted to wing span load theory. It is the author's feeling that engineering analysts frequently solve an aircraft dynamics problem by stability augmentation principles without realizing that options may exist to obtain similar results by an airframe configuration modification.

Both longitudinal and lateral-directional flight modes are considered in Chapter 3 during a review of static stability and control concepts. Linear analysis methods are used with emphasis on matrix approaches in setting up the problems as a prelude to the more demanding considerations involved with time-dependent, coupled, differential equations. Another reason for starting the material coverage in this manner is that static stability and control results can be recognized as corresponding to either the zero frequency portion of a frequency-response dynamic analysis or the steady-state answer in a time-history response analysis after all transients have dissipated.

The governing equations of motion are developed in Chapter 4 with a heavy dependence on the use of linear algebra as applied to first principles of engineering mechanics (i.e., conservation of linear and angular momentum). Subsequently, the governing equations are linearized by small perturbation assumptions when establishing the uncoupled longitudinal and lateral-directional dynamic relations. Dimensional stability derivatives are employed so that results take place in real time. Techniques for solving the equations of motion are described in Chapter 5 starting with a review of Laplace transform methods and ending with an introduction to state-space fundamentals.

The longitudinal- and lateral-directional dynamics of the airframe are treated separately in Chapters 6 and 7 with the goals being 1) to identify modal behaviors, 2) to develop simplifying approximations for describing the individual modes, and 3) to determine the airframe's response to control inputs. Harmonic control inputs are viewed in the frequency domain whereas other control inputs such as pulses and doublets are viewed in the time domain. Modern control theory concepts makes analysis convenient in either the frequency or the time domain. The use of airframe controls for stability augmentation purposes are illustrated for a few select situations following an introduction to the concept of controllability.

An insight into the nonlinear form of airframe dynamics is provided by considering the full set of coupled equations of motion. Analysis techniques applicable to describe airframe pitch–yaw inertial coupling, airframe wing rock, and stall dynamics are discussed in Chapter 8. Examples are provided to illustrate that a limit cycle (e.g., wing rock and/or stall porpoising) represents a stable system in the sense of Lyapunov.

Finally, flight is not always an ideal situation in smooth air. Atmospheric turbulence effects on airframe dynamics are treated in Chapter 9. An introduction to random variables and stochastic processes is included in the presentation because turbulence is best viewed as a random process acting as input to the motion of an aircraft vehicle.

Appendices are provided dealing with selected mathematical concepts that support the text material. It is expected that the presence of the material will act as a refresher for some readers and as a stimulus for further consideration by other readers. In addition, the stability derivatives for a varied group of aircraft are available as appendices for use in support of problem exercises.

Many examples are presented to illustrate basic linear algebra procedures applicable to modern control theory principles. The software subroutines employed are based on MATLAB™. MATLAB is a registered trademark of The Math-Works, Inc.

Acknowledgment

The material in this text is an outgrowth of lessons learned by the author from friends in the academic community, from fellow engineers while employed in the airframe industry, and from military aviators while teaching the subject matter at the Naval Postgraduate School (NPS). The probing questions by the students at the NPS during their studies of airframe dynamics has provided much of the motivation for the text. In particular, the author desires to both acknowledge and thank the following individuals for the insights they provided, which is hopefully reflected in the outlook contained in this book: Y. C. "Burt" Fung, University of California at San Diego, who has been a research advisor, mentor, and friend; Richard W. Bell, Russell W. Duren, and Richard M. Howard, Naval Postgraduate School, faculty colleagues, who gave both encouragement and advice for this book; and Joseph F. Sutter, Boeing Commercial Airplane Group, who introduced the author to the fine points of engineering tradeoffs during the design of the first U.S. jet transport.

Louis V. Schmidt
November 1997

1
Introduction

1.1 Background

The subject of aircraft flight dynamics relies on an understanding of numerous physical disciplines including aerodynamics, propulsion, structures, and engineering mechanics. As one becomes more familiar with the subject, a realization comes that the kinship of flight dynamics to the vibration of mechanical bodies also applies to other multi-degree-of-freedom physical systems such as ships, submarines, automobiles, missiles, and spacecraft. Both stability and control are concerns in a dynamic system. Although many researchers had attempted to solve the design problem of a crew-carrying, powered aircraft in the latter part of the 19th and early part of the 20th centuries, it was the Wright brothers who first succeeded in December 1903 at Kitty Hawk. An analysis by Culick and Jex[1] describes that although the first Wright Flyer was an unstable vehicle longitudinally, the brothers overcame this factor by their selection of a canard lifting surface for longitudinal control, wing warping for lateral control, a vertical tail to compensate for adverse yaw, and equally important, their piloting skills developed from glider experience as preparation for powered flight. It is only in recent years that an experimental aircraft, the X-29, has had a higher degree of longitudinal instability. Success in handling and flying qualities today is in a large measure due to the integration of compact, reliable digital computers with robust control laws.

When considering the equations of motion governing flight, mention must be made of the pioneering studies by Lanchester[2] and Bryan.[3] Lanchester's approach involved basic physical concepts whereas Bryan used principles of theoretical mechanics to describe aircraft motion. Bryan's approach forms the genesis for contemporary analysis methods. In more recent times, the understanding of flight dynamics has matured due to the efforts of many, including Perkins and Hage,[4] Babister,[5] and Etkin.[6]

The analysis of flight dynamics can be either in the frequency or the time domain, depending on the intent of the studies. The reasons for considering either of these domains will become clearer during subsequent studies using control theory methods. Currently, it is acceptable to categorize control theory as either classical or modern. The classical approach, used primarily in the era of 1930–1960, solved sets of ordinary differential equations using Laplace transform techniques, which was an outgrowth of the Heaviside operational calculus methods in vogue during the early 1900s. The word modern, when used in conjunction with control theory, is a contemporary description involving the use of state-space techniques, based on the orderly world of linear algebra, for the analysis of physical system dynamic properties. There is a danger that what we view as (or call) modern today will be considered differently in the future.

1.2 Definitions

One of the difficulties encountered by students of flight mechanics relates to sign conventions, physical units, and nomenclature involving multiple subscripting.

2 INTRODUCTION TO AIRCRAFT FLIGHT DYNAMICS

The conventions used in the following material will follow closely to the guidelines established jointly by the American National Standards Institute (ANSI) and AIAA.[7] These conventions are used in a large part by industry and researchers in the United States.

1.2.1 Coordinate System

A right-hand orthogonal coordinate set will be used with the origin located at the vehicle's center of gravity (c.g.). In general, x will be forward, y will be a lateral direction (toward the right wing tip) and z will be downward (normal to the x–y plane). Figure 1.1 illustrates the coordinate system. It should be recognized that, in general, the x–z plane is a plane of symmetry and the actual position of the x and z axes may be rotated about the y axis for reasons of convenience, such as having the x axis aligned with the relative absolute velocity vector as an initial condition. This latter body-axis orientation is known as the airframe's stability axes.

1.2.2 Forces

The forces considered when developing the equations of motion will be denoted by uppercase letters in the x–y–z coordinate system. These forces, X, Y, and Z, will have positive values acting in the positive directions of the respective x–y–z reference frame. The X, Y, and Z terms can be viewed as components of

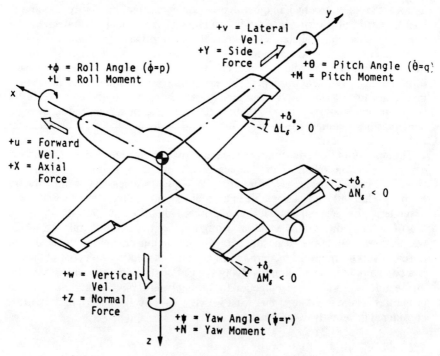

Fig. 1.1   Coordinate sign convention.

a total force vector acting on the vehicle at its c.g. If the forces were not in static equilibrium, then linear momentum conservation principles would prevail to describe the velocity changes.

1.2.3 Moments

Similarly, the moment vector acting on the vehicle at its c.g. will have components L, M, and N acting along the x–y–z coordinate system, as shown in Fig. 1.1. The L, M, and N moment terms are described as the roll, pitch, and yaw moments, respectively, and are generally due to inputs derived from either airframe aerodynamics or propulsion systems. Again, if the moments were not in static equilibrium, conservation principles of angular momentum would prevail to yield angular motion changes.

1.2.4 Velocities

The total linear velocity of the airframe may be considered as a vector V having components of u, v, and w acting in the positive x, y, and z directions, respectively. The angular velocity vector Ω will be described by the p, q, and r components with definitions relative to the body axes, as shown in Fig. 1.1. In matrix notation, the vector terms become

$$V = u\boldsymbol{e}_x + v\boldsymbol{e}_y + w\boldsymbol{e}_z = \begin{Bmatrix} u \\ v \\ w \end{Bmatrix}$$

$$\Omega = p\boldsymbol{e}_x + q\boldsymbol{e}_y + r\boldsymbol{e}_z = \begin{Bmatrix} p \\ q \\ r \end{Bmatrix} \tag{1.1}$$

where $\boldsymbol{e}_x, \boldsymbol{e}_y$, and $\boldsymbol{e}_z$ are the unit vectors in the x–y–z coordinate system attached to the airframe at the c.g.

The velocities defined by Eq. (1.1) will be either specified as initial conditions or determined analytically in accord with methods developed in subsequent chapters. It is appropriate, though, to recognize that these velocities are real physical quantities that can be measured by actual sensors on the airframe.

The angular rates, p, q, and r, can be sensed by rate gyros for stability augmentation usage in either analog or digital format. A rate gyro display instrument is usually found in an airplane cockpit to provide turn-rate information to the pilot, as indicated by the deflection of a needlelike pointer. The magnitude of the velocity vector $|V|$ is normally shown by an instrument (airspeed indicator) that relates airframe velocity to the pressure difference between the total pressure and the static pressure, which are commonly sensed by a pitot tube and flush orifice(s) located on the airframe's exterior. Indicated airspeed can be corrected to a true airspeed once account is taken for instrument calibration error, pressure source position error, compressibility effects, and air density (provided by an independent altitude measurement).

Velocity may be expressed as either true, V_t, or equivalent, V_e; the relationship to each other is from the dynamic pressure expression

$$Q = \tfrac{1}{2}\rho V_t^2 = \tfrac{1}{2}\rho_0 V_e^2 \qquad (1.2)$$

where

ρ = atmospheric density at altitude above sea level
ρ_0 = ideal sea-level atmospheric density
V_t = true airspeed (TAS)
V_e = equivalent airspeed (EAS)

From Eq. (1.2), one finds that

$$V_t = \sqrt{\rho_0/\rho}\, V_e$$

A listing of atmospheric properties for an ideal atmosphere as a function of altitude is given in Appendix A.

It should be noted that a letter sequence is used fairly consistently when defining the coordinates, forces, moments, and velocities. The technique, traditionally used by aeronautical engineers, provides a degree of uniformity in notation. It will soon become apparent that the nomenclature will need expanding; otherwise, the normal alphabet use would need repeating. Symbol augmentation will be achieved by introducing Greek symbols such as $\alpha, \beta, \delta, \phi, \theta$, and ψ.

1.2.5 Control Systems

In addition to conventional control systems such as ailerons, elevators, and rudders, external control of the forces and moments acting on the vehicle can also occur from a variety of other physical devices including 1) thrust vectoring by a jet engine's exhaust, 2) speedbrakes/spoilers mounted on the fuselage or wing, 3) horizontal tail mounted forward of the wing (canard surface), 4) vee tail to replace horizontal and vertical tail(s) aft of the main wing lifting surface, 5) vertical fin on the underside of the fuselage for direct side-force control, and 6) variable wing camber.

For the most part, the material described in this text will assume a conventional control system approach with the recognition that once the principle of control sensitivities are understood, other control approaches may be introduced and considered as alternative controllers of airframe dynamics. The conventional controls shown in Fig. 1.1 include the aileron (mounted on the wing trailing edge), the elevator (mounted on the horizontal tail), and the rudder (mounted on the vertical tail). These control surfaces produce their aerodynamic force and moment inputs due to a change in camber of a lifting surface.

Longitudinal control deflection δ_e is shown in Fig. 1.1 as positive with the trailing edge down. If the longitudinal control were behind the aircraft c.g., one would expect the resulting pitch moment M about the c.g. to be negative in sign. The rate of change of pitching moment with elevator deflection is described by the sensitivity term of $\Delta M_\delta (= \partial M/\partial \delta_e)$. The two force sensitivity terms, ΔX_δ and ΔZ_δ, are important in dynamics but are overshadowed by the importance of ΔM_δ.

Lateral control deflection δ_a is shown as positive with right-hand aileron trailing edge up and left-hand aileron trailing edge down. It will be assumed that the deflection pattern is antisymmetric, which results in wing span loads producing a rolling moment, i.e., positive aileron control tends to lower the right-hand wing in the direction of increasing roll (bank) angle ϕ. Although lateral control produces yawing moments and side forces on the airframe, the dominant term is the rate of change of rolling moment with aileron deflection, described by the sensitivity term $\Delta L_\delta (= \partial L/\partial \delta_a)$.

Directional control deflection δ_r is shown in Fig. 1.1 as positive with the rudder's trailing edge rotating to the left and will normally produce a positive side force with a corresponding negative yawing moment about the c.g. The prime influence of rudder control is the production of a yawing moment as described by the sensitivity term $\Delta N_\delta (= \partial N/\partial \delta_r)$. This stability derivative, in addition to the associated rolling moment and side-force sensitivities, will vary in value (and sign) depending on airplane angle of attack.

It should be recognized in the preceding statements of control sensitivities that linearity is assumed. For the majority of the material in this text, this assumption will be made while developing the basic concepts of airframe dynamics and dynamic response. However, one should realize that linearity will not apply at high angles of attack and sideslip, and the linearized equations of motion will not adequately describe the airframe dynamics.

Control surface angles are customarily defined relative to the local hinge line (or pivot axis). Consequently, the control deflection angles do not, in general, form an orthogonal set. However, the moments and forces produced by their respective deflections will be considered as acting in the airframe's orthogonal frame of reference. Consider an elevator control with a swept-back hinge line mounted on a horizontal tail having a dihedral angle. In spite of this situation, the capability of elevator control will be evaluated by its production of X and Z forces and M pitching moments, as expressed in the airframe's orthogonal set of aerodynamic influence coefficients.

1.2.6 Dimensionless Aerodynamic Coefficients

The dimensional stability derivatives to be used in the equations of motion have as a basis the dimensionless coefficients defined as follows:

$$
\begin{aligned}
C_X &= \frac{X \text{ force}}{QS} = (\text{positive}, +x \text{ body-axis direction}) \\
C_Y &= \frac{Y \text{ force}}{QS} = (\text{positive}, +y \text{ body-axis direction}) \\
C_Z &= \frac{Z \text{ force}}{QS} = (\text{positive}, +z \text{ body-axis direction}) \\
C_\ell &= \frac{\text{roll moment}, L}{QSb} = (\text{positive, right-wing-down direction}) \\
C_m &= \frac{\text{pitch moment}, M}{QSc} = (\text{positive, nose-up direction}) \\
C_n &= \frac{\text{yaw moment}, N}{QSb} = (\text{positive, nose-right direction})
\end{aligned}
\quad (1.3)
$$

where

Q = dynamic pressure = $0.5\rho(u^2 + v^2 + w^2) = 0.5\gamma p M^2$
ρ = air density at specified altitude
p = static pressure at specified altitude
M = Mach number of aircraft
γ = ratio of specific heats, normally 1.4 for air
S = reference wing area
b = reference wing span
c = reference wing chord

The subscript of c.g. to indicate the reference axis for moment coefficients has been omitted for sake of clarity.

Two other dimensionless force coefficients, which are commonly used in aerodynamic (e.g., wing theory) and performance analyses, are the airplane lift and drag coefficients. The use of L as a subscript for lift force should not be confused with the notation of L to indicate a rolling moment,

$$C_L = \frac{\text{lift force}}{QS} = \text{(positive up and normal to the velocity vector)}$$

$$C_D = \frac{\text{drag force}}{QS} = \text{(positive aft and parallel to the velocity vector)}$$

(1.4)

1.2.7 Angles

The orientation of the velocity vector with respect to the body-axis coordinate system establishes two angles of significance in the production of aerodynamic forces and moments, namely, angle of attack α and sideslip angle β. Consider the velocity of the airframe's c.g. as being given by V as defined in Eq. (1.1). From Fig. 1.2, it may be seen that the u, v, and w velocity components along the x, y, and z body axes are given by

$$u = V \cos\beta \cos\alpha$$
$$v = V \sin\beta$$
$$w = V \cos\beta \sin\alpha$$

(1.5)

where

$$V = (u^2 + v^2 + w^2)^{\frac{1}{2}}$$

The angles α and β may be identified from Eq. (1.5) as

$$\alpha = \tan^{-1}(w/u)$$
$$\beta = \sin^{-1}(v/V)$$

(1.6)

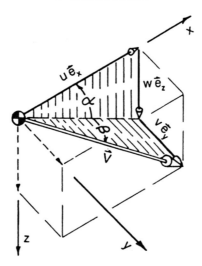

Fig. 1.2 Airframe velocity components.

1.3 Coordinate Transformations

This section will describe coordinate system transformations using the force vector represented by the sum of the aerodynamic lift and drag force components as an example. The resultant force vector will then be re-expressed as the X and Z force components in the body-axis coordinate system. For simplicity, let us assume that the aircraft is in level flight equilibrium, which implies that the velocity vector is horizontal. A sketch of the system is shown in Fig. 1.3.

It will be noted in Fig. 1.3 that the body's x and z reference axes are inclined at an angle of α relative to the level-flight reference plane. Recall from Fig. 1.1, the body-axis coordinate x is defined positive in a forward direction with the origin at the c.g. whereas z is defined positive from the origin extending downwards, normal to the x axis. The X and Z force components are positive with respect to the positive x and z axis directions, respectively.

The lift and drag forces combine into a resultant force, which may be considered as a vector that is described by components from the assumed coordinate system.

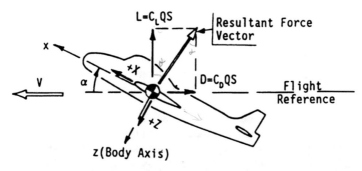

Fig. 1.3 Airframe force components.

Changes in the coordinate system will affect only the values of the components whereas the resultant remains invariant. It may be seen from Fig. 1.3 that the X and Z force components may be expressed in terms of lift L and drag D by

$$-X = D \cos \alpha - L \sin \alpha$$
$$-Z = D \sin \alpha + L \cos \alpha \quad (1.7)$$

or in matrix notation

$$-\begin{Bmatrix} X \\ Z \end{Bmatrix} = \begin{bmatrix} \cos \alpha & -\sin \alpha \\ \sin \alpha & \cos \alpha \end{bmatrix} \begin{Bmatrix} D \\ L \end{Bmatrix} = T_\alpha \begin{Bmatrix} D \\ L \end{Bmatrix} \quad (1.8)$$

where T_α represents the α angular rotation transformation matrix.

Next consider the reverse direction to resolve components from the x–z coordinates to the velocity (stability axis) reference frame,

$$D = -X \cos \alpha - Z \sin \alpha$$
$$L = X \sin \alpha - Z \cos \alpha$$

which may be expressed in matrix notation similar to Eq. (1.8),

$$\begin{Bmatrix} D \\ L \end{Bmatrix} = -\begin{bmatrix} \cos \alpha & \sin \alpha \\ -\sin \alpha & \cos \alpha \end{bmatrix} \begin{Bmatrix} X \\ Z \end{Bmatrix} = (-1) T_\alpha^{-1} \begin{Bmatrix} X \\ Z \end{Bmatrix} \quad (1.9)$$

Restating the transformation matrix, T_α, and its inverse, one notes that the matrix inverse is equal to the transpose of the original matrix, i.e., $T_\alpha^{-1} = T_\alpha^T$,

$$T_\alpha = \begin{bmatrix} \cos \alpha & -\sin \alpha \\ \sin \alpha & \cos \alpha \end{bmatrix} \quad \text{and} \quad T_\alpha^{-1} = \begin{bmatrix} \cos \alpha & \sin \alpha \\ -\sin \alpha & \cos \alpha \end{bmatrix}$$

Definition: A matrix with the property that its transpose is the same as its inverse is an orthogonal matrix.

For a more general discussion on angular transformations and other linear algebra properties, refer to Appendix D.

1.4 Wing Properties

It will prove useful to define a few geometric and aerodynamic properties of a wing inasmuch as the wing usually is the main lifting surface of an aircraft. Wing geometry also is used as reference quantities when defining both dimensionless and dimensional stability derivatives.

Consider a straight tapered wing as shown in Fig. 1.4 having a wing span b. The root and tip chords are expressed as c_r and c_t, respectively, which yields a wing taper ratio λ, where $\lambda = c_t/c_r$. It is frequently found convenient to restate the wing lateral coordinate y in a dimensionless form (fraction of wing semispan) of η where $\eta = 2y/b$. Other significant geometric properties include the spanwise

INTRODUCTION

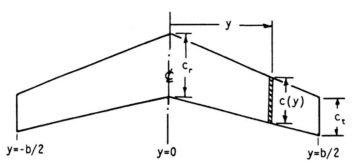

Fig. 1.4 Sketch of wing planform.

variations of built-in twist and airfoil thickness and the sweepback angle of a reference chord line, such as the wing leading edge or quarter chord.

Wing area S for a general wing planform is

$$S = 2\int_0^{b/2} c(y)\,dy = b\int_0^1 c(\eta)\,d\eta \tag{1.10}$$

Airframe manufacturers may list reference wing areas that differ from the ideal situation of Eq. (1.10) due to their procedures in accounting for areas hidden in the fuselage carry-through region, planform irregularities such as wing root leading- or trailing-edge extensions, etc. Once a reference area is established for an airframe, all dimensionless coefficients are standardized to that value.

Wing chord, $c(y)$ or $c(\eta)$ for a straight tapered wing is

$$\begin{aligned} c(y) &= c_r + (2y/b)(c_t - c_r) \\ c(\eta) &= c_r + \eta(c_t - c_r) = c_r[1 + \eta(\lambda - 1)] \end{aligned} \tag{1.11}$$

Integration of Eq. (1.10) for a straight tapered wing yields

$$S = 0.5b(c_r + c_t) = 0.5bc_r(\lambda + 1)$$

The reference chord length used in relating pitch moment to a dimensionless moment coefficient, although arbitrary, frequently is the mean aerodynamic chord (MAC) which is denoted symbolically as $\bar{c}$. The MAC, which is not the average geometric chord, is given by

$$\bar{c} = \int_0^1 c^2(\eta)\,d\eta \bigg/ \int_0^1 c(\eta)\,d\eta \tag{1.12}$$

For a wing with a single straight-tapered planform, the MAC is given by

$$\bar{c} = \frac{2}{3}\left[(c_r + c_t) - \frac{c_r c_t}{(c_r + c_t)}\right] \tag{1.13}$$

with a spanwise location $\bar{y}$ where $c(\bar{y})$ equals $\bar{c}$ given by

$$\bar{y} = \frac{b(c_r + 2c_t)}{6(c_r + c_t)}$$

Wing aspect ratio (AR) is described as the ratio of the wing span, b, to the wing's geometric average chord, $c_{ave} = S/b$, i.e.,

$$AR = b^2/S \qquad (1.14)$$

In aircraft performance analyses, wing AR serves an important role for estimating the wing's induced drag and lift curve slope.

Aerodynamic properties of a wing are frequently described by the total wing lift and pitching moment coefficients, C_L and C_m, respectively. It is useful to realize that the total lift and pitching moments represent a spanwise integration of the sectional lift and pitching moments over the surface. A symmetrical span load variation as a function of wing span station is shown in Fig. 1.5. The running span load, lift per unit span, is denoted by ℓ, which should not be confused with the roll moment coefficient subscript of Eq. (1.3).

For a symmetrically loaded wing, i.e., $\ell(+y) = \ell(-y)$ and $m(+y) = m(-y)$, the wing lift and pitching moments become

$$L = 2 \int_0^{b/2} \ell(y) dy = b \int_0^1 \ell(\eta) d\eta$$

$$M = 2 \int_0^{b/2} m(y) dy = b \int_0^1 m(\eta) d\eta \qquad (1.15)$$

It should be noted that the pitching moments are referenced to the same axial x station for both the section and total moments. The sectional lift and pitching moment coefficients (using lower case c for the coefficients) become

$$c_\ell = \ell/Qc \quad \text{and} \quad c_m = m/Qc^2 \qquad (1.16)$$

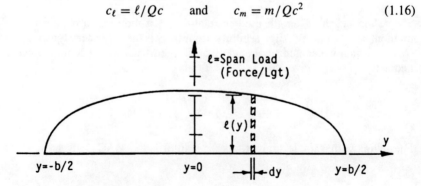

Fig. 1.5 Symmetric wing span load distribution.

1.5 Dimensions and Units

The stability derivatives used in describing the equations of motion will be based on the following English unit system:

b = wing span, ft
c = wing chord, ft
S = wing area, ft^2
V = airspeed, ft/s (ft · s^{-1})
W = weight, lb
g = gravitational constant = 32.174 ft/s^2 (ft · s^{-2})
m = mass (W/g), slugs (lb · s^2 · ft^{-1})
ρ = atmospheric density, slugs/ft^3 (lb · s^2 · ft^{-4})
F = force, lb
M = moment, ft · lb
Q = dynamic pressure, lb/ft^2 (lb · ft^{-2})

The metric unit system, also known as the Système Internationale (SI), is related to the English unit system (used in this text) by the conversion factors given in Table 1.1.

The influence of compressibility is associated with the Mach number M, which is defined as the ratio of vehicle velocity to the speed of sound, i.e.,

$$M = V/a \qquad (1.17)$$

where the speed of sound a is

$$a = \sqrt{\gamma RT} = \sqrt{\gamma p/\rho} \qquad (1.18)$$

The two forms for the sound speed in Eq. (1.18) may be recognized as a result of applying the equation of state for a gas (i.e., considered here as the atmosphere) as

$$p = \rho RT \qquad (1.19)$$

where R is the gas constant (cf. Appendix A), which is invariant with altitude for the normal atmosphere, and T is the temperature.

Table 1.1 Conversion of English units to metric units

Property	English units	Metric units
Length	1 ft	0.30480 m
Mass	1 slug	14.5930 kg
Pressure	1 lb/ft^2	47.8803 N/m^2
		0.478803 bar
Force	1 lb	4.44822 N
Density	1 slug/ft^3	515.379 kg/m^3
Temperature	1 °R	(5/9) K

Table 1.2 Atmospheric properties, sea-level standard day

Property	English units	Metric units
Pressure p	2116.2 lb/ft^2	1013.25 N/m^2
	29.92 in. Hg	760 mm Hg
Density ρ	0.002377 slug/ft^3	1.225 kg/m^3
Temperature T	518.67°R	288.15 K
	(59.0°F)	(15.0°C)
Sound speed a	1116.4 ft/s	340.3 m/s

The properties of the standard atmosphere as a function of geometric altitude are presented in Appendix A for the English unit system. A comparison of the sea-level properties of the standard atmosphere between the English and metric unit system is shown in Table 1.2.

References

[1] Culick, F. E., and Jex, H. R., "Aerodynamics, Stability and Control of the 1903 Wright Flyer," *The Wright Flyer, An Engineering Perspective,* Smithsonian Institution Press, Washington, DC, 1987, pp. 19–43.

[2] Lanchester, F. W., *Aerodonetics,* Constable and Co., Ltd., London, 1908.

[3] Bryan, G. H., *Stability in Aviation,* Macmillan, London, 1911.

[4] Perkins, C. D., and Hage, R. E., *Airplane Performance, Stability, and Control,* Wiley, New York, 1949, Chaps. 5–11.

[5] Babister, A. W., *Aircraft Stability and Control,* Pergamon, New York, 1961.

[6] Etkin, B., *Dynamics of Flight, Stability and Control,* 2nd ed., Wiley, New York, 1982, Chaps. 4–7.

[7] Anon., "Recommended Practice for Atmospheric and Space Flight Vehicle Coordinate Systems," ANSI/AIAA Rept. R-004-1992, Washington, DC, Feb. 1992.

Problems

1.1. Verify that the product of the transformation matrix T_α from Eq. (1.8) with its transpose, T_α^T, yields the unit diagonal identity matrix denoted as $[I]$. Hint: Two matrices A and B, each of order $(n \times n)$ can be multiplied to obtain the matrix product C where the c_{ij} element of C is given by

$$c_{ij} = \sum_{k=1}^{n} a_{ik} b_{kj}$$

1.2. Show that if a wing had a constant sectional lift coefficient ($c_\ell = \text{const} = k$), then the total wing lift coefficient C_L also would be k.

1.3. Verify Eq. (1.12) for the MAC definition, $\bar{c}$. Hint: Assume that the wing has an unswept quarter-chord line and the sectional moment coefficient about the quarter-chord line is constant (c_m = const = k). Determine the reference chord such that the total wing pitching moment coefficient C_m also will be k.

Remark: Based on the result of Problem 1.2, it is reasonable to expect that a similar result would apply to the wing's pitching moment coefficient.

1.4. Verify Eq. (1.13) for the value of $\bar{c}$ for a wing with a straight tapered planform.

1.5. Given a wing with an elliptical planform (e.g., World War II Spitfire fighter), show that the MAC is given by

$$\bar{c} = (8/3\pi)c_r$$

Hint: Define the wing planform as a quadratic relation having a root chord value of $c(0) = c_r$ and a tip chord value of $c(y/2) = 0$,

$$(c/c_r)^2 + \eta^2 = 1$$

where $\eta = 2y/b$.

2
Aerodynamic Principles

2.1 Background

Many published sources are available that provide airframe stability derivatives for a wide range of aircraft and flight conditions, e.g., McRuer et al.,[1] Heffley and Jewell,[2] Teper,[3] Blakelock,[4] and Nelson.[5] The first three sources provide listings in dimensional derivative form whereas the latter two sources tabulate dimensionless derivatives. The relationship between dimensional and dimensionless stability derivatives will be developed in Chapter 4 in conjunction with the linearization of the equations of motion. Stability derivatives in the dimensional form are convenient for the application of modern control theory methods when solving flight dynamic problems in either the real-time or frequency domains. Appendix B provides a collection of dimensional stability derivatives for a number of aircraft using available NASA reports.[2,3]

Estimation methods for the stability derivatives will vary in reliability primarily due to configuration dependence. The U.S. Air Force DATCOM[6] is frequently cited as a data source for estimating dimensionless stability derivatives. In actual design, experimental techniques (both wind-tunnel and flight testing) are used to determine the derivatives with predictive support provided by computational methods, such as described by Katz and Plotkin.[7] Discussions on the aerodynamic sources of the stability derivatives may be found in many classic stability and control texts (e.g., Etkin[8] and Perkins and Hage[9]), and the following material has a heavy dependence on such references.

It is the purpose of this chapter to use elementary aerodynamic principles to clarify the physical meaning of stability derivatives. In particular, it is hoped that the reader's understanding of damping terms such as pitch, roll, and yaw damping will be clarified by this chapter. The first damping term (pitch) is due primarily to the influence of the horizontal tail, the second damping term (roll) is strongly dependent on the wing, whereas the last damping term (yaw) is largely due to the action of the vertical tail. Elements of wing theory are introduced as a reinforcement to the concepts being described in order to distinguish between various forms of aerodynamic modeling when making estimates of stability derivatives.

The Taylor series formula for a function of several variables may be applied to aerodynamic coefficients in a general sense as the following illustrates.

Consider a function $f(a + x_1, b + x_2)$ that is continuous in the neighborhood of the location $x_1 = x_2 = 0$ and has continuous partial derivatives, up to and including those of order n, in the vicinity of the assumed location. The function may be then expressed as

$$f(a + x_1, b + x_2) = f_0 + f_{x_1} x_1 + f_{x_2} x_2$$
$$+ \tfrac{1}{2!}\left[f_{x_1 x_1} x_1^2 + 2 f_{x_1 x_2} x_1 x_2 + f_{x_2 x_2} x_2^2 \right] + \cdots + R_n \qquad (2.1)$$

where

$$f_0 = f(a, b)$$

at $x_1 = x_2 = 0$ and

$$f_{x_1} = \frac{\partial f}{\partial x_1}$$

evaluated at $x_1 = x_2 = 0$, also

$$f_{x_2} = \frac{\partial f}{\partial x_2} \qquad f_{x_1 x_1} = \frac{\partial^2 f}{\partial x_1^2}, \ldots$$

For the situation of aerodynamic coefficients considered in this text, the linear form of the Taylor's series formula will be considered, which implies that second- and higher-order partial derivatives will not be considered. Wind-tunnel data on many aircraft support this assumption providing regions such as aircraft stall are not being considered.

The term stability derivatives will be recognized as a representation of first partial derivatives in the appropriate linearized Taylor's series expansion. It is in this application that one must be careful in the choice of subscript notation. As an example, consider an aerodynamic variable, such as aircraft pitching moment coefficient about the aircraft's c.g., as representing a trim value plus a linear sum of perturbation terms from trim. This approach provides an estimate of the change in the aerodynamic variable due to changes in the state and input control variables (e.g., angle of attack, control deflection, angle-of-attack rate, pitch-attitude rate, etc.). These words may be expressed in a mathematical form by

$$C_{m_{cg}} = C_{m_0} + C_{m_\alpha}(\alpha + \Delta\alpha) + C_{m_\delta}(\delta + \Delta\delta)$$
$$+ C_{m_{\dot\alpha}}\left(\Delta\dot\alpha \frac{c}{2V}\right) + C_{m_q}\left(\Delta\dot\theta \frac{c}{2V}\right) \qquad (2.2)$$

where trim in steady flight implies that

$$0 = C_{m_0} + C_{m_\alpha}\alpha + C_{m_\delta}\delta$$

and α and δ represent steady values of angle of attack and control deflection, respectively, for a condition of static equilibrium. Subtracting the trim condition from the total expression for pitching moment, Eq. (2.2), yields

$$\Delta C_m = C_{m_\alpha}\Delta\alpha + (c/2V)\left[C_{m_{\dot\alpha}}\Delta\dot\alpha + C_{m_q}\Delta\dot\theta\right] + C_{m_\delta}\Delta\delta \qquad (2.3)$$

where

C_{m_0} = aircraft pitching moment coefficient at $C_L = 0$, $\delta = \Delta\delta = 0$
C_{m_α} = sensitivity of pitch moment to $\Delta\alpha$ change, $\partial C_m/\partial\alpha$
C_{m_δ} = control effectiveness derivative, $\partial C_m/\partial\delta$
$C_{m_{\dot\alpha}}$ = pitch moment damping due to $\Delta\dot\alpha$, $\partial C_m/\partial(\dot\alpha c/2V)$

C_{m_q} = pitch moment damping due to $\Delta\dot{\theta}$, $\partial C_m/\partial(\dot{\theta}c/2V)$
ΔC_m = unbalance in pitch moment coefficient for aircraft in a state different than static trim
Δ = value of perturbation variable
c = reference chord length
V = freestream velocity

Note that the terms $(\dot{\alpha}c/2V)$ and $(\dot{\theta}c/2V)$ represent dimensionless rate terms, which implies that the corresponding damping derivatives remain dimensionless.

2.2 Longitudinal Stability Derivatives

The airframe, considered as a plant in modern control system terminology, will be described in Chapter 4 using dimensional stability derivatives. The stability derivatives applicable to the airframe's longitudinal motion will include functions dependent on small perturbations of the axial velocity, $u(t)/V$; the normal velocity, $\alpha(t) = w(t)/V$; the damping terms $\dot{\alpha}$ and $\dot{\theta}(=q)$; and the longitudinal control deflection δ. Longitudinal control deflection δ will often be subscripted by either the letter e or s depending on whether the control is provided by a plain flap (elevator) or a complete surface (stabilizer), respectively. Also it is convenient typically to consider time-dependent perturbations such as $\alpha = \alpha(t)$ vice the initial form of $\Delta\alpha$, as used earlier in Sec. 2.1.

2.2.1 Derivatives Due to u/V

The presence of the dynamic pressure term Q in the axial velocity dimensional stability derivative expressions makes it convenient to consider these derivatives in the form of X_u, Z_u, and M_u as shown next using the definitions derived in Chapter 4,

$$X_u = \frac{S}{mV}\frac{\partial(QC_x)}{\partial(u/V)}$$

$$Z_u = \frac{S}{mV}\frac{\partial(QC_z)}{\partial(u/V)} \qquad (2.4)$$

$$M_u = \frac{Sc}{I_yV}\frac{\partial(QC_m)}{\partial(u/V)}$$

where

S = reference wing area, ft^2
c = reference wing chord, ft
m = airframe mass (W/g), slugs
I_y = pitch axis mass moment of inertia, slug · ft^2
u = axial velocity perturbation, ft · s^{-1}
V = freestream air velocity, ft · s^{-1}
Q = dynamic pressure, lb · ft^{-2}

C_x = dimensionless axial force coefficient
C_z = dimensionless normal force coefficient
C_m = dimensionless pitching moment coefficient, referenced to c.g.

The dependence of the dynamic pressure on the $u(t)$ velocity perturbation can be expressed as

$$Q = 0.5\rho V^2 \left[1 + \left(\frac{u}{V}\right)\right]^2 = Q_0\left[1 + 2\left(\frac{u}{V}\right) + \left(\frac{u}{V}\right)^2\right]$$

where Q_0 is the initial trim aircraft value of dynamic pressure. In a linearized, small perturbation form, Q becomes

$$Q = Q_0[1 + 2(u/V)] \qquad (2.5)$$

To a first-order approximation, the initial stability axis assumption allows one to state that

$$C_x = -C_D$$

Therefore, following through on the linearization and recognizing that X_u is dependent on both Q and C_x, one finds

$$X_u = -\frac{Q_0 S}{mV}\left[2C_D + \frac{\partial C_D}{\partial(u/V)}\right]$$

The second term in the preceding expression reflects the partial derivative of dimensionless drag coefficient with respect to airspeed. Without compressibility influences, this term would vanish. However, because airframe Mach number is given by $M = V/a$, where a is the sound speed, the second expression becomes

$$\left[\frac{\partial C_D}{\partial(u/V)}\right]\left(\frac{a}{a}\right) = M\frac{\partial C_D}{\partial M}$$

In low subsonic flight, the rate of change of aircraft C_D at constant α is quite small. As shown in Fig. 2.1, an airframe critical Mach number corresponds to $M = M_{cr}$

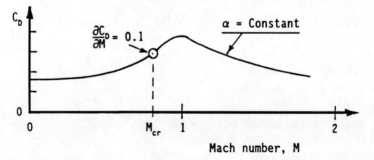

Fig. 2.1 Sketch of C_D variation with Mach number.

when $\partial C_D/\partial M = +0.1$ (i.e., a common definition). Making this substitution, one obtains that

$$X_u = -\frac{QS}{mV}\left[2C_D + M\frac{\partial C_D}{\partial M}\right] \quad (2.6)$$

where the 0 subscript has been dropped from the Q_0 expression after the partial derivatives have been obtained. Note that the use of the symbol M should not be confused with the airframe pitching moment when considered in this context. As will become more evident later, the notation in aircraft flight mechanics involves the use of many symbols and subscripts.

It may be concluded from looking at Fig. 2.1 that the X_u derivative of Eq. (2.6) will be dominated at low subsonic Mach numbers by the trim value of airframe drag coefficient C_D, whereas flight in or beyond the transonic region will bring the compressibility induced term into dominance. Also, the form of X_u in Eq. (2.6) assumes that there is no variation in propulsive system thrust with velocity perturbation, an assumption that is reasonable for a jet-engine powered vehicle.

One can deduce by similar reasoning, when the assumption is made that $C_z = -C_L$, the Z_u derivative is

$$Z_u = -\frac{QS}{mV}\left[2C_L + M\frac{\partial C_L}{\partial M}\right] \quad (2.7)$$

and the M_u derivative is

$$M_u = +\frac{QSc}{I_y V}M\frac{\partial C_m}{\partial M} \quad (2.8)$$

The absence of a C_m term in Eq. (2.8) is consistent with the assumption of the aircraft being in initial trim about the c.g. at the beginning of the dynamic maneuver. The $\partial C_m/\partial M$ derivative is associated with the term "tuck under," a Mach-dependent trim change experienced by many aircraft during transonic flight. This effect was first encountered by piston-powered fighters during World War II when making high-speed dives. Modern airfoil design combined with stability augmentation makes the presence of tuck under relatively unknown to today's pilot.

2.2.2 Derivatives Due to α

The dimensional derivatives X_α, Z_α, and M_α arise from taking partial derivatives of the corresponding dimensionless coefficients as

$$X_\alpha = \frac{QS}{m}\frac{\partial C_x}{\partial \alpha}$$

$$Z_\alpha = \frac{QS}{m}\frac{\partial C_z}{\partial \alpha} \quad (2.9)$$

$$M_\alpha = \frac{QSc}{I_y}\frac{\partial C_m}{\partial \alpha}$$

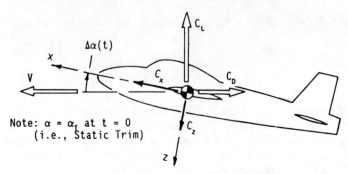

Fig. 2.2   Sketch of aircraft in $\Delta\alpha(t)$ perturbation.

Figure 2.2 illustrates an aircraft encountering an angle-of-attack perturbation, $\Delta\alpha(t)$, about an initially trimmed angle α_T. The velocity vector is initially aligned with the body axes (defined as stability axes) before the start of the $\Delta\alpha(t)$ perturbation. In other words, $\alpha = \alpha_T + \Delta\alpha(t)$, where α_T is measured relative to an arbitrary aircraft datum line.

The axial and normal force coefficients, C_x and C_z, respectively, remain attached to the airframe body axes during the motion and C_x is aligned with the wind vector V, before the onset of motion from the initial trim condition. The airframe C_L and C_D values apply at $[\alpha_T + \Delta\alpha(t)]$ using quasi-steady aerodynamic assumptions. In addition, it is convenient to state trigonometric functions as

$$C_\alpha = \cos \Delta\alpha(t) \quad \text{and} \quad S_\alpha = \sin \Delta\alpha(t)$$

Using the principles described in Sec. 1.3, and considering $\Delta\alpha$ as a perturbation term, one may define a matrix transformation in the form of

$$\begin{Bmatrix} C_x \\ C_z \end{Bmatrix} = -\begin{bmatrix} C_\alpha & -S_\alpha \\ S_\alpha & C_\alpha \end{bmatrix} \begin{Bmatrix} C_D \\ C_L \end{Bmatrix} = -[T_\alpha] \begin{Bmatrix} C_D \\ C_L \end{Bmatrix} \qquad (2.10)$$

From Eq. (2.10), C_x may be identified as

$$C_x = -C_D C_\alpha + C_L S_\alpha$$

Taking the partial derivative of C_x with respect to $\Delta\alpha$ yields

$$\frac{\partial C_x}{\partial \alpha} = +C_D S_\alpha - \left(\frac{\partial C_D}{\partial \alpha}\right) C_\alpha + C_L C_\alpha + \left(\frac{\partial C_L}{\partial \alpha}\right) S_\alpha$$

Linearizing for small $\Delta\alpha(t)$ changes plus use of the stability axis assumption provides

$$\frac{\partial C_x}{\partial \alpha} = \left[C_D + \left(\frac{\partial C_L}{\partial \alpha}\right)\right]\Delta\alpha + \left[C_L - \left(\frac{\partial C_D}{\partial \alpha}\right)\right]$$

which simplifies further by an order of magnitude consideration to

$$\frac{\partial C_x}{\partial \alpha} = \left[C_L - \left(\frac{\partial C_D}{\partial \alpha}\right)\right]$$

and finally

$$X_\alpha = \frac{QS}{m}\left[C_L - \frac{\partial C_D}{\partial \alpha}\right] \quad (2.11)$$

The second row in the matrix relation of Eq. (2.10) also provides an expression for C_z as an orthogonal transformation from C_D and C_L,

$$C_z = -C_D S_\alpha - C_L C_\alpha$$

Taking the partial derivative of C_z with respect to $\Delta\alpha$ followed by a similar linearization process as just used yields $\partial C_z/\partial \alpha$ as

$$\frac{\partial C_z}{\partial \alpha} = -\left[C_D + \left(\frac{\partial C_L}{\partial \alpha}\right)\right]$$

and the corresponding dimensional stability derivative as

$$Z_\alpha = -\frac{QS}{m}\left[C_D + \frac{\partial C_L}{\partial \alpha}\right] \quad (2.12)$$

The airframe's lift coefficient, as in Eq. (2.11), consists primarily of wing and horizontal tail contributions because the fuselage in a conventional aircraft configuration contributes little to the lift force. An approximate expression for the complete aircraft lift coefficient is given by

$$C_L = \left(\frac{\partial C_L}{\partial \alpha}\right)_{\text{wing}} (\alpha - \alpha_0)$$
$$+ \left(\frac{\partial C_L}{\partial \alpha}\right)_H [(1 - \epsilon_\alpha)(\alpha - \alpha_0) - \epsilon_0 + \delta_H]\left(\frac{S_H}{S}\right)\eta_H$$

where

$(\partial C_L/\partial \alpha)_{\text{wing}}$ = wing–body lift-curve slope, rad^{-1}
$(\partial C_L/\partial \alpha)_H$ = horizontal tail lift-curve slope referenced to tail area (S_H), rad^{-1}
α = wing angle of attack, rad
ϵ_α = average change in downwash angle at the tail due to wing α change, $\partial\epsilon/\partial\alpha$
ϵ_0 = tail downwash angle when α is α_0, rad
δ_H = tail incidence angle, rad
η_H = ratio of dynamic pressure at the tail relative to the freestream value, Q_H/Q

The dominant term in Eq. (2.12) is the complete aircraft lift-curve slope, $\partial C_L/\partial\alpha$, which may be obtained from the aircraft's C_L expression as

$$\left(\frac{\partial C_L}{\partial \alpha}\right)_{a/c} = \left(\frac{\partial C_L}{\partial \alpha}\right)_{\text{wing}} + \left(\frac{\partial C_L}{\partial \alpha}\right)_H (1 - \epsilon_\alpha)\left(\frac{S_H}{S}\right)\eta_H \quad (2.13)$$

The terms in Eq. (2.13) may be corrected separately for the influence of airframe elastic deformation including body bending to provide an estimate for the influence of static aeroelasticity. Accounting for such effects is important when considering airframe stability derivatives on relatively elastic airframes such as large jet transports.

The term contributing to the M_α derivative may be estimated directly from the airframe's dimensionless static stability derivative. Because the complete aircraft lift from α acts at an x location along the fuselage axis of symmetry described as the neutral point, the static stability derivative can be expressed relative to the complete airframe lift-curve slope as

$$\frac{\partial C_m}{\partial \alpha} = \left(\frac{\partial C_L}{\partial \alpha}\right)_{a/c} \left[\left(\frac{x}{c}\right)_{np} - \left(\frac{x}{c}\right)_{cg}\right] = \left(\frac{\partial C_L}{\partial \alpha}\right)_{a/c} \left(\frac{\Delta x}{c}\right)_{np} \quad (2.14)$$

where it is assumed in Eq. (2.14) that 1) both c.g. and neutral point locations are relative to the leading edge of the reference chord and 2) the body-axis coordinate system is used for x, cf. Fig. 1.1.

A longitudinally stable aircraft corresponds to $(\Delta x/c)_{np}$ having a negative sign, which implies that the c.g. is forward of the neutral point. In this case, a positive angle-of-attack perturbation results in a restorative (stable) pitching moment about the c.g. from the lift change that acts at the neutral point. A related term, the static margin, is described[1] as

$$\text{static margin} = -(\Delta x/c)_{np}$$

with the obvious benefit that a measure of the aircraft's static stability is provided by a positive number. Because the neutral point is invariant for a given flight condition, the static margin increases as the c.g. is moved forward.

In case the airframe has a manual elevator control system, a distinction must be made between stick-free and stick-fixed static stability in order to reflect the pitch moments induced by elevator floating during angle-of-attack changes. In a strict sense, Eq. (2.14) represents stick-fixed static stability for a manual longitudinal control system. No such distinction is needed if the control system were irreversible, which is the case for most high-performance aircraft.

The location of the (stick-fixed) neutral point relative to the aircraft c.g. may be estimated based on developing the complete aircraft static stability term $(\partial C_m/\partial \alpha)$ directly from the wing and tail contributions to pitch moment. Consider Eq. (2.14) in an alternate form as

$$\frac{\partial C_m}{\partial \alpha} = \left(\frac{\partial C_L}{\partial \alpha}\right)_{wing} \left(\frac{\Delta x}{c}\right)_{ac} - \left(\frac{\partial C_L}{\partial \alpha}\right)_H (1 - \epsilon_\alpha) V_H \eta_H \quad (2.15)$$

where

$(\Delta x/c)_{ac}$ = location of the wing–body aerodynamic center (a.c.) relative to the c.g. and normalized with respect to the wing reference chord length

V_H = horizontal tail volume coefficient, $(S_H \ell_H / Sc)$

ℓ_H = distance between wing–body a.c. and horizontal tail center of lift from aircraft α, $(x_{ac} - x_H)$; positive value for a tail located aft of the wing surface.

AERODYNAMIC PRINCIPLES

The neutral point location can be found from Eqs. (2.14) and (2.15) as

$$\left(\frac{\Delta x}{c}\right)_{np} = \frac{\left(\frac{\partial C_L}{\partial \alpha}\right)_{wing}\left(\frac{\Delta x}{c}\right)_{ac} - \left(\frac{\partial C_L}{\partial \alpha}\right)_{H}(1-\epsilon_\alpha)V_H \eta_H}{\left(\frac{\partial C_L}{\partial \alpha}\right)_{a/c}} \quad (2.16)$$

The horizontal tail's contribution to airframe static stability, as shown by Eq. (2.16), is influenced by the quantity $(1 - \epsilon_\alpha)V_H$. The tail volume coefficient V_H must be in balance with the wing downwash term ϵ_α when an airframe neutral point location is a design constraint. Tradeoff studies made during the airframe design development frequently disclose that raising the horizontal tail location onto the vertical tail (e.g., tee tail) places the tail in a reduced downwash field with the benefit that the tail volume coefficient may be reduced while maintaining the same desired airframe stability margin. However, it is often found in the tradeoff studies that the reduced tail size for a tee-tail configuration introduces added weight to the total tail structure for reasons of strength and avoidance of tail flutter problems. Other factors considered in design when establishing the horizontal tail placement includes influences such as tail size on aircraft performance and tail vertical location on aircraft stall behavior.

2.2.3 Derivatives Due to Pitch Rate q

The rate of change of pitch attitude, $q = d\theta/dt$, produces both a force and a moment on the airframe. The dominant effect of pitch rate is on pitching moment and to a lesser effect on the normal force. The axial force due to pitch rate normally is negligible and, consequently, does not appear in listings of airframe stability derivatives. Rate-dependent derivatives are labeled damping because these terms in the equations of motion are dissipative in nature and reduce motion oscillations by their energy absorbing capability. The pitch-rate dimensional derivatives are defined by

$$X_q = -\frac{QS}{m}\left(\frac{c}{2V}\right)C_{D_q} = 0 \quad \text{(by assumption)}$$

$$Z_q = -\frac{QS}{m}\left(\frac{c}{2V}\right)C_{L_q} \quad (2.17)$$

$$M_q = \frac{QSc}{I_y}\left(\frac{c}{2V}\right)C_{m_q}$$

By convention, the dimensionless damping partial derivatives are

$$C_{D_q} = \frac{\partial C_D}{\partial(qc/2V)}$$

$$C_{L_q} = \frac{\partial C_L}{\partial(qc/2V)} \quad (2.18)$$

$$C_{m_q} = \frac{\partial C_m}{\partial(qc/2V)}$$

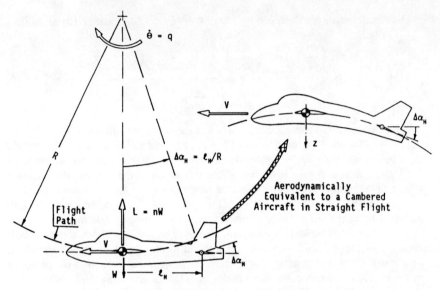

Fig. 2.3 Airframe in curvilinear flight.

where the dimensionless pitch rate $qc/2V$ represents q being normalized by $2V/c$. The reference chord c is the same as used in establishing the pitch moment coefficient C_m. The use of $c/2$ vs c is a convention that may be attributed to thin-airfoil theory where the distance between the a.c. and the tangential flow control point (at the rearward a.c.) is $c/2$.

The partial derivatives with respect to pitch rate q may be visualized by considering an aircraft at constant velocity in curvilinear flight about a center of rotation, as shown in Fig. 2.3. The radius of curvature for the flight path has been exaggerated in order to clarify the aerodynamic sources for the force and moment changes as a result of pitch rate. It is well known in aerodynamics that a vehicle in curvilinear flight is equivalent aerodynamically to a like-curved vehicle in rectilinear flight, also shown in Fig. 2.3. The validity of this concept can be recognized by noting that the surface boundary conditions for the airflow over both bodies are identical, thereby yielding identical aerodynamic solutions.

The main source in providing pitch damping for the described model arises from the horizontal tail's induced flow angle causing a change in aircraft lift and pitching moment. It can be noted from Fig. 2.3 that the freestream velocity is related to pitch rate by

$$V = qR$$

and the tail incidence change from a positive value of q is

$$\Delta\alpha_H = \ell_H/R = \ell_H q/V \tag{2.19}$$

Example 2.1: Estimation of Flight Radius of Curvature

To appreciate the magnitude of the induced flight curvature, consider an aircraft in a 2-g pullup maneuver while traveling at an airspeed of $V = 500$ ft/s.

This implies that the airframe lift is $L = 2W$ where W is the airframe weight ($W = mg$). An estimate for the radius of curvature, R, comes from a consideration of the centrifugal force yielding

$$R = V^2/(n-1)g = 7770 \text{ ft}$$

for this example. Also observe that in unaccelerated level flight, where $n = 1$, the radius of curvature would be infinite in value.

It is also convenient to estimate the induced angle change for an airframe for the example flight condition. Consider an assumed aircraft with a tail length of 20 ft. Then from Eq. (2.19), one estimates that

$$\Delta \alpha_H = (20 \text{ ft})/(7770 \text{ ft}) = 0.00257 \text{ rad} (= 0.15 \text{ deg})$$

A small induced tail incidence contributes significantly to the damping of a longitudinal flight oscillation.

The production of aircraft lift coefficient due to the angle induced at the horizontal tail by pitch rate may be estimated using Eq. (2.19),

$$\Delta C_L = \left(\frac{S_H}{S}\right)\left(\frac{\partial C_L}{\partial \alpha}\right)_H \left(\frac{\ell_H q}{V}\right)$$

$$= 2V_H \left(\frac{\partial C_L}{\partial \alpha}\right)_H \left(\frac{qc}{2V}\right) \quad (2.20)$$

The partial derivative of Eq. (2.20) with respect to dimensionless pitch rate yields an estimate of the tail's contribution to airframe pitch damping,

$$\left(\Delta C_{L_q}\right)_H = 2V_H \left(\frac{\partial C_L}{\partial \alpha}\right)_H$$

$$\left(\Delta C_{m_q}\right)_H = -2V_H \left(\frac{\ell_H}{c}\right)\left(\frac{\partial C_L}{\partial \alpha}\right)_H \quad (2.21)$$

There also is a contribution to pitch damping from the wing and body, and the amount will, of course, be configuration dependent. A frequent approximation to the estimation of the total airframe damping derivative is to increase the estimated tail contribution by 10%. Therefore, an estimate of the complete aircraft pitch damping would be

$$C_{L_q} = (2.20) V_H \left(\frac{\partial C_L}{\partial \alpha}\right)_H$$

$$C_{m_q} = -(2.20) V_H \left(\frac{\ell_H}{c}\right)\left(\frac{\partial C_L}{\partial \alpha}\right)_H \quad (2.22)$$

Although it is not readily apparent by comparing Eqs. (2.21) and (2.22) where the 10% estimation factor comes from, an example will be given as an illustration of a potential source.

Example 2.2: Estimation of Pitch Moment Damping

Consider a representative, conventional airframe with the following assumed characteristics:

$$\left(\frac{\partial C_L}{\partial \alpha}\right)_H = 4.0 \text{ rad}^{-1}$$

$$(\ell_H/c) = 2.0$$

$$(S_H/S) = 40 \text{ ft}^2/200 \text{ ft}^2 = 0.20$$

$$V_H = 0.40$$

Estimate pitch damping using Eq. (2.22) as

$$C_{m_q} = -(2.20)(0.4)(2.0)(4.0 \text{ rad}^{-1}) = -7.04 \text{ rad}^{-1}$$

It can be shown that a two-dimensional airfoil (not a wing) of chord c in incompressible flow would have a damping derivative of

$$c_{m_q} = -\pi/4 \text{ rad}^{-1}$$

If this sectional property were assumed to translate as a total wing value (using a wing strip theory approach), then the assumption of a 10% increase in damping due to the presence of the wing could be rationalized from an order of magnitude basis.

In actual airframe design, more accurate span load methods would be used to predict pitch damping using both available wind-tunnel test data sources and theoretical wing span load analysis methods. In some situations, account also would be made of airframe elastic deformations when estimating the pitch damping terms.

2.2.4 Derivatives Due to Angle-of-Attack Rate $\dot{\alpha}$

The rate of change of the vertical velocity perturbation, $d\alpha/dt$, produces aerodynamic damping with the primary effect being the development of normal force and pitching moment terms. The alpha-rate dimensional derivatives are defined by

$$X_{\dot{\alpha}} = -\frac{QS}{m}\left(\frac{c}{2V}\right)C_{D_{\dot{\alpha}}} = 0 \quad \text{(by assumption)}$$

$$Z_{\dot{\alpha}} = -\frac{QS}{m}\left(\frac{c}{2V}\right)C_{L_{\dot{\alpha}}} \quad (2.23)$$

$$M_{\dot{\alpha}} = \frac{QSc}{I_y}\left(\frac{c}{2V}\right)C_{m_{\dot{\alpha}}}$$

By convention, the dimensionless damping partial derivatives are

$$C_{D_{\dot{\alpha}}} = \frac{\partial C_D}{\partial(\dot{\alpha}c/2V)}$$

$$C_{L_{\dot{\alpha}}} = \frac{\partial C_L}{\partial(\dot{\alpha}c/2V)} \quad (2.24)$$

$$C_{m_{\dot{\alpha}}} = \frac{\partial C_m}{\partial(\dot{\alpha}c/2V)}$$

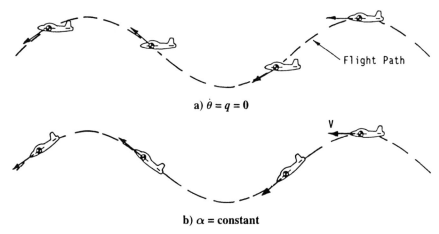

Fig. 2.4 Comparison of airframe flight trajectories.

where the dimensionless angle-of-attack rate $\dot{\alpha}c/2V$ represents $\dot{\alpha}$ being normalized by $2V/c$.

It is helpful to visualize airframe motion that produces either pure pitch- or alpha-rate damping inasmuch as the stability derivatives from these motions relate to the aerodynamic forces and moments that act on the airframe by principles of linear superposition. Figure 2.4a illustrates an airframe trajectory where the pitch orientation relative to the horizon is constant (i.e., $q = 0$) while the angle between the vehicle and the trajectory's tangent (velocity vector) is changing to give a variable $\alpha(t)$ perturbation. In contrast, Fig. 2.4b maintains the aircraft's position relative to the velocity vector (i.e., $\dot{\alpha} = 0$) while allowing the pitch orientation to vary with time. Under the assumed trajectories, there would be aerodynamic terms arising from the $d\alpha/dt$ influence in Fig. 2.4a whereas the motion depicted in Fig. 2.4b would introduce q related aerodynamic damping terms. Of course, an actual flight trajectory would be a linear combination of the two motion types.

Damping due to $d\alpha/dt$ will arise due to contributions from both the wing and the horizontal tail. However, the wing lift and moment response to $d\alpha/dt$ is difficult to quantify. Consequently, the tail contribution is most frequently considered as the major portion of $d\alpha/dt$ related damping under a quasi-steady flow assumption.

Figure 2.5a is a simplified representation of a lifting wing by a single vortex horseshoe consisting of a bound element on the wing and trailing portions extending downstream to infinity. The horizontal tail is shown as immersed in the downwash field of the idealized vortex system. Figure 2.5b depicts the flowfield a short instant after the wing angle of attack is increased by $\Delta\alpha$. The resulting increase in bound vorticity $\Delta\Gamma$ due to $\Delta\alpha$ results in the downstream convection of a line vortex by consideration of the classical Helmholtz theorem[10] governing the creation of vorticity. The modeling of the $d\alpha/dt$ influence includes the following.

1) Downwash at the tail does not respond instantaneously to changes in wing lift caused by the angle-of-attack perturbation $\Delta\alpha$.

2) Tail response to the flowfield depends on the downwash in the neighborhood of the tail that is generated by the wing's trailing vortex system.

3) Because the vorticity change $\Delta\Gamma$ is convected by the freestream velocity V the change in circulation of the wing will not be apparent at the tail as a downwash angle change until a (transport) time, $\Delta t = \ell_H/V$, has elapsed.

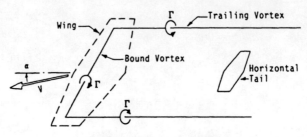

a) Initial steady-state condition

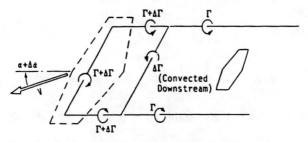

b) Unsteady condition due to $\Delta\alpha$

Fig. 2.5 Time lag of downwash at the tail.

4) Therefore, downwash at the tail, $\epsilon_H(t)$, corresponds to the wing lift production at an earlier time of $(t - \Delta t)$.

Based on the described modeling, an estimate of the correction to the quasi-steady tail downwash angle due to the time lag in downwash production would be

$$\Delta\epsilon = -\epsilon_\alpha \dot{\alpha} \Delta t = -\epsilon_\alpha \dot{\alpha}(\ell_H/V) = -\Delta\alpha_H \quad (2.25)$$

The production of aircraft lift coefficient due to the angle induced at the horizontal tail by the time-lag-of-downwash effect may be estimated using Eq. (2.25),

$$\Delta C_L = \left(\frac{S_H}{S}\right)\left(\frac{\partial C_L}{\partial \alpha}\right)_H \epsilon_\alpha \dot{\alpha}\left(\frac{\ell_H}{V}\right)$$

$$\Delta C_L = 2V_H \left(\frac{\partial C_L}{\partial \alpha}\right)_H \epsilon_\alpha \left(\frac{\dot{\alpha} c}{2V}\right) \quad (2.26)$$

The partial derivative of Eq. (2.26) with respect to dimensionless angle-of-attack rate yields an estimate of the tail's contribution to damping,

$$(\Delta C_{L_{\dot{\alpha}}})_H = 2V_H \epsilon_\alpha \left(\frac{\partial C_L}{\partial \alpha}\right)_H$$

$$(\Delta C_{m_{\dot{\alpha}}})_H = -2V_H \epsilon_\alpha \left(\frac{\ell_H}{c}\right)\left(\frac{\partial C_L}{\partial \alpha}\right)_H \quad (2.27)$$

AERODYNAMIC PRINCIPLES

Table 2.1 Airframe pitch damping derivative

Airframe	Mach no.	Alt.	C_{m_q}, rad^{-1}	$C_{m_{\dot{\alpha}}}$, rad^{-1}
Boeing 747	0.25	S.L.[a]	−20.8	−3.2
Convair 880	0.25	S.L.	−12.1	−4.1
Lockheed Jetstar	0.25	S.L.	−8.0	−3.0
Lockheed F-104A	0.25	S.L.	−5.8	−1.6
LTV A-7A	0.30	S.L.	−3.8	−0.7
Douglas A-4D	0.40	All	−3.7	−1.1

[a] Sea level.

As with the pitch-damping derivatives, a 10% increase is applied to represent the complete aircraft estimate for damping due to $d\alpha/dt$, i.e.,

$$C_{L_{\dot{\alpha}}} = (2.20) V_H \epsilon_\alpha \left(\frac{\partial C_L}{\partial \alpha} \right)_H$$

$$C_{m_{\dot{\alpha}}} = (-2.20) V_H \epsilon_\alpha \left(\frac{\ell_H}{c} \right) \left(\frac{\partial C_L}{\partial \alpha} \right)_H \quad (2.28)$$

Because the value of ϵ_α on a conventional aircraft frequently ranges from 0.2 to 0.4, comparison of Eq. (2.28) with Eq. (2.22) leads one to expect that a listing of $M_{\dot{\alpha}}$ in a tabulation will, in general, be less than M_q by approximately the same scaling value. Available airplane values of C_{m_q} and $C_{m_{\dot{\alpha}}}$ (Refs. 2 and 3) are shown in Table 2.1 both to compare and to show the approximate magnitude of the respective dimensionless derivatives.

It will be seen in Chapter 6 that the short-period approximation includes the damping moment in the form of $(M_q + M_{\dot{\alpha}})$ as a single term under the assumption that $\alpha(t)$ and $\theta(t)$ perturbations are approximately equivalent in both amplitude and phase. It is for this reason that parameter identification methods applied to flight-test derived short-period time histories have difficulty in distinguishing between the two forms of damping. Parameter identification results usually express damping by the sum.

2.3 Lateral-Directional Stability Derivatives

The stability derivatives applicable to the airframe's lateral-directional motion will include functions dependent on the normalized lateral velocity $\beta(t) = v(t)/V$, damping perturbation terms due to yaw rate $r = d\psi/dt$, and roll rate $p = d\phi/dt$ along with derivatives due to lateral and directional control. The damping derivative due to $\dot{\beta}$ is usually considered as negligible and will not be presented because modeling of it for purposes of estimation is not well understood. The dimensional force and moments to be considered include Y (side force), L (rolling moment), and N (yawing moment).

2.3.1 Derivatives Due to β

The dimensional derivatives Y_β, L_β, and N_β arise from taking partial derivatives of the corresponding dimensionless coefficients as

$$Y_\beta = \frac{QS}{m} \frac{\partial C_y}{\partial \beta}$$

$$L_\beta = \frac{QSb}{I_x} \frac{\partial C_\ell}{\partial \beta} \quad (2.29)$$

$$N_\beta = \frac{QSb}{I_z} \frac{\partial C_n}{\partial \beta}$$

where

b = reference wing span, ft
I_x = roll axis mass moment of inertia, slug · ft^2
I_z = yaw axis mass moment of inertia, slug · ft^2

The reference values of Q, S, and m were described earlier in Sec. 2.2.

The Y_β and N_β stability derivatives are dominated by contributions from the vertical tail and fuselage if linear superposition approaches were used in modeling estimates. The wing influence is usually considered as minor. Fig. 2.6 depicts a plan view for a typical airframe in body-axis orientation when exposed to a positive sideslip velocity.

The dimensionless side-force derivative due to β is given approximately by

$$\frac{\partial C_y}{\partial \beta} = \left(\frac{\Delta \partial C_y}{\partial \beta}\right)_{body} - \left(\frac{\partial C_L}{\partial \alpha_V}\right)_V (1 - \sigma_\beta)\left(\frac{S_V}{S}\right)\eta_V \quad (2.30)$$

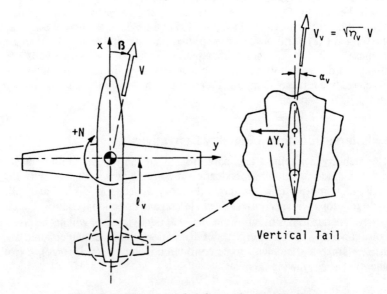

Fig. 2.6 Plan view of airframe during sideslip.

where

$(\Delta \partial C_y/\partial \beta)_{\text{body}}$ = body contribution to side-force derivative, rad^{-1}
$(\partial C_L/\partial \alpha_V)_V$ = vertical tail lift-curve slope referenced to vertical tail area (S_V), rad^{-1}
σ_β = average change in sidewash at the vertical tail per unit sideslip β angle, $\partial \sigma/\partial \beta$
η_V = ratio of dynamic pressure at the tail to the freestream value, Q_V/Q

The minus sign on the vertical tail contribution can be realized from Fig. 2.6 because positive β produces vertical tail lateral force in the negative y direction.

The body contribution is highly configuration dependent and difficult to predict. The vertical tail contribution is usually established from wind-tunnel tests by comparing tail-off with tail-on experiments during sideslip sweeps while holding α constant. The sidewash term σ_β varies with airframe angle of attack. Although σ_β is normally on the order of 0.3, it can attain a value on the order of 1 at angles of attack in the stall region. Under these circumstances, the vertical tail could be immersed in separated flow from the wing and fuselage and, consequently, be unable to develop any side force due to changes in sideslip angle.

The yawing moment stability derivative C_{n_β} can be approximated by considering it as a sum of the individual components

$$\frac{\partial C_n}{\partial \beta} = \left[\frac{\Delta \partial C_n}{\partial \beta}\right]_{\text{body}} + \left[\frac{\Delta \partial C_n}{\partial \beta}\right]_{\text{wing}} + \left[\frac{\Delta \partial C_n}{\partial \beta}\right]_V \quad (2.31)$$

C_{n_β} is frequently described as the weathercock stability derivative because it is this term that develops a restoring moment in the presence of sideslip tending to return the airframe to a zero β condition, much like a weathervane.

The $[\Delta \partial C_n/\partial \beta]_{\text{body}}$ term is usually destabilizing whereas a small stabilizing contribution is normally provided to the weathercock stability by the $[\Delta \partial C_n/\partial \beta]_{\text{wing}}$ term. Predictions of these terms are difficult to make. The vertical tail's contribution to directional stability can be estimated as a dimensionless moment developed by the tail's side force coefficient, i.e.,

$$\left[\frac{\Delta \partial C_n}{\partial \beta}\right]_V = +\left(\frac{\partial C_L}{\partial \alpha_V}\right)_V (1 - \sigma_\beta)\left(\frac{\ell_V}{b}\right)\left(\frac{S_V}{S}\right)\eta_V$$

$$= +\left(\frac{\partial C_L}{\partial \alpha_V}\right)_V (1 - \sigma_\beta) V_V \eta_V \quad (2.32)$$

where $V_V = (S_V \ell_V/Sb)$, is the vertical tail volume coefficient.

The C_{n_β} derivative will have a positive sign when providing a stabilizing influence to directional stability. At high angles of attack, C_{n_β} can become quite small and even change sign ($-$ is unstable). The literature is rich with concerns on the influence of C_{n_β} on the stall departure behavior of high-performance aircraft. Because C_{n_β} is dependent on the nature of the sidewash variation with β, it is frequently necessary to tabulate ΔC_n as a function of sideslip β and angle of attack α when describing airframe directional characteristics in a flight simulator environment.

The importance of a rolling-moment term due to sideslip angle in aircraft handling comes from the recognition that, unlike a surface ship with roll coupling between the c.g. and the center of buoyancy, an aircraft platform does not directly develop a restoring roll moment due to a roll angle change from trim. Consequently, when a roll angle disturbance occurs during flight (e.g., a 5-deg-right roll angle change due to a sudden gust) the airframe would initially experience a positive body-axis side-force input from the y component of weight, which in turn would result in a positive sideslip angle as the airframe attempts to equilibrate aerodynamically. The sideslip angle then interacts with the weathercock stability, C_{n_β}, to remove the sideslip, and a rolling moment, also due to sideslip, would occur concurrently tending to remove the roll angle disturbance (i.e., raise the right wing). The events described actually are dynamic in scope with the lateral and directional influences being continually coupled. Weathercock stability requires C_{n_β} to be positive in sign whereas roll moment due to sideslip, C_{ℓ_β}, should be negative in sign.

The roll moment due to sideslip, C_{ℓ_β}, depends to a first order on four factors, i.e., wing dihedral angle, wing–body interference, wing sweep angle, and vertical tail roll contribution. Although all four factors are key to establishing the C_{ℓ_β} stability derivative, the first item listed has led to the colloquial usage of dihedral effect to denote the C_{ℓ_β} term.

C_{ℓ_β} **due to wing dihedral angle.** Consider a wing with an unswept $c/4$ line having a geometric dihedral angle of Γ, as shown in Fig. 2.7. It is assumed that the x body axis is aligned with the initial velocity vector before the introduction of α and β, cf. Fig. 2.7a. By considering a view of the right-wing panel looking forward (Fig. 2.7b), a coordinate transformation can be made to identify the flow angle

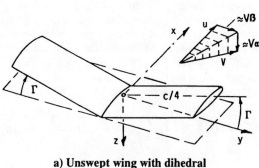

a) Unswept wing with dihedral

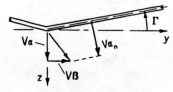

b) Rear view, right wing

Fig. 2.7 Wing dihedral effect.

components normal and tangential to the wing panel, α_n and β_t, respectively. The transformation for the right-wing panel yields

$$\begin{Bmatrix} V\alpha_n \\ V\beta_t \end{Bmatrix} = \begin{bmatrix} C_\Gamma & S_\Gamma \\ -S_\Gamma & C_\Gamma \end{bmatrix} \begin{Bmatrix} V\alpha \\ V\beta \end{Bmatrix} \quad \text{where} \quad \begin{cases} C_\Gamma = \cos\Gamma \\ S_\Gamma = \sin\Gamma \end{cases}$$

The velocity normal to the right-wing panel, which is influential in establishing the span loading, can be seen from the first row of the preceding matrix transformation relation as

$$V\alpha_n = VC_\Gamma\alpha + VS_\Gamma\beta$$

and α_n is approximated by applying the small angle assumption as

$$\alpha_n = \alpha + \Gamma\beta = \alpha + \Delta\alpha_\Gamma$$

On the left-wing panel, $\Delta\alpha_\Gamma$ becomes $-\Gamma\beta$. The effect of sideslip angle on dihedral effect is to introduce an antisymmetric angle input $\Delta\alpha_\Gamma$ into the wing span load determination. The resulting antisymmetric span load will produce a negative roll moment (i.e., right wing up, left wing down). The C_{ℓ_β} term due to geometric dihedral effect on this wing model will be negative in sign corresponding to a statically stabilizing influence.

The antisymmetric wing span load due to β will superimpose on the symmetric load to α and thereby develop a yawing moment increment from the antisymmetric induced drag distribution. This yawing moment derivative increment enters into the $[\Delta C_{n_\beta}]_{\text{wing}}$ term mentioned in Eq. (2.31), and although small in magnitude, will normally have a positive sign (stable contribution) because the antisymmetric load due to plus β is positive on the right-wing panel.

C_{ℓ_β} *due to wing–body interference.* The wing–body interference on dihedral effect can be visualized by considering a cylindrical fuselage experiencing crossflow laterally due to the sideslip velocity component $V\beta$. The flow patterns over the bodies in crossflow are depicted by the streamlines, as shown in Fig. 2.8, for two wing positions.

The high-wing configuration, as shown in Fig. 2.8a, encounters upflow on the inboard portion of the right wing and downflow on the left wing when exposed to positive sideslip β. The induced flow angles produce an antisymmetric span load distribution. The resulting roll moment acts counter-clockwise about the x body axis (negative in sign), which tends to increase the dihedral effect derivative. Conversely, a low-wing configuration results in a roll moment due to β that is

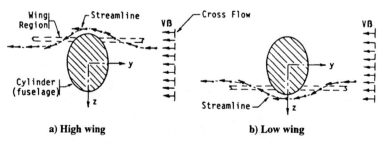

a) High wing   b) Low wing

Fig. 2.8 Wing–body interference.

positive in sign, thereby reducing dihedral effect. Design implications are that a low-wing configuration will require more geometric dihedral angle to maintain the same total dihedral effect as a comparable high-wing configuration. Differences in dihedral angles can amount to between 5–10 deg due to extremes of wing placement for the same value of the total dihedral effect derivative.

As a historical note, a low-wing, twin-engined, straight-tapered-wing transport built after World War II had undesirable lateral-directional flying qualities because the designers neglected to use wing dihedral. These aircraft were not produced in any quantity.

From a practical standpoint, dihedral in a low-wing airframe is also advantageous for maintaining ground clearance during crosswind landings.

C_{ℓ_β} *due to wing sweep angle.* Modern jet aircraft with wing sweepback have considerable induced dihedral effect due to sweep. Conversely, forward sweep acts to reduce the induced dihedral effect. Although a wing span load analysis could be made to estimate the effect, it is more fitting to remain intuitive for concept purposes. Basically, the logic starts by recognizing that the lift-curve slope on a straight, two-dimensional swept wing varies inversely with the cosine squared of sweep angle.[10] Figure 2.9 depicts a three-dimensional wing with sweep angle Λ operating at a sideslip angle β.

If one were to assume a form of strip theory for estimating the effect of sweep on wing lift, an approximate aerodynamic model would distinguish between the lift produced on the right- and left-wing panels as

$$(\text{Lift})_{\text{RH}} = C_L Q(S/2) \cos^2(\Lambda - \beta)$$

$$(\text{Lift})_{\text{LH}} = C_L Q(S/2) \cos^2(\Lambda + \beta)$$

The asymmetric lift load results in an increment to the dihedral effect from wing sweepback having the form of

$$\left[\Delta C_{\ell_\beta}\right]_\Lambda = -(\text{const})C_L \sin 2\Lambda$$

with the constant having a value on the order of 0.1–0.3 depending on the configuration. The following lists the impact of this estimation.

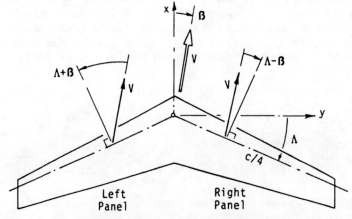

Fig. 2.9 Swept-back wing in sideslip.

1) Wing sweep results in dihedral effect, C_{ℓ_β}, being C_L dependent.
2) Forward sweep introduces a destablizing dihedral effect.
3) A 35-deg swept-back wing at high values of C_L can have added dihedral effect equivalent to 5–7 deg of geometric dihedral.
4) Flap deflection, as during landing approach, increases dihedral effect due to the C_L influence on a swept-back wing.

C_{ℓ_β} *due to vertical tail.* As noted earlier, the vertical tail is dominant in the production of side force and yawing moment due to sideslip angle β. In addition, the side force due to β will induce a rolling moment as a consequence of the z coordinate of the tail-force centroid. The contribution to dihedral effect, $[\Delta C_{\ell_\beta}]_V$, although generally small compared to the wing term, will usually be stabilizing ($-$sign) at low angles of attack, change with angle of attack, and become either neutral or destabilizing at or near to stall angles. The actual contribution is highly configuration dependent.

In summary, the aerodynamic mechanisms giving rise to dihedral effect have been illustrated by approximate concepts. The wing configuration is considered as the dominant factor in establishing dihedral effect, C_{ℓ_β}, with geometric items such as dihedral angle, sweep angle, and body placement all being important.

Representative airplane values of C_{y_β}, C_{n_β}, and C_{ℓ_β} from Refs. 2 and 3 are shown in Table 2.2 to indicate the order of magnitude and sign for stable, dimensionless stability derivatives due to β.

2.3.2 Derivatives Due to Yaw Rate r

The rate of change of yaw (heading) angle, $r = d\psi/dt$, results in the production of side forces as well as roll and yaw moments. The dimensional roll damping stability derivatives Y_r, L_r, and N_r are defined by

$$Y_r = \frac{QS}{m}\left(\frac{b}{2V}\right)C_{y_r}$$

$$L_r = \frac{QSb}{I_x}\left(\frac{b}{2V}\right)C_{\ell_r} \qquad (2.33)$$

$$N_r = \frac{QSb}{I_z}\left(\frac{b}{2V}\right)C_{n_r}$$

Table 2.2 Airframe derivatives due to sideslip

Airframe	Mach no.	Alt.	C_{y_β}, rad^{-1}	C_{n_β}, rad^{-1}	C_{ℓ_β}, rad^{-1}
Boeing 747	0.25	S.L.[a]	−0.96	0.15	−0.22
Convair 880	0.25	S.L.	−0.88	0.14	−0.20
Lockheed Jetstar	0.25	S.L.	−0.72	0.12	−0.16
Lockheed F-104A	0.25	S.L.	−1.17	0.50	−0.15
LTV A-7A	0.30	S.L.	−0.87	0.11	−0.10
Douglas A-4D	0.40	All	−0.98	0.25	−0.12

[a]Sea level.

By convention, the dimensionless yaw damping derivatives used in Eq. (2.33) are

$$C_{y_r} = \frac{\partial C_y}{\partial (rb/2V)}$$

$$C_{\ell_r} = \frac{\partial C_\ell}{\partial (rb/2V)} \quad (2.34)$$

$$C_{n_r} = \frac{\partial C_n}{\partial (rb/2V)}$$

The dimensionless stability derivatives are expressed with respect to dimensionless yaw rate where the normalizing factor is $b/2V$, which compares to the $c/2V$ factor used with pitch rate. The wing span b is significant from a scaling standpoint when considering the lateral-directional aerodynamic coefficients.

Figure 2.10 depicts the plan view of an airframe when experiencing pure positive yaw rate with the center of rotation shown at a lateral distance R from the airplane's c.g., on the y body-axis coordinate.

The dimensionless C_{y_r} derivative can be considered as being made up of contributions from the vertical tail and wing as

$$\frac{\partial C_y}{\partial (rb/2V)} = \left[\frac{\Delta \partial C_y}{\partial (rb/2V)}\right]_V + \left[\frac{\Delta \partial C_y}{\partial (rb/2V)}\right]_{\text{wing}} \quad (2.35)$$

The vertical tail contribution, which is analogous to the horizontal tail contribution to pitch damping [cf. Eq. (2.21)], is dependent on the vertical tail incidence induced by rotary flight, as shown in Fig. 2.10, i.e.,

$$\Delta \alpha_V = \ell_V/R = \ell_V r/V \quad (2.36)$$

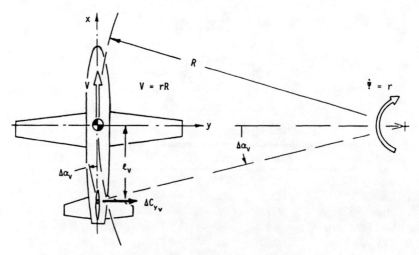

Fig. 2.10 Airframe in turning flight.

which yields an estimate for the vertical tail contribution to the stability derivative as

$$\left[\frac{\Delta \partial C_y}{\partial (rb/2V)}\right]_V = \left(\frac{S_V}{S}\right)\left(\frac{\partial C_L}{\partial \alpha}\right)_V \left(\frac{2\ell_V}{b}\right) = 2V_V \left(\frac{\partial C_L}{\partial \alpha}\right)_V \quad (2.37)$$

The $[\Delta \partial C_y / \partial (rb/2V)]_{\text{wing}}$ contribution is usually considered as negligible, which leaves Eq. (2.37) as the main term considered when making an estimate of the C_{y_r} derivative. However, the C_{y_r} derivative will frequently be neglected when establishing the airframe equations of motion because of its minor influence on the airframe's lateral-directional dynamics.

The roll moment due to yaw rate arises primarily from the vertical tail and wing contributions. The induced curvature on the fuselage is normally considered as a negligible influence when estimating C_{ℓ_r}. Therefore,

$$\frac{\partial C_\ell}{\partial (rb/2V)} = \left[\frac{\Delta \partial C_\ell}{\partial (rb/2V)}\right]_V + \left[\frac{\Delta \partial C_\ell}{\partial (rb/2V)}\right]_{\text{wing}}$$

$$= (\Delta C_{\ell_r})_V + (\Delta C_{\ell_r})_{\text{wing}} \quad (2.38)$$

The $(\Delta C_{\ell_r})_V$ term may be estimated by considering the rolling moment produced by the vertical tail's generated side force due to $rb/2V$ acting at a center of pressure that is a distance z_V (body-axis notation) vertically from the x axis. Plus z_V corresponds to the centroid being below the x axis. This latter statement is based on the use of stability axes as a reference system, which means that the x body axis is aligned with the velocity vector at the start of the dynamic maneuver. At low angles of attack, the centroid of vertical tail lift (due to rotary flight) will usually be above the x axis, which implies that the $(\Delta C_{\ell_r})_V$ contribution would then be positive in value. An approximation for the vertical tail's contribution is

$$(\Delta C_{\ell_r})_V = -\left(\frac{z_V}{b}\right)(\Delta C_{y_r})_V$$

The $(\Delta C_{\ell_r})_{\text{wing}}$ term is an important contributor to the cross derivative of roll due to yaw rate. It is modeled by considering the differentially induced dynamic pressures between the left- and right-wing panels. The increased and decreased dynamic pressures on the left- and right-wing panels, respectively, produce an antisymmetric span load, which gives a positive rolling moment that is a function of yaw rate. When modeling this term, wing aspect ratio, taper ratio, sweep angle, and Mach number all play an important role. An initial approximation, based on elementary wing strip theory approach, is

$$(\Delta C_{\ell_r})_{\text{wing}} = +\left(\frac{C_L}{4}\right)$$

In summary, C_{ℓ_r} is usually positive in value, is dependent on wing lift coefficient, and is slightly influenced by the vertical tail contribution. An approximate estimate for C_{ℓ_r} is given by

$$C_{\ell_r} = +\frac{C_L}{4} - \left(\frac{z_V}{b}\right)(\Delta C_{y_r})_V \quad (2.39)$$

The third yaw rate derivative, C_{n_r}, is known as the yaw damping term and is an important factor in establishing the damping level of the Dutch-roll mode described in Chapter 6. This derivative is primarily made up of contributions from the vertical tail and wing with the vertical tail portion being dominant,

$$\frac{\partial C_n}{\partial (rb/2V)} = \left[\frac{\Delta \partial C_n}{\partial (rb/2V)}\right]_V + \left[\frac{\Delta \partial C_n}{\partial (rb/2V)}\right]_{wing}$$

$$= (\Delta C_{n_r})_V + (\Delta C_{n_r})_{wing} \quad (2.40)$$

An estimate of $(\Delta C_{n_r})_V$ may be made by considering the yawing moment produced by the vertical tail during yaw rate motion. The stabilizing influence (negative sign) arises when the positive side force developed due to yaw rate r is aft of the c.g., which results in a negative yawing moment. An approximation for the vertical tail's contribution is

$$(\Delta C_{n_r})_V = -(\ell_V/b)(\Delta C_{y_r})_V$$

The wing contribution may be visualized by considering the alteration to both induced and profile wing drag distributions due to aircraft yaw rate. A positive yaw rate results in a drag increase on the left-wing panel and a corresponding decrease on the right-wing panel. The antisymmetric drag distribution equates to a negative yawing moment, which in turn provides a stable, yaw rate damping action. However, the dominant term in overall yaw damping comes from the vertical tail, and initial airframe derivative estimations would use such an approach.

Representative values of the rotary derivatives, C_{ℓ_r} and C_{n_r}, are shown in Table 2.3 based on information from Refs. 2 and 3. The C_{y_r} stability derivative is considered as negligible and is not listed.

2.3.3 Derivatives Due to Roll Rate p

The rate of change of bank angle, $p = d\phi/dt$, primarily results in the production of roll and yaw moments. The dimensional roll damping stability derivatives

Table 2.3 Airframe derivatives due to yaw rate

Airframe	Mach no.	Alt.	C_{n_r}, rad^{-1}	C_{ℓ_r}, rad^{-1}
Boeing 747	0.25	S.L.[a]	−0.30	0.10
Convair 880	0.25	S.L.	−0.19	0.20
Lockheed Jetstar	0.25	S.L.	−0.14	0.10
Lockheed F-104A	0.25	S.L.	−0.73	0.25
LTV A-7A	0.30	S.L.	−0.31	0.15
Douglas A-4D	0.40	All	−0.35	0.14

[a]Sea level.

Y_p, L_p, and N_p, are defined by

$$Y_p = \frac{QS}{m}\left(\frac{b}{2V}\right)C_{y_p} = 0 \quad \text{(by assumption)}$$

$$L_p = \frac{QSb}{I_x}\left(\frac{b}{2V}\right)C_{\ell_p} \qquad (2.41)$$

$$N_p = \frac{QSb}{I_z}\left(\frac{b}{2V}\right)C_{n_p}$$

By convention, the dimensionless roll damping derivatives used in Eq. (2.41) are

$$C_{y_p} = \frac{\partial C_y}{\partial(pb/2V)} = 0 \quad \text{(by assumption)}$$

$$C_{\ell_p} = \frac{\partial C_\ell}{\partial(pb/2V)} \qquad (2.42)$$

$$C_{n_p} = \frac{\partial C_n}{\partial(pb/2V)}$$

These derivatives are with respect to the dimensionless roll rate, $pb/2V$, which is commonly referred to as the airplane's roll helix angle. This term is a measure of the wing tip's helical trajectory during a pure roll maneuver. From an aerodynamic standpoint, $pb/2V$ represents the induced value of antisymmetric angle of attack at the $y = b/2$ wing tip. The Y_p derivative, like the Y_r derivative, is usually neglected. Because the predominant item in the preceding derivatives is due to the induced aerodynamics on the wing from the antisymmetric roll helix angle influence, the ensuing discussion will focus on the wing contribution to L_p and N_p.

The roll rate is assumed about the x body axis as shown in Fig. 2.11. From a practical standpoint, this assumption implies that angle of attack (which is producing C_L) is below stall, and the aerodynamics is governed by principles of linear superposition. As indicated in Fig. 2.11, the positive roll rate induces a positive angle of attack at the right-wing tip of $\Delta\alpha(b/2) = pb/2V$. Because roll rate is occurring about the x body axis, the induced angle of attack varies linearly from $\Delta\alpha = +pb/2V$ at the right-wing tip to $-pb/2V$ at the left-wing tip. The antisymmetric induced angle distribution will be considered for its influence on the derivatives.

The dimensionless derivative, C_{n_p}, is a cross derivative that acts to couple yaw moment with roll rate. Another cross derivative mentioned earlier, C_{ℓ_r}, serves in the opposite manner by cross-coupling roll moment with yaw rate. These two terms provide a rationale for lateral-directional airframe dynamics being a coupled interaction, e.g., Dutch-roll motion.

The C_{n_p} derivative is modeled by considering the antisymmetric drag developed by the wing during pure rolling. Both the profile and induced drag distributions are altered by positive roll rate. The angle increase on the right-wing panel during a positive roll maneuver will increase the profile drag on that panel thereby creating a positive yawing moment. Likewise, the left-wing panel will experience a decrease

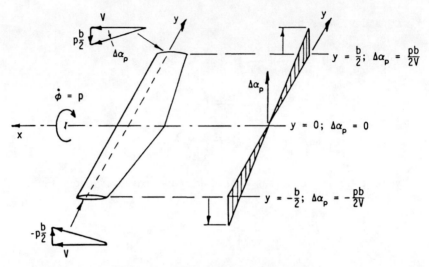

Fig. 2.11 Airplane wing in rolling flight.

in profile drag, which also adds to the positive yawing moment. However, the positive contribution to C_{n_p} from profile drag is normally overshadowed by the induced drag influence and, consequently, is frequently neglected during derivative estimation.

The yawing moment due to induced drag changes may be estimated by considering the wing lift as being inclined forward on the right-wing panel and aft on the left panel due to the roll induced $\Delta\alpha$, which results in a negative value. The yawing moment contribution may be expressed as a function of wing (airplane) lift coefficient. An approximation, based on strip-loading theory applied to an untapered wing with an elliptical span loading is given by

$$C_{n_p} = -(C_L/8) \tag{2.43}$$

The C_{ℓ_p} derivative, known as roll damping, is extremely important when considering pure roll maneuvering because a steady roll rate due to lateral control occurs when the moment produced by roll damping balances the applied roll moment due to lateral control input from ailerons or spoilers. A roll helix angle capability, $pb/2V$, on the order of 0.07–0.09 is frequently considered as a design goal by the aerodynamicist. The upper bound on desired roll helix angle (0.09 rad = 5.2 deg) lends credence to the use of linear superposition methods when modeling the influence of roll damping during flight at low angles of attack, i.e., unstalled flight.

The spanwise antisymmetric wing angle-of-attack variation induced by roll helix angle $\Delta\alpha_p$ results in the development of an antisymmetric section lift coefficient perturbation Δc_ℓ on the wing, as can be seen in Fig. 2.12. The antisymmetric loading produces no change in total wing lift coefficient ΔC_L, while producing a wing roll moment coefficient ΔC_ℓ.

Note: The use of a lower case c to describe a section lift coefficient Δc_ℓ, should not be confused with the use of an upper case C to describe a total wing rolling-moment coefficient ΔC_ℓ.

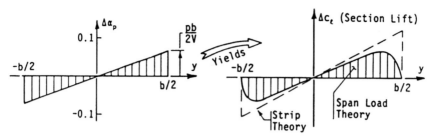

Fig. 2.12 Span load due to rolling flight.

The C_{l_p} derivative may be estimated using available span load methods that provide the section lift coefficient distribution by an evaluation of the integral expression

$$C_{l_p} = -\frac{b}{2S} \int_0^1 \Delta c_l(\eta) c(\eta) \eta \, d\eta \qquad (2.44)$$

where

$\Delta c_l(\eta)$ = wing sectional lift coefficient spanwise variation due to $\Delta \alpha_p$
$\Delta \alpha_p$ = induced antisymmetric angle of attack due to unit value of $pb/2V$
c = local wing chord, $c(\eta)$, ft
b = wing span, ft
S = wing area, ft^2
η = dimensionless lateral dimension, $(2y/b)$

Equation (2.44) can be approximated using strip theory, which assumes that the local Δc_l is determined solely by the local wing lift-curve slope and angle of attack. This assumption neglects the three-dimensional influence from span loadings on other portions of the wing. The method is convenient for providing an order of magnitude value for the derivative, but is not considered as an accurate approach. In addition, the technique does not account for effects due to wing aspect ratio, wing sweep angle, and Mach number. The strip-theory approximation alters Eq. (2.44) to

$$C_{l_p} = -\frac{b}{2S} \int_0^1 \left(\frac{\partial c_l}{\partial \alpha}\right)_{\text{wing}} c(\eta) \eta^2 \, d\eta \qquad (2.45)$$

where Δc_l has been replaced by $[\eta(\partial c_l/\partial \alpha)_{\text{wing}}]$. Figure 2.12 also illustrates the distinction between a strip-theory solution and more exact span load models. The strip-theory approach would show sectional loadings at the wing tips, which does not agree with the physical boundary condition of no pressure difference at the wing tips.

Example 2.3: Estimation of Wing Roll Damping Derivative

Elementary wing strip theory will be used to estimate C_{l_p} on a straight tapered wing in incompressible flow. Assumptions and definitions include the following.

1) Section lift-curve slope is $(\partial c_\ell/\partial \alpha)_{\text{wing}} = 2\pi$ rad^{-1}.
2) Wing taper ratio $\lambda = c_t/c_r$ where c_t and c_r are the wing tip and root chords, respectively.
3) Expressions for wing area and chord, from Chapter 1, are

$$S = 0.5bc_r(\lambda + 1)$$

$$c = c_r[1 + \eta(\lambda - 1)]$$

Substitution into Eq. (2.45) gives

$$C_{\ell_p} = -\frac{1}{(1+\lambda)} \int_0^1 (2\pi)\eta^2[1 + \eta(\lambda - 1)]d\eta$$

which simplifies to

$$C_{\ell_p} = -\frac{\pi(1 + 3\lambda)}{6(1 + \lambda)} \tag{2.46}$$

An approximate estimate for the roll damping derivative of a constant chord wing is

$$C_{\ell_p} = -\pi/3$$

The preceding estimates of the roll damping derivative provide an order of magnitude value for the term. Normally, the strip-theory approach overestimates the value. Initial estimations for the damping derivative during preliminary design stages would employ empirical methods such as those available from the U.S. Air Force DATCOM[6] or the literature. The determination of airframe air loads by computer modeling is another approach for estimating the damping derivative. Experimental determinations require special wind-tunnel configurations, a technique usually reserved for evaluations at high angles of attack when the behavior can be highly nonlinear.

Representative airplane values for the roll-related damping derivatives, C_{n_p} and C_{ℓ_p}, are shown in Table 2.4 based on information from Refs. 2 and 3. The C_{y_p} is considered as negligible and is not listed.

Table 2.4 Airframe derivatives due to roll rate

Airframe	Mach no.	Alt.	C_{n_p}, rad^{-1}	C_{ℓ_p}, rad^{-1}
Boeing 747	0.25	S.L.[a]	−0.12	−0.45
Convair 880	0.25	S.L.	−0.05	−0.38
Lockheed Jetstar	0.25	S.L.	−0.12	−0.37
Lockheed F-104A	0.25	S.L.	−0.14	−0.29
LTV A-7A	0.30	S.L.	−0.01[b]	−0.25
Douglas A-4D	0.40	All	+0.02[b]	−0.26

[a] Sea level.
[b] C_{n_p} derivative is nearly zero or slightly positive in value; presumably the profile drag and tail contributions overcame the lift coefficient dependence for these configurations.

2.3.4 Derivatives Due to Lateral Control

This section will treat the lateral control derivatives because of their relationship to a subsequent span load theory discussion in Sec. 2.5. The dimensional lateral control derivatives are defined relative to the lateral control deflection δ as

$$Y_\delta = (QS/m)C_{y_\delta}$$
$$L_\delta = (QSb/I_x)C_{\ell_\delta} \qquad (2.47)$$
$$N_\delta = (QSb/I_z)C_{n_\delta}$$

By convention, the dimensionless lateral control derivatives used in Eq. (2.47) are

$$C_{y_\delta} = \frac{\partial C_y}{\partial \delta}$$
$$C_{\ell_\delta} = \frac{\partial C_\ell}{\partial \delta} \qquad (2.48)$$
$$C_{n_\delta} = \frac{\partial C_n}{\partial \delta}$$

Lateral control may be achieved by many methods including wing warping as used by the Wright brothers during their first flight demonstration at Kitty Hawk, North Carolina, in December 1903. Contemporary lateral control configurations include the following.

1) Ailerons located on the wing trailing edge act as antisymmetric deflected plain flaps. They develop lift due to a form of variable camber.

2) Spoilers located on the wing upper surface are deflected upward on one side to produce a rolling moment by the local lift reduction. Like the aileron, they act aerodynamically as a variable camber device. Spoilers have the advantage of not introducting an excessive torsional moment into the wing structure when deflected, which makes their consideration desirable as a design option for a relatively elastic wing. Symmetric deflection of wing spoilers has an added benefit by their action as speedbrakes. It is possible to reduce the change in aircraft pitching moment due to spoiler deflection, considered as an undesirable cross-coupling effect, by the proper sizing and lateral placement of the spoilers on the wing during the design process.

Positive aileron deflection is defined as trailing edge up on the right wing in order to develop a positive rolling moment, cf. Fig. 1.1. The sign convention used to describe lateral control deflections is not standardized in the literature.

The Y_δ lateral control derivative will normally be neglected when the control is solely by aileron deflection. However, spoilers on swept-back wing aircraft often have a Y_δ derivative. This derivative is usually not considered of great import when conducting flight dynamic analyses.

The N_δ derivative may be visualized as a yawing moment reaction due to the development of an antisymmetric induced drag on the wing. For positive aileron deflection, the right- and left-wing panels experience a decrease and increase, respectively, in both span load and induced drag. The antisymmetric induced drag change would normally result in a negative yawing moment, which is described

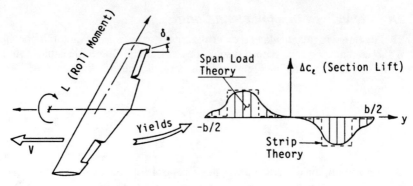

Fig. 2.13   Span load due to aileron control.

as an adverse-yaw effect. With spoiler roll control, the change in profile drag can overshadow the induced drag influence and provide a pro-yaw effect. Configuration alterations to the ailerons may be made to counter the adverse-yaw tendency of the N_δ derivative by creating an antisymmetric profile drag using features such as vertically offset hinge lines, control nose reshaping, etc. In practice, the experienced pilot adjusts to the actual aircraft's lateral control yaw moment behavior by intuitively introducing slight compensating rudder deflections. Alternatively, an aileron–rudder interconnect (either mechanical or electrical) can be introduced into the control system design if the N_δ derivative were ill behaved. Many pilots of high-performance jet, combat aircraft fly feet-on-the-floor except when needing to develop sideslip for crosswind landings or yawing moment for engine-out flight.

Unlike elevator control where a deflection results in a change to the flight trim condition, lateral control is primarily a rate producing device. Figure 2.13 illustrates the roll moment due to an antisymmetric span load caused by aileron control deflection. The dashed line represents a span load estimation using a strip-theory approach, where as stated earlier, every wing chordwise section is assumed to behave two dimensionally. A representative portrayal of a span load distribution using three-dimensional theory is shown as the curved, solid line in order to emphasize that the span load distribution involves aerodynamic cross talk from all regions of the wing. The result of wing theory is to smooth the load distribution and provide section air loads at the wing tip and wing root regions even though the plain flap device (i.e., aileron) is located elsewhere. It is for these reasons that the design of an optimum spanwise location for an aileron system to produce a given amount of roll control usually does not include the aileron extending to the wing tip.

The following example is given to illustrate the application of strip theory for estimating roll control effectiveness. In general, estimations by this approach will be high in magnitude as was the case in Example 2.3 when estimating roll damping.

Example 2.4:   Estimation of Aileron Roll Control

Like Example 2.3, strip theory will be used to make an estimate of C_{ℓ_δ} on an unswept, straight tapered wing in incompressible flow. The aileron will be assumed as a plain flap device where wing taper ratio $\lambda = c_t/c_r = 0.4$, $E = c_F/c = 0.20$ (hinge line on 80% chord), $\eta_i = 0.6$ (inboard end of aileron), and

AERODYNAMIC PRINCIPLES

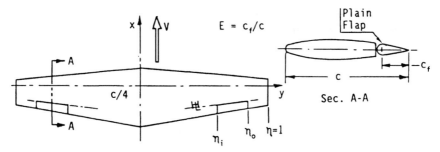

Fig. 2.14 Aileron configuration, Example 2.4.

$\eta_o = 0.8$ (outboard end of aileron). A sketch of the configuration is shown in Fig. 2.14.

Using thin airfoil theory (e.g., Glauert[11]), the section lift coefficient due to plain flap deflection is given by

$$\Delta c_\ell = \left(\frac{\partial c_\ell}{\partial \delta}\right)\delta$$

where

$$\left(\frac{\partial c_\ell}{\partial \delta}\right) = 2[\cos^{-1}(1 - 2E) + 2\sqrt{E(1-E)}]$$

A slight variation of Eq. (2.44) allows the definition of the lateral control roll derivative to be stated as

$$C_{\ell_\delta} = -\frac{b}{2S}\int_0^1 (\Delta c_\ell)_{\text{ail}} c(\eta)\eta \, d\eta$$

The section lift coefficient in this integral expression would be obtained by using a three-dimensional span load theory for a unit aileron deflection imposed as an initial condition. When strip theory is used for an approximate estimate, the preceding integral simplifies for the tapered wing to

$$C_{\ell_\delta} = -\frac{b}{2S}\int_{\eta_i}^{\eta_o} \left(\frac{\partial c_\ell}{\partial \delta}\right)_{\text{ail}} c(\eta)\eta \, d\eta$$

For the configuration of this example, evaluation of the lateral-control derivative using the positive deflection sign convention yields

$$C_{\ell_\delta} = +0.199 \text{ rad}^{-1}$$

Likewise, an estimate of the roll damping derivative for this example using Eq. (2.46) is

$$C_{\ell_p} = -0.823 \text{ rad}^{-1}$$

If maximum aileron deflection (for linear control) were ±15 deg (±0.26 rad), then an estimate of the maximum roll moment coefficient due to aileron control would be

$$(C_\ell)_{\text{max ail}} = +0.052$$

If a pure roll rate motion were considered, the equilibrium roll helix angle after all startup transients had decayed would be given by

$$0 = (C_\ell)_{\text{max ail}} + (C_{\ell_p})(pb/2V)$$

which states that by a strip-theory approach an estimate of the maximum roll helix angle for the assumed configuration would be

$$(pb/2V)_{\text{max}} = -(0.052)/(-0.823) = 0.063 \text{ rad}$$

In actual practice, the lateral-control derivatives would be angle-of-attack dependent when α was beyond the linear range. Roll control can become negligible or reverse when the aircraft approaches or is in the stall region. Also the presence of deflected wing flaps will alter the aircraft's roll capability. Wind-tunnel tests are normally conducted to evaluate the described control sensitivity terms.

2.4 Wing Theory

The wing theory to be described, will be oriented to the evaluation of wing dominated influences on selected stability and control derivatives. Thin-airfoil theories will be considered because the effects of thickness distributions on the flowfields will be of secondary import. The prime concern will be directed at the production of lift from vortices located on the wing chord plane. An initial review of two-dimensional, incompressible, thin-airfoil theory will be followed by modeling of a three-dimensional wing for span load determinations. Typical span load solutions will include additional loading due to α, i.e., lift-curve slope; wing contribution to pitching moment; roll damping from the wing; and aileron control effectiveness.

Subsonic compressibility effects at Mach numbers below M_{cr} will be modeled by use of the Prandtl–Glauert–Goethert transformation.[10] Inasmuch as the material of this text is introductory in nature, discussions relating to transonic, supersonic, and hypersonic flight regions are left as specialized applications and are not treated here. A very complete discussion on numerical modeling techniqes for span load evaluations is presented by Katz and Plotkin.[7]

2.4.1 Thin-Airfoil Theory

The airfoil will be assumed as a thin lifting surface representing a two-dimensional section in incompressible flow. Classical methods for finding the pressure distributions, section lift, and section pitching moments use techniques such as conformal transformations and integral relations, cf. Glauert.[11] The approach presented here is focused toward numerical methods using panel modeling in conjunction with control points that permit matching of tangential flow boundary conditions. These concepts will be applied later when considering three-dimensional wing theory.

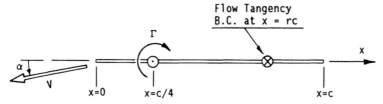

Fig. 2.15 Two-dimensional thin airfoil at angle of attack.

Consider a thin, uncambered airfoil of chord c represented by a vortex located at the aerodynamic center as shown in Fig. 2.15. The airfoil is assumed to be at an angle of attack α. Under these circumstances, thin-airfoil theory states that the section lift coefficient is

$$c_\ell = \frac{\ell}{Qc} = \left(\frac{\partial c_\ell}{\partial \alpha}\right)\alpha = (2\pi)\alpha \qquad (2.49)$$

where

$\partial c_\ell/\partial \alpha$ = ideal incompressible lift-curve slope, 2π, rad^{-1}
ℓ = running span load, lb-ft^{-1}
Q = dynamic pressure = $0.5\rho V^2$, lb-ft^{-2}
c = section chord, ft

The thin airfoil at angle of attack has been represented by a single vortex of strength Γ located at the aerodynamic center ($x = c/4$) as an approximation. The vortex strength Γ is related to the total air load (per unit span) by the Kutta–Joukowsky law[10] as

$$\ell = \rho V \Gamma \qquad (2.50)$$

The theoretical value of section lift depends on finding the vorticity distribution along the airfoil's chord that satisfies the tangential flow boundary condition everywhere along the airfoil and the Kutta condition at the trailing edge. For the situation as shown in Fig. 2.15, where the vorticity distribution is replaced by a concentrated vortex at the aerodynamic center, tangential flow can be satisfied at only one place on the airfoil. Consider that there is a location, $x = rc$, where tangential flow is satisfied. Then at that point on the airfoil,

$$\alpha = (\Delta w/V) \qquad (2.51)$$

where

Δw = induced downwash at $x = rc$ due to vortex Γ
$(\Delta w/V)$ = induced downwash angle due to vortex Γ

Application of the Biot–Savart law[10] for a two-dimensional vortex combined with the flow tangency boundary condition leads to

$$\left(\frac{\Delta w}{V}\right) = \frac{\Gamma}{2\pi V c(r - 0.25)} \qquad (2.52)$$

Substitution of the information from Eqs. (2.49–2.51) into Eq. (2.52) permits satisfaction of the tangential flow boundary condition to be satisfied at $r = 0.75$.

Note: The rearward aerodynamic center at the 0.75 wing chord is a control point for flow tangency when lift due to angle of attack is modeled by a line vortex located at the 0.25 wing chord (a.c.).

The term rearward a.c. is described in terms of both steady and unsteady aerodynamics by Fung.[12] It should be carefully noted that representation of a continuous vorticity function on an airfoil by a single bound line vortex can have the flow tangency boundary condition satisfied at only one place along the chord.

The airfoil model could be improved by considering n chordwise panels each of length (c/n). The discrete vortex elements located at the a.c. for each of the n panels would have their strengths determined by satisfying flow tangency at each of the n control points. The final result of using the panel method as a numerical approximation for the additional loading case will be that the total loading continues to act at the quarter chord (a.c.) and the section lift-curve slope remains in accord with Eq. (2.49).

Example 2.5: Model a Thin Airfoil by Two Panels

Assume that the thin airfoil is represented by two adjacent panels, each of chord $c/2$. A line vortex will be considered as acting at the local a.c. of each panel. Figure 2.16 shows the arrangement.

The two vortices, Γ_1 and Γ_2, are located at $x = c/8$ and $5c/8$, respectively. Two control points are located at the rearward a.c. of each panel, i.e., $x = 3c/8$ and $7c/8$. At the first rearward center (control point 1), the total induced downwash is related to the angle of attack at point 1 by

$$\alpha_1 = (\Delta w/V)_{11} + (\Delta w/V)_{12}$$

where $(\Delta w/V)_{ij}$ is the downwash at i due to the vortex at j. Typically, $(\Delta w/V)_{11}$ can be expressed as

$$\left(\frac{\Delta w}{V}\right)_{11} = \left(\frac{1}{\pi c}\right)\left(\frac{2\Gamma_1}{V}\right) = \left(\frac{1}{\pi c}\right)\left(\frac{\ell}{Q}\right)_1$$

If $(1/\pi c)$ were defined as an influence coefficient a_{11}, and the remaining influence coefficients determined, the resulting two simultaneous equations with two unknowns could then be expressed as

$$[A]\{\ell/Q\} = \{\alpha\} \qquad (2.53)$$

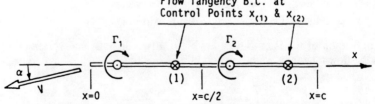

Fig. 2.16 Two panel thin-airfoil model at angle of attack.

or

$$(1/\pi c)\begin{bmatrix} 1 & -1 \\ 1/3 & 1 \end{bmatrix}\begin{Bmatrix} (\ell/Q)_1 \\ (\ell/Q)_2 \end{Bmatrix} = \begin{Bmatrix} \alpha_1 \\ \alpha_2 \end{Bmatrix}$$

For the case of additional loading, $\alpha_1 = \alpha_2 = \alpha = 1.0$ rad, the air loadings can be found by premultiplying Eq. (2.53) by $[A]^{-1}$, the inverse of matrix $[A]$, to give

$$(\ell/Q)_1 = (3/2)\pi c$$

$$(\ell/Q)_2 = (1/2)\pi c$$

The total section loading is given by

$$c_\ell c = (\ell/Q)_1 + (\ell/Q)_2 = 2\pi c$$

Therefore, the section lift-curve slope, $(\partial c_\ell/\partial \alpha)$, remains at the familiar value of 2π when using a two-panel modeling approach. The added information obtained from the example is in the form of more detail on the distribution of section loading. Moments taken about the leading edge due to the air loadings $(\ell/Q)_1$ and $(\ell/Q)_2$ considered acting at $x = c/8$ and $5c/8$, respectively, will confirm that the centroid of air loading due to angle of attack, the a.c., remains at the $c/4$ location.

Key factors to be noted from the two-dimensional thin-airfoil example include the following.

1) An airfoil experiencing pure angle of attack as an input can be approximated by placing a single vortex at the $c/4$ a.c. location and adjusting the vortex strength to satisfy the tangential flow boundary condition at the $3c/4$ rearward a.c.

2) Panel modeling can provide more information as to the variation in chordwise aerodynamic loading by the numerical approximation of Eq. (2.53), where each panel's control point is at the panel's rearward a.c.

3) Increasing the number of chordwise panels improves the accuracy of the chordwise loading, and in the limit as the number of panels tends to infinity, the loading will be recognized as the pressure distribution.

4) The effects of airfoil camber can be modeled by using the slopes of the camber line at each of the control points as input values for the α column vector in Eq. (2.53). Numerical accuracy will be improved by increasing the number of panels, as one would expect.

2.4.2 Three-Dimensional Wing Theory

Three-dimensional wing span load determinations will be modeled by placing horseshoe vortices over the wing surface in contrast to the thin-airfoil (two-dimensional) model where a line vortex on the airfoil surface extended laterally from plus to minus infinity. Both modeling concepts are in accord with the Helmholtz theorem,[10] which states the following.

1) The strength of a vortex filament is constant everywhere along its length.

2) A vortex filament cannot end in a fluid; it must extend to the boundaries of the fluid, go to infinity, or form a closed path.

3) In the absence of rotational external forces, a fluid that is initially irrotational remains irrotational.

In the three-dimensional wing situation, the vortex will be considered as starting from infinity downstream, proceeding upstream as a free vortex aligned with the

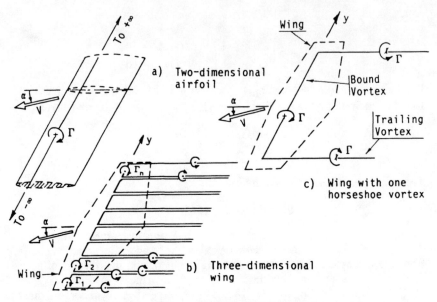

Fig. 2.17 Vortex models for airfoils and wings.

airstream, becoming attached to the wing surface in the form of a bound vortex, then departing from the wing and proceeding downstream aligned again with airstream as a free vortex. The term horseshoe vortex is used to describe the form of the vortex configuration.

Figure 2.17 illustrates the difference in modeling concepts between an airfoil and a wing. All vortex arrangements agree with the Helmholtz theorem as just stated. In Fig. 2.17a, the airfoil section has the same aerodynamic properties everywhere along its lateral dimension and the vortex extends to the fluid boundaries at plus and minus infinity. In Fig. 2.17b, the wing lift development is shown as due to a series of horseshoe vortices laid adjacent to each other in a pattern between the wing tips. The vortex portions of the horseshoe array individually develop lift as a consequence of the Kutta–Joukowsky law [Eq. (2.50)], yielding a total wing lift as the sum of the individual contributions.

Figure 2.17c is an approximate representation of the wing by a single horseshoe vortex where the span of the bound portion is less than the wing span. It should be apparent that a single horseshoe vortex, although satisfactory for an elementary understanding of the development of wing lift, does not allow for spanwise variations in load distribution. The single set of trailing vortices may be viewed as a result of the numerous free vortices, as shown in Fig. 2.17b, coalescing into a pair of trailing vortices. The single, trailing vortex pair model can be seen in nature when observing the vapor trails behind a high-flying aircraft.

The downwash field induced by the elements of the horseshoe vortices may be determined for the purpose of matching tangential flow boundary conditions on the wing in much the same manner as was shown in the thin-airfoil modeling. The downwash can be found using the Biot–Savart law.[10] Figure 2.18 illustrates the concept for an application of the Biot–Savart law to a finite length vortex, i.e., a vortex element of strength Γ induces a velocity Δw_p at a point p located at a

AERODYNAMIC PRINCIPLES

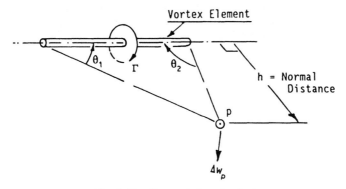

Fig. 2.18 Vortex induced velocity.

distance h from the axis of the vortex element by

$$\Delta w_p = \Gamma(\cos\theta_1 + \cos\theta_2)/(4\pi h) \quad (2.54)$$

where the induced velocity acts normal to the plane defined by the axis of the vortex and the normal distance h.

The width of the trailing vortices in the single horseshoe is usually considered to be between 0.8 and 0.9 of the wing span. However, for purposes of modeling, the downwash flow induced by the single horseshoe vortex representation will be assumed to satisfy flow tangency at a rearward a.c. (control point) located at the $0.75c$ position on the wing centerline whereas the bound vortex will be assumed as positioned on the wing's unswept $0.25c$ line. The span of the trailing vortices can be estimated by applying the Biot–Savart law to calculate the downwash velocity at the control point and relating the vortex strength to the total wing lift coefficient C_L that has been found by more accurate modeling procedures. Example 2.6 will illustrate the concept.

Example 2.6: Estimate Horseshoe Vortex Width

Consider an unswept, constant chord wing of AR = 8 in incompressible flow that is experiencing additional type of lift due to constant angle of attack α. More accurate wing theory[11] provides a lift-curve slope estimate for this wing of $\partial C_L/\partial \alpha = 4.84$ rad^{-1}. The wing will be represented by a horseshoe vortex system, cf. Fig. 2.19.

Total lift due to the bound vortex, using Eq. (2.50), is

$$L = \rho V \Gamma(\eta b)$$

which establishes the wing lift coefficient as

$$C_L = C_{L_\alpha}\alpha = L/(QS) = (2\Gamma/V)(\eta/c) \quad (2.55)$$

because wing area for the constant chord wing is $S = bc$.

Next, relate angle of attack α to $\Sigma \Delta w/V$ at the control point to satisfy the flow-tangency boundary condition. The control point is located at the rearward a.c. on the wing centerline by a symmetry assumption. The Biot–Savart law is then applied to find the contributions to vortex induced downwash from the three elements of the horseshoe vortex.

$$\left(\frac{\Delta w}{V}\right)_{bd} = \left(\frac{\Gamma}{4\pi V}\right)\left(\frac{1}{x}\right)\left(\frac{2y}{\sqrt{x^2+y^2}}\right)$$

$$\left(\frac{\Delta w}{V}\right)_{tr} = \left(\frac{\Gamma}{4\pi V}\right)\left(\frac{1}{y}\right)\left(1+\frac{x}{\sqrt{x^2+y^2}}\right)$$

and

$$\frac{\Sigma \Delta w}{V} = \left(\frac{2\Gamma}{V}\right)\left(\frac{1}{4\pi}\right)\left(\frac{x+\sqrt{x^2+y^2}}{xy}\right)$$

After replacing $\Sigma \Delta w/V$ by α, using Eq. (2.55) to find $(2\Gamma/V)$ and simplifying, one finds that

$$1 = \left(\frac{C_{L_\alpha}}{4\pi}\right)\left(\frac{c}{\eta}\right)\left(\frac{x+\sqrt{x^2+y^2}}{xy}\right) \tag{2.56}$$

Recognizing from Fig. 2.19 that $x = c/2$, $y = \eta b/2$, and $b/c = \text{AR}$, Eq. (2.56) may be solved for the vortex width on the example wing using $C_{L_\alpha} = 4.84 \text{ rad}^{-1}$ and AR = 8. The solution for the trailing vortex width is $0.887b$.

Conclude from this example that a single horseshoe vortex model provides a first approximation for describing the development of wing lift due to angle of attack. The principle used in this example can be extended to more detailed wing span load determinations involving the matching of tangential flow boundary conditions at control points to establish the strengths of vortex elements.

A modified Weissinger approach will be used to improve the modeling of a straight tapered wing having sweepback. As shown in Fig. 2.20, the wing lift production is represented by many horseshoe vortex elements with the bound portions of the horseshoes aligned with the local $c/4$ line. Control points are located at the local $3c/4$ points midway between each of the trailing vortices.

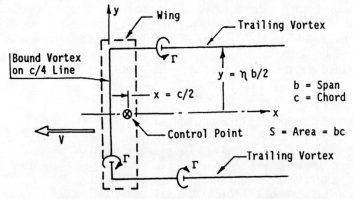

Fig. 2.19 Horseshoe vortex model, Example 2.6.

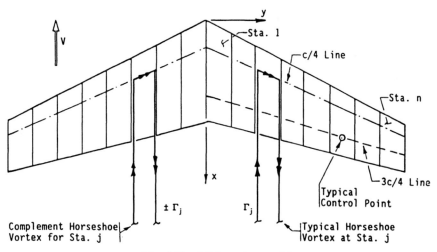

Fig. 2.20 Lifting wing model.

The strength of the individual vortices extending over the wing span is established by satisfying the flow-tangency requirement at each of the control points. Equation (2.53) is applicable to the three-dimensional wing when suitable modifications are made to the aerodynamic influence coefficients. Consider the following relation:

$$a_{i1}(\ell/Q)_1 + \cdots + a_{ij}(\ell/Q)_j + \cdots = \alpha_i$$

for i and $j = 1, 2, \ldots, n$. The n simultaneous equations can be stated in matrix form as

$$[A]\{\ell/Q\} = \{\alpha\} \qquad (2.57)$$

where

a_{ij} = element (ij) in the $[A]$ matrix, the downwash induced angle at control point i due to (ℓ/Q) from vortex j
$(\ell/Q)_j$ = span load due to vortex $j = 2\Gamma_j/V$
α_i = angle of attack at control point for spanwise station i

The off-diagonal terms in the aerodynamic influence coefficient matrix provide a downwash input at one station from lift developed at another station. If the cross talk influence were neglected, the $[A]$ matrix would become diagonal and the resulting span load solution would be considered as the result of strip theory. The strip-theory method for determining wing lift production is not accurate and is mainly used for illustrative purposes.

The vortex model shown in Fig. 2.20 includes n equal-spaced elements covering the complete wing span. Although the actual wing shape and $\{\alpha\}$ distribution determines a practical value for n, experience has shown that wings with aspect ratios of 5 or more can be modeled reasonably well when n is 20 (10 per wing panel). If one had a concern for load accuracy in wing regions where spanwise load gradients were steep, such as the wing tip, vortex elements with shorter spans could be included in that region of the model. However, it is practical in numerical modeling to reduce the number of vortex elements by a factor of two based on the recognition that linear superposition prevails and loads can be due to either symmetric or antisymmetric $\{\alpha\}$ column vectors in Eq. (2.57). Symmetric $\{\alpha\}$ inputs would correspond to situations such as additional loading (i.e., constant angle α across the wing span) and geometric (built-in) twist distributions. Antisymmetric $\{\alpha\}$ inputs occur when solving span loads due to roll helix angle and aileron control inputs.

A continuous function, such as the wing lift distribution extending between $y = b/2$ and $-b/2$, can be represented as the sum of a symmetric and antisymmetric lift function. Consider $\ell(y)$ in the spanwise region of $-b/2 \leq y \leq b/2$. It can be restated as

$$\ell(y) = [0.5\ell(y) + 0.5\ell(y)] + [0.5\ell(-y) - 0.5\ell(-y)]$$
$$= 0.5[\ell(y) + \ell(-y)] + 0.5[\ell(y) - \ell(-y)]$$
$$= f_s(y) + f_a(y)$$

where

$f_s(y)$ = symmetric function, i.e., $f_s(-y)$
$f_a(y)$ = antisymmetric function, i.e., $-f_s(-y)$

The separation of the span load problem into symmetric and antisymmetric solutions can be seen by reconsidering Eq. (2.57) in a partitioned form that represents right- and left-hand wing panels. Then

$$\begin{bmatrix} [A_{11}] & [A_{12}] \\ [A_{21}] & [A_{22}] \end{bmatrix} \begin{Bmatrix} \{\ell/Q\}_{\text{RH}} \\ \{\ell/Q\}_{\text{LH}} \end{Bmatrix} = \begin{Bmatrix} \{\alpha\}_{\text{RH}} \\ \{\alpha\}_{\text{LH}} \end{Bmatrix}$$

where typically

$[A_{11}]$ = influence of $\{\ell/Q\}_{\text{RH}}$ on $\{\alpha\}_{\text{RH}}$
$[A_{12}]$ = influence of $\{\ell/Q\}_{\text{LH}}$ on $\{\alpha\}_{\text{RH}}$

It can be shown that for symmetric span load determinations, when

$$\{\alpha\}_{\text{RH}} = \{\alpha\}_{\text{LH}}$$

then $[A_{11}] = [A_{22}]$ and $[A_{12}] = [A_{21}]$. And for antisymmetric span load determinations, when

$$\{\alpha\}_{\text{LH}} = -\{\alpha\}_{\text{RH}}$$

then $[A_{22}] = -[A_{11}]$ and $[A_{21}] = -[A_{12}]$.

Based on these considerations, it is numerically convenient to solve the span load distribution on one wing panel for either a symmetric or antisymmetric condition. The aerodynamic matrix relation [Eq. (2.57)] will have an $[A]$ matrix for either the symmetric or antisymmetric form, i.e.,

$$[A]_s = [A]_{\text{RH}} + [A]_{\text{LH}}$$

$$[A]_a = [A]_{\text{RH}} - [A]_{\text{LH}}$$

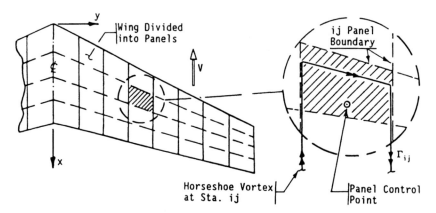

Fig. 2.21 Wing panel model.

The symmetric form of Eq. (2.57) will apply to span load solutions involving the determination of wing lift coefficient C_L and pitching moment coefficient C_m for use in longitudinal related flight analyses. Similarly, the antisymmetric form will apply to determinations of rolling moment coefficients C_ℓ in lateral-directional flight analyses.

The modified Weissinger approach may be altered to a panel method by dividing each spanwise section into a number of separate chordwise panels as shown in Fig. 2.21. The main benefit from a panel method is that information is obtained concerning the load distribution in both the spanwise and chordwise directions. The panel modeling approach has a direct analogy to the concept described previously in Example 2.5 for an airfoil modeled by multiple chordwise segments.

Subsonic compressibility effects for Mach numbers less than M_{cr} can be accommodated in the wing models described in this chapter by applying the Prandtl–Glauert–Goethert transformation.[10] The incompressible form of the aerodynamic influence coefficient matrix may be used as equivalent to the subsonic form if the x coordinate is altered by the transformation

$$x_M = \frac{x_0}{\sqrt{1 - M^2}} \qquad (2.58)$$

where

x_0 = physical x coordinate, as defined in Fig. 2.21
x_M = transformed x coordinate
M = freestream Mach number, $M < M_{cr} < 1$

The resulting air loads may be numerically integrated into proper dimensionless coefficient form using the original physical geometry based on the use of the x_0 coordinate system.

The modeling principles described in this section may be modified to include the presence of other physical features; e.g., the fuselage may be modeled by adding a three-dimensional vortex lattice panel array representation for the body's surface and prescribing flow tangency on the body as well as on the wing. Adding more control points increases the order of the matrix equation, a trade that increases solution complexity while in turn providing more information concerning the flowfield. However, the depth of panel modeling described in this chapter is adequate for the reader to develop an appreciation of the aerodynamic basics when estimating stability derivatives. The reader interested in more complete airframe modeling concepts will find the treatment by Katz and Plotkin[7] as informative.

2.5 Stability Derivatives Using Wing Theory

The subsonic wing theory described in Sec. 2.4 has been programmed using Basic as a computer language for application to a straight tapered wing. A listing of a program entitled SPANLD.BAS, which is suitable for use on a desktop computer, is provided in Appendix C. The program is menu driven for the input of wing geometry, subsonic Mach number, and selection of span load type. The following wing related stability derivatives will be briefly investigated to illustrate a few modeling principles: 1) symmetric span loading, including additional loading due to angle of attack and pitch damping, C_{m_q}, due to the wing, and 2) antisymmetric span loading, including roll damping, C_{ℓ_p}, due to the wing and roll control, C_{ℓ_δ}, due to a plain flap (aileron) deflection.

2.5.1 Additional Loading

The symmetric span load due solely to angle of attack is denoted as additional loading. The significant items of interest with respect to additional loading include the lift-curve slope and the location of the wing-alone neutral point. Total wing lift also arises due to contributions from geometric (built-in) twist, camber, and dead-weight induced twists.

The effect of including chordwise panels on the determination of lift-curve slope and wing neutral point location is shown in Fig. 2.22 for a series of similar wings having quarter-chord sweep angles of 0, 15, and 30 deg. Increasing the number of chordwise panels had little influence on the estimate of lift-curve slope (cf. Fig. 2.22a). The neutral point estimate showed convergence when two or more chordwise panels were considered in the model (cf. Fig. 2.22b). Note that the unswept wing had a neutral point that differed from $0.25\bar{c}$ (i.e., an airfoil's a.c. and a wing-alone neutral point are different). The effect of sweepback was both to decrease the lift-curve slope and to move the wing's neutral point aft relative to the unswept wing.

Subsonic compressibility effects upon wings with quarter-chord sweep angles of 0, 15, and 30 deg, using the Prandtl–Glauert–Goethert transformation rules, are shown in Fig. 2.23 by the variation of lift-curve slope with Mach number. The effect of sweepback is to reduce the influence of compressibility on the relative increase of lift-curve slope with Mach number. An approximate estimate for the variation of subsonic lift-curve slope is given by

$$(C_{L_\alpha})_M = (C_{L_\alpha})_{M=0} / \left[1 - (M \cos \Lambda_{c/4})^2\right]^{\frac{1}{2}} \quad (2.59)$$

The model used for a span load solution (cf. Appendix C) does not account for local areas of supersonic flow on the wing. Consequently, the values of C_{L_α} shown in Fig. 2.23 do not reflect proper behavior in the transonic region as would normally be realized from wind-tunnel testing of an aircraft model or found from flight tests of an aircraft.

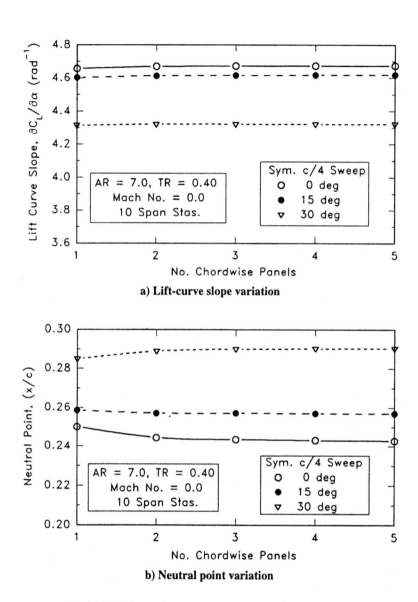

a) Lift-curve slope variation

b) Neutral point variation

Fig. 2.22   Effects of panel modeling on additional loading.

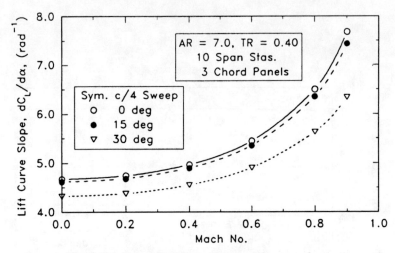

Fig. 2.23 Compressibility effects on lift-curve slope.

The effect of sweepback on the spanwise distribution of additional loading is to increase the sectional loading in the outboard regions. Conversely, a swept-forward wing will tend to have increased wing loading due to angle of attack in the inboard wing regions (cf. Fig. 2.24). The ordinate in Fig. 2.24 expresses the section loading ($c_\ell c$) normalized with respect to $C_L c_{ave}$, which implies that the area under the curve is unity. Use of the normalized sectional loading is a common method for displaying loading trends on a wing.

Without aerodynamic modifications by choice of airfoil, camber, thickness, and built-in twist, the swept-back wing frequently tends initially to develop separated (stalled) flow in the outboard wing regions, which leads to a pitch-up stability change by the wing at high angles of attack. However, placement of the horizontal

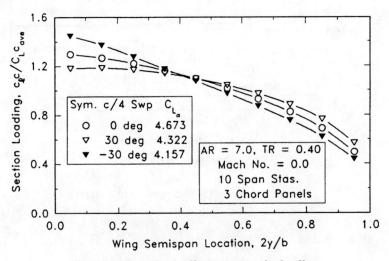

Fig. 2.24 Sweepback effects on spanwise loading.

tail relative to the wing often has a marked influence on the stall behavior of the complete airframe. Aerodynamics of the complete airframe in the neighborhood of stalled flight is highly nonlinear and span load modeling can at best only provide trends.

2.5.2 Pitch Damping Due to the Wing

In Sec. 2.2, the pitch damping that arose from an aircraft in circular flight was described as aerodynamically equivalent to a curved aircraft in rectilinear flight where the radius of the circular arc was related to pitch rate and airframe velocity by $R = V/q$. The wing contribution to pitch damping may be modeled by considering the wing as having a circular arc camber distribution centered about the 0.25 MAC. When modeling the wing-induced aerodynamic contribution, a panel method is essential in order to reflect the chordwise variation of the tangential flow boundary condition. It will be found (cf. Appendix C) that the $\{\Delta\alpha\}$ input matrix terms for the span load problem per unit value of $(qc/2V)$ are given by

$$\Delta\alpha(i) = [X3(i) - X_{c/4}]/(c/2)$$

where

$X3(i)$ = location of the ith control point
$X_{c/4}$ = location of the wing's 0.25 MAC
c = wing MAC

At zero sweepback, it will be observed from Fig. 2.25 that the estimated value of wing contribution to C_{m_q} is not significantly different from the strip-theory assumption of $\Delta C_{m_q} = -\pi/4$, as mentioned in Example 2.2. The effect of wing sweep in either direction is to increase the wing pitch damping due to the wing panel control points being displaced farther fore and aft from the 0.25 MAC. But as pointed out in Sec. 2.2, the horizontal tail remains as the prime contributor to airframe pitch damping.

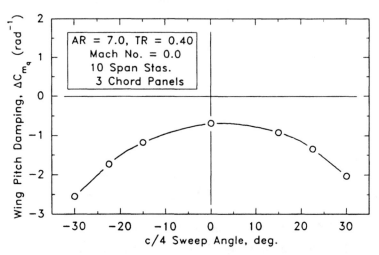

Fig. 2.25 Sweepback effects on wing pitch damping.

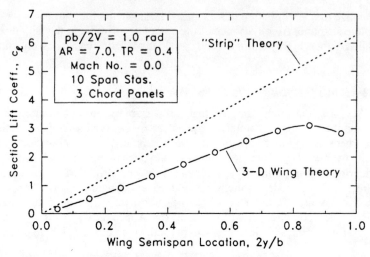

Fig. 2.26 Spanwise loading for roll damping.

2.5.3 Roll Damping Due to the Wing

The span load program of Appendix C was used to estimate the roll damping derivative for a number of cases. An incompressible solution for an unswept wing with aspect ratio of 7 and taper ratio of 0.4 provided a roll damping estimate of

$$C_{\ell_p} = -0.503 \text{ rad}^{-1}$$

The strip-theory estimate in Example 2.4 for C_{ℓ_p} was -0.823 rad^{-1}, an estimate that was 64% higher. However, strip theory does not consider wing AR, and its estimate would be closer to wing-theory results for higher ARs. Figure 2.26 presents a comparison of the spanwise variation of section lift coefficient between strip theory and the modified Weissinger span load model. It is especially evident in the wing tip region that the three-dimensional nature of span loading is not represented by the strip-theory approach and the use of strip theory would provide a nonconservative estimate of roll damping.

The modified Weissinger span load model makes possible estimates of subsonic compressibility effects on wing roll damping and allows consideration of variations in taper ratio, AR, and sweepback angle.

Figure 2.27 shows the effect of Mach number upon C_{ℓ_p} for wings with quarter-chord sweepback angles of 0 and 30 deg. The subsonic compressibility effect on roll damping is similar to that for additional loading. Although three chordwise panels were used in the evaluation shown in Fig. 2.27, the nature of the wing flow during pure roll makes possible reasonably good roll damping estimations using only a swept lifting vortex (modified Weissinger) approach.

2.5.4 Roll Control Due to Ailerons

The wing roll moment due to aileron control requires the use of panel modeling techniques. For instance, a 20% chord aileron could be modeled by a panel approach that involved five chordwise elements. The aileron, which may be

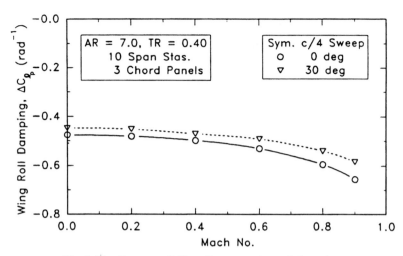

Fig. 2.27 Compressibility effects on wing roll damping.

considered as a plain flap control, can be modeled by satisfying tangential flow at the appropriate trailing-edge panels.

The program listed in Appendix C was used to investigate an optimum location for the spanwise placement of a 20% chord aileron on an unswept wing for the cases of ailerons having spans of 10 and 20% semispan, respectively. The design study is seen in Fig. 2.28 where the abscissa is the aileron's outboard tip location. It is important to note that, for the assumed wing-aileron examples, the aileron configuration with the outboard tip located at about 85% of the wing semispan provided the maximum amount of ideal roll control effectiveness. This observation implies 1) the aerodynamic influence of the control surface extends both inboard and outboard of its wing location and 2) an aileron extending to the tip cannot influence air loads where no wing surface exists.

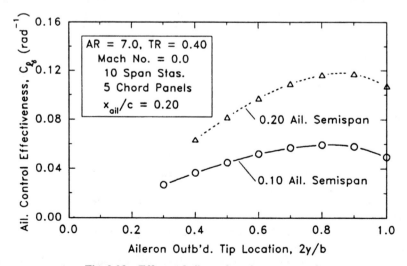

Fig. 2.28 Effects of aileron location on control.

2.5.5 Concluding Remarks

The illustrations showing the use of a wing-theory model to estimate wing dominated stability and control derivatives were applicable to a rigid airframe. In reality, the wing is an elastic structure that deflects in both bending and torsion under the presence of air loads. The influence of wing deflections on the aerodynamic solutions may be treated by modifying the span load model so as to include the effect of the structurally induced angles. One can describe the structural deflection angles due to air loads in a matrix operator sense by

$$\{\Delta\alpha\}_s = Q[S]\{\ell/Q\}$$

where

$\{\Delta\alpha\}_s$ = column vector of structural deflection angles
$[S]$ = matrix array of structural influence coefficients

It can be shown[13] that an equilibrium solution for the span load distribution where static aeroelastic influences are considered may be represented by altering Eq. (2.57) to include the structural deflections, i.e.,

$$[[A] - Q[S]]\{\ell/Q\} = \{\alpha\} \qquad (2.60)$$

The span load solution, $\{\ell/Q\}$, for the elastic wing may be found using Eq. (2.60) by premultiplying the rigid-wing input angle column vector $\{\alpha\}$ by the inverse of the $[[A] - Q[S]]$ matrix. Symmetric and antisymmetric solutions may be obtained by using the appropriate form of the aerodynamic influence coefficient matrix $[A]$. If the inverse of the matrix were singular, no span load solution would exist. Such a situation corresponds to a wing torsional divergence condition (cf. Fung[12]).

It should be noted that the rigid wing span load solution would be obtained by setting the dynamic pressure Q equal to zero in Eq. (2.60). Again, as in modeling the aerodynamic matrix $[A]$, a choice of modeling techniques may be used when establishing the structural influence matrix $[S]$. A metal wing with a spar-box structure can be modeled by a beam with an elastic axis whereas a composite wing structure might involve a more complex representation to include chordwise influences.

When making an accurate analysis of the complete airframe, the influence of body bending and torsion should be included in addition to accounting for the static aeroelastic nature of the horizontal and vertical tails. Airframes with relatively elastic airframes such as transports and jet bombers experience significant influences from static aeroelastic effects on the airframe's stability derivatives, such as the lift-curve slope and neutral point; damping in pitch, yaw, and roll; and control effectiveness terms.

References

[1]McRuer, D., Ashkenas, I., and Graham, D., *Aircraft Dynamics and Automatic Control*, Princeton Univ. Press, Princeton, NJ, 1973, pp. 273, 687–743.

[2]Heffley, R. K., and Jewell, W. F., "Aircraft Handling Qualities Data," NASA CR-2144, Dec. 1972.

[3]Teper, G. L., "Aircraft Stability and Control Data," Systems Control Technology, Inc., Rept. STI-TR 176-1, Hawthorne, CA, April 1969.

[4]Blakelock, J. H., *Automatic Control of Aircraft and Missiles*, 2nd ed., Wiley, New York, 1991, pp. 601–603.

[5]Nelson, R. C., *Flight Stability and Automatic Control*, McGraw–Hill, New York, 1989, pp. 250–260.

[6]Anon., "USAF Stability and Control DATCOM," Flight Control Div., Air Force Flight Dynamics Laboratory, Wright-Patterson AFB, OH, 1978.

[7]Katz, J., and Plotkin, A., *Low-Speed Aerodynamics, from Wing Theory to Panel Methods*, McGraw–Hill, New York, 1991, Chap. 12.

[8]Etkin, B., *Dynamics of Flight, Stability and Control*, 2nd ed., Wiley, New York, 1982, Chap. 5.

[9]Perkins, C., and Hage, R., *Airframe Performance, Stability and Control*, Wiley, New York, 1949, Chaps. 5–9.

[10]Kuethe, A. M., and Chow, C.-Y., *Foundations of Aerodynamics: Bases of Aerodynamic Design*, 4th ed., Wiley, New York, 1986, pp. 51–55, 275–283.

[11]Glauert, H., *The Elements of Aerofoil and Airscrew Theory*, 2nd ed., Cambridge Univ. Press, New York, 1959, pp. 87–93.

[12]Fung, Y. C., *The Theory of Aeroelasticity*, Wiley, New York, 1955, pp. 82, 126, 407.

[13]Schmidt, L. V., "Static Aeroelastic Effects upon Wing Span Loads," *AIAA 9th Applied Aerodynamics Conf.*, CP918, AIAA, Washington, DC, Sept. 1991.

Problems

2.1. Consider an aircraft with a straight-tapered wing defined by: S = wing area = 500 ft^2; S_H = horizontal tail area = 100 ft^2; AR = wing aspect ratio = 7.0; λ = wing taper ratio = 0.40; $(C_{L_\alpha})_W$ = wing lift-curve slope = 4.67 rad^{-1}; $(C_{L_\alpha})_H$ = tail lift-curve slope; = 3.00 rad^{-1}; ℓ_H = tail length = 19.0 ft; $\eta_H = Q_H/Q = 1.0$ (e.g., ideal jet aircraft); ϵ_α = wing downwash influence at tail = 0.30; and $(\Delta x/c)_{ac}$ = wing a.c. relative to the c.g. = +0.01. Find:

a) Complete aircraft (controls fixed) lift-curve slope.

b) $(\Delta x/c)_{np}$ for complete aircraft (controls fixed).

Remarks: Here $(\Delta x/c)_{np}$ is known as the static stability margin and defines the axial distance between the c.g. and the aircraft's neutral point, normalized with respect to the reference wing chord (assume $c = \bar{c}$ = MAC). The stability margin is frequently multiplied by 100 and then expressed as a percent margin.

2.2. Consider the aircraft of Problem 2.1 flying at $M = 0.7$ and at an altitude of 20,000 ft. The fuselage aft of the wing acts as a cantilevered beam with a body bending stiffness constant of $K = 1.4 \times 10^{-6}$ rad-lb^{-1} that describes the structurally induced change in horizontal tail incidence due to horizontal tail lift, i.e., $\Delta\alpha_H = -KL_H$ where L_H = horizontal tail lift force, lb. Find the influence of body bending on the static stability margin. Hint: Let $\alpha_H = (1 - \epsilon_\alpha)\alpha_W + \Delta\alpha_H$.

Remarks: In actual design practice, many terms that are present in the basic equations are considered relative to their being modified by the influence of static aeroelasticity when estimating stability margins. Some structurally dependent influences include: 1) wing lift-curve slope change, 2) wing aerodynamic center shift, 3) horizontal tail lift-curve slope change, and 4) body bending due to tail load.

2.3. Consider an aircraft with a constant chord wing flying in turning flight. Assume that the section lift coefficient is constant across the wing span. Using wing strip theory, show that the wing contribution to the C_{ℓ_r} stability derivative may be expressed as

$$(\Delta C_{\ell_r})_{\text{wing}} = +(C_L/3)$$

2.4. Consider the aircraft of Problem 2.3 when the section lift loading is elliptical. Using wing strip theory, show that the wing contribution to the C_{ℓ_r} stability derivative may be expressed as

$$(\Delta C_{\ell_r})_{\text{wing}} = +(C_L/4)$$

2.5. Consider the aircraft of Problem 2.3 when the section lift loading is elliptical. Using wing strip theory, show that the wing contribution to the C_{n_p} stability derivative may be expressed as

$$(\Delta C_{n_p})_{\text{wing}} = -(C_L/8)$$

2.6. Consider a thin airfoil of chord c, moving at a velocity V, while at a distance h above the ground. If the thin airfoil had a sectional lift-curve slope of 2π rad^{-1} when in free air, find a relation to describe the change in sectional lift-curve slope as the airfoil nears the ground, i.e., as a function of (c/h). Hints: Assume that the thin airfoil may be represented by a concentrated vortex of strength Γ located at $0.25c$ with flow tangency maintained at the $0.75c$ location on the airfoil. The presence of the ground requires an image airfoil with vortex strength of $-\Gamma$ located a distance $-h$ beneath the surface in order to maintain flow tangency at the ground surface.

2.7. Reconsider an airfoil in ground effect similar to Problem 2.6 using the two-panel approach of Example 2.5 to represent the airfoil. The image of the airfoil should also be a two-panel representation. Find:
   a) The increase in incompressible airfoil lift-curve slope relative to the free-air value when the airfoil is a distance $h = c/2$ above the ground.
   b) Compare results with those obtained in Problem 2.6.
   c) The corresponding a.c. location and compare with the location assumed in Problem 2.6.

Remarks: A logical extension of Problems 2.6 and 2.7 is to consider a complete aircraft in ground effect for both takeoff and landing configurations. From a historical perspective, it is interesting to note that there were accidents during the early 1900s when heavily laden test aircraft could take off but did not have sufficient propulsive power to accelerate and rise above ground effect. During an aircraft design cycle, it is quite common to conduct low-speed wind-tunnel tests on a model located near to a simulated ground plane in order to identify ground effects on lift and drag forces and on pitching moments.

2.8. A 10,000-lb executive jet with an unswept, straight-tapered wing made a low-altitude (50-ft) flyby over an idealized sea-level airport at an equivalent airspeed of $V_e = 100$ kn (168 ft-s^{-1}). The wing loading (W/S) was 40 lb-ft^{-2} and the wing AR was 7.0. Assume that the wing lift may be approximated by a single horseshoe vortex whose span is 0.90 of the wing span. Estimate the vertical trajectory of the trailing vortices at a fixed location along the runway. Hints: 1) Assume that the trailing vortices are acting as infinitely long vortices; i.e., the aircraft is at least 170 ft down the runway 1 s after passing by the fixed location. 2) Images of the trailing vortices are required in order to satisfy the boundary condition of no vertical velocity at ground level. 3) It is suggested that the vertical trajectory of the trailing vortices be estimated in 1-s intervals for 4 s.

2.9. Verify the following relation:

$$\int_0^1 \frac{c_\ell c}{C_L c_{\text{ave}}} \eta = 1$$

where η is the wing semispan coordinate $= 2y/b$.

3
Static Stability and Control

3.1 Background

Airframe static stability and control will be considered in this chapter. In addition to increasing an understanding of flight mechanics, some aspects of this topic will be seen to correlate with the zero-frequency case when flight dynamics problems are investigated (e.g., load factor due to longitudinal control). The airframe will normally be considered as being in static trim, a term implying zero acceleration in all six body axes (i.e., x, y, z, ϕ, θ, and ψ). Although a body may be in static equilibrium, the question still remains whether the trim condition is stable. Stability about a trim point is generally viewed relative to the body's response when a small perturbation displaces it from the equilibrium point. If the body returned to the trim point following a perturbation, then the trim condition would be described as a stable static equilibrium. Conversely, unstable static equilibrium corresponds to the body diverging from the trim point when slightly perturbed.

The following discussions on static stability and control will treat the longitudinal and lateral-directional axes as two separate topics because they are usually uncoupled in practice. The relations to be developed will use static stability derivatives on the assumption that static equilibrium of the airframe exists. Damping derivatives will be used in the developments when the control determination involves establishing steady motion rates such as in pitch-up or banked-turn maneuvers. Although dimensional stability derivatives may be used in the following derivations when the airframe is at a constant velocity, it is both customary and more convenient to use dimensionless coefficients when considering airframe static stability and control.

3.2 Longitudinal Stability and Control

Static stability associated with an aircraft's longitudinal axis normally is addressed by the sign of the C_{m_α} stability derivative. A minus sign implies that a perturbation increase in angle of attack from the flight trim setting will result in a negative pitching moment being developed, where $\Delta C_m = C_{m_\alpha} \Delta \alpha$ will be in a direction to restore the aircraft α back to α_{trim}. The degree of longitudinal static stability for an aircraft (without stability augmentation) will depend on the location of the aircraft c.g. relative to the neutral point, as described in Sec. 2.2.2 by Eq. (2.14).

There have been several high-performance aircraft whose designs included having their c.g.s located behind rather than ahead of the neutral point, e.g., the General Dynamics (now Lockheed–Martin) F-16 fighter and the Grumman (now Northrop–Grumman) X-29 swept-forward wing experimental aircraft. These aircraft were deliberately designed statically unstable in the classical sense and were influential in introducing the term of relaxed static stability into the stability and control vocabulary. Both aircraft proved acceptable with respect to flying qualities because their control systems included the use of multiply redundant, stability augmentation systems. This text will consider aircraft having the classical form of static stability, i.e., the c.g. forward of the neutral point.

A key design consideration is the fore and aft movement of the c.g. relative to the neutral point during all possible portions of the flight profile under various locations of crew, passengers, fuel, cargo, and ordnance. The forward c.g. limit typically places a requirement on the maximum longitudinal control available for a safe landing flare maneuver from an approach trim setting while the aft c.g. limit involves control sensitivity concerns, which may affect safe flying qualities. A rule of thumb frequently used is that the aft c.g. limit should have a 5% static margin relative to the neutral point.

Design problems associated with the c.g. location are legendary; two early examples are as follows.

1) An early U.S. Air Force jet fighter/trainer had many internal fuel tanks, which required careful fuel management by the pilot throughout the flight mission in order that the aircraft remain within acceptable c.g. boundaries.

2) An early supersonic jet bomber required careful fuel management during transition between subsonic and supersonic flight because of the marked shift in neutral point between the two flight regimes; i.e., a clue as to this effect is obtained from thin-airfoil theory, which states that the aerodynamic center is at the $c/4$ and $c/2$ locations for sub- and supersonic flow, respectively.

Aircraft longitudinal control brings many requirements into play. These requirements include the following.

1) The aircraft should be trimmable at a given value of C_L, c.g. location, and airframe configuration (e.g., flaps up, flaps down, flaps down with gear extended, etc.).

2) The flight velocity should increase by pilot application of nose-down ($+\delta$) longitudinal control and decrease by a reversal of the control direction. This requirement corresponds to the control needed to change C_L in steady flight. Whereas thrust changes affect rate of climb or descent, they usually are weakly coupled to aircraft C_L (or velocity) changes.

3) The aircraft must have a maneuvering capability to do either pull-ups or turns subject to staying within the maneuvering ($V-n$) envelope.

4) Recovery from transient maneuvers, such as incurred by aircraft wake and turbulent gust encounters, must be possible using longitudinal control.

Most general aviation aircraft (many Piper aircraft are exceptions) have a fixed horizontal tail surface with both trim and control being provided by the elevator. In contrast, high-performance aircraft frequently have a movable horizontal tail that can be used for both trim and control or, alternatively, solely for trim with control provided by the elevator surface. One reason for using a movable horizontal tail for trim purposes is to reduce the aircraft trim drag and thereby improve aircraft performance. The complexities of a movable horizontal tail are usually found to be unwarranted for use by general aviation aircraft.

3.2.1 Longitudinal Control for Velocity Change

The trim velocity of an aircraft determines the equilibrium lift coefficient, which in turn is established by the use of longitudinal control. There are two independent

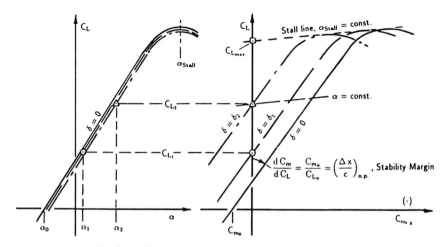

Fig. 3.1 Sketch of the C_m and α variation with C_L.

variables, α and δ, which are determined by satisfying equilibrium in lift and pitching moment for a given aircraft configuration. A graphical sketch (Fig. 3.1) shows the C_m and α variation with C_L for fixed values of longitudinal control as might be obtained from wind-tunnel tests on a scale model of the aircraft. The zero-lift pitching-moment coefficient, when control $\delta = 0$, is denoted as C_{m_0} in Fig. 3.1. The zero-lift intercept for a C_m–C_L curve corresponding to a positive trim C_L value must be a positive value for a statically stable aircraft. The $\alpha = $ const line on the C_m–C_L plot is shown for various control angles. It is not at a constant value of airplane C_L due to a ΔC_L term arising from $(C_{L_\delta}\Delta\delta)$. A stable stall break also is shown in Fig. 3.1, which implies that when the airplane α reaches and exceeds α_{stall}, a negative change in pitching moment occurs to lower the aircraft's nose for the initiation of stall recovery. A stall demonstration is a dynamic maneuver used in the determination of the minimum flying speed for a given airplane weight, configuration, and c.g. position.

It is customary in data plots of C_m vs C_L to show $(-)C_m$ toward the right, as in Fig. 3.1, a practice probably based on the convenience offered when graphically interpreting aircraft stability and control. Although not shown in Fig. 3.1, tail-off curves as obtained from wind-tunnel tests are frequently presented in order to provide information of the horizontal tail's contribution to the aircraft's stability and control.

The initial trim point at C_{L_1} may be found by considering the static equilibrium relations in C_m and C_L using Fig. 3.1 to clarify the expressions.

Moment equilibrium is given by

$$0 = C_{m_{cg}} = C_{m_0} + C_{m_\alpha}(\alpha_1 - \alpha_0) + C_{m_\delta}\delta_1$$

Lift equilibrium is given by

$$C_{L_1} = C_{L_\alpha}(\alpha_1 - \alpha_0) + C_{L_\delta}\delta_1$$

where $\alpha_0 = \alpha$ for zero lift when control $\delta = 0$ and C_{m_0} is the pitching-moment coefficient when $\alpha = \alpha_0$ and $\delta = 0$. Note that C_{m_0} is invariant with respect to c.g. position.

The two equilibrium relations may be combined in matrix form as

$$\begin{bmatrix} C_{m_\alpha} & C_{m_\delta} \\ C_{L_\alpha} & C_{L_\delta} \end{bmatrix} \begin{Bmatrix} \alpha_1 \\ \delta_1 \end{Bmatrix} = \begin{Bmatrix} C_{m_\alpha}\alpha_0 - C_{m_0} \\ C_{L_1} + C_{L_\alpha}\alpha_0 \end{Bmatrix}$$

or symbolically as

$$[A] \begin{Bmatrix} \alpha_1 \\ \delta_1 \end{Bmatrix} = \begin{Bmatrix} 0 \\ C_{L_1} \end{Bmatrix} + \begin{Bmatrix} C_{m_\alpha}\alpha_0 - C_{m_0} \\ C_{L_\alpha}\alpha_0 \end{Bmatrix} \qquad (3.1)$$

A solution for α_1 and δ_1 at the trim value of C_{L_1} is found by premultiplying both sides of Eq. (3.1) by the inverse of the matrix $[A]$, i.e.,

$$\begin{Bmatrix} \alpha_1 \\ \delta_1 \end{Bmatrix} = [A]^{-1} \begin{Bmatrix} 0 \\ C_{L_1} \end{Bmatrix} + [A]^{-1} \begin{Bmatrix} C_{m_\alpha}\alpha_0 - C_{m_0} \\ C_{L_\alpha}\alpha_0 \end{Bmatrix} \qquad (3.2)$$

where

$$[A]^{-1} = \frac{1}{\Delta} \begin{bmatrix} C_{L_\delta} & -C_{m_\delta} \\ -C_{L_\alpha} & C_{m_\alpha} \end{bmatrix}$$

and

$$\Delta = \det[A] = C_{m_\alpha}C_{L_\delta} - C_{m_\delta}C_{L_\alpha}$$

The second term in Eq. (3.2) represents the aerodynamic constants due to the C_{m_0}, α_0, and C_{m_α} values for a prescribed c.g. location. It should be noted that the inversion of a nonsingular, 2×2 matrix was done by inspection. The inversion of higher-order, nonsingular matrices are better left in symbolic form and then solved by numerical methods.

It can be seen from Eq. (3.2) that trim at a second value of lift coefficient, C_{L_2}, will require α and δ to satisfy Eq. (3.2) by the following relation:

$$\begin{Bmatrix} \alpha_2 \\ \delta_2 \end{Bmatrix} = [A]^{-1} \begin{Bmatrix} 0 \\ C_{L_2} \end{Bmatrix} + [A]^{-1} \begin{Bmatrix} C_{m_\alpha}\alpha_0 - C_{m_0} \\ C_{L_\alpha}\alpha_0 \end{Bmatrix}$$

The determination of the α and δ values at a given trim C_L assumed that thrust induced lift force and pitching moments were negligible. Coupling of the trim relation [Eq. (3.2)] would be handled by introducing thrust-related stability derivatives. The thrust values used with these derivatives would primarily depend on maintaining x-axis force equilibrium and would reflect whether the aircraft was in level flight or in a climb/descent condition. Static x-axis equilibrium in the low-subsonic case corresponds approximately to

$$0 = T_{\text{prop}} - QS\left(C_{D_0} + \frac{C_L^2}{\pi \text{AR}e}\right) - mg\sin\Theta \qquad (3.3)$$

where

T_{prop} = propulsion thrust force
C_{D_0} = zero lift drag coefficient
e = Oswald's airplane efficiency factor[1]
Θ = aircraft pitch attitude relative to the horizon

It should be noted that the linearized form of the x-axis momentum conservation relation introduced the X_α stability derivative (cf. Sec. 2.2.2) and applied to small perturbations about a trim C_L value whereas Eq. (3.3) relates to the X_0 term in Eq. (4.35) with a quadratic dependence on C_L.

Example 3.1

This example will illustrate, using Eq. (3.2) for an assumed aircraft, the steps required to determine 1) α and δ for a given initial trim velocity and c.g. position and 2) the variation of δ required to change air speed at a given c.g. position.

Consider an aircraft described by W/S = wing loading = 50 lb-ft^{-2}, stick-fixed neutral point at $0.35c$, $C_{m_0} = +0.020$, $C_{m_\delta} = -0.75$ rad^{-1}, $C_{L_\alpha} = 5.00$ rad^{-1}, $C_{L_\delta} = +0.25$ rad^{-1}, $\alpha_0 = -1.0$ deg ($= -0.01745$ rad), and $V_{\text{trim}} = 250$ kn EAS (KEAS). Assume that linearity prevails and that both compressibility effects and propulsion system interactions are negligible. The dynamic pressure at trim is given by Eq. (1.2),

$$Q_1 = \tfrac{1}{2}\rho_0 V_e^2 = (250 \text{ KEAS})^2/295.37 = 211.6 \text{ psf}$$

The trim lift coefficient is

$$C_{L_1} = \frac{(W/S)}{Q_1} = \frac{50 \text{ lb-ft}^2}{211.6 \text{ psf}} = 0.2363$$

Assume that the initial c.g. position corresponds to a 5.0% static margin, and then obtain C_{m_α} by using Eq. (2.14). Here

$$C_{m_\alpha} = C_{L_\alpha}(\Delta x/c)_{\text{np}} = (5.00 \text{ rad}^{-1})(-0.05) = -0.25 \text{ rad}^{-1}$$

Expanding the terms in accord with Eq. (3.2) for the assumed aircraft trim condition yields

$$\begin{Bmatrix} \alpha_1 \\ \delta_1 \end{Bmatrix} = [A]^{-1} \begin{Bmatrix} 0 \\ 0.2363 \end{Bmatrix} + [A]^{-1} \begin{Bmatrix} -0.0156 \\ -0.0873 \end{Bmatrix}$$

where

$$[A]^{-1} = \begin{bmatrix} 0.0678 & 0.2034 \\ -1.3559 & -0.0678 \end{bmatrix} \text{(rad)}$$

Completion of the matrix operations gives

$$\left\{\begin{array}{c}\alpha_1\\\delta_1\end{array}\right\} = \left\{\begin{array}{c}0.0293\\0.0111\end{array}\right\} \text{(rad)} = \left\{\begin{array}{c}1.68\\0.64\end{array}\right\} \text{(deg)}$$

Reconsider the previous steps for $V_e = 180$ KEAS and find α_2 and δ_2 for $(x/c)_{cg} = 0.30$ from Eq. (3.2),

$$Q_2 = 109.7 \text{ psf} \qquad C_{L_2} = 0.4558$$

By using Eq. (3.2) for the revised C_L, one finds that

$$\left\{\begin{array}{c}\alpha_2\\\delta_2\end{array}\right\} = [A]^{-1}\left\{\begin{array}{c}0\\0.4558\end{array}\right\} + [A]^{-1}\left\{\begin{array}{c}-0.0156\\-0.0873\end{array}\right\}$$

$$\left\{\begin{array}{c}\alpha_2\\\delta_2\end{array}\right\} = \left\{\begin{array}{c}0.0739\\-0.0038\end{array}\right\} \text{(rad)} = \left\{\begin{array}{c}4.23\\-0.22\end{array}\right\} \text{(deg)}$$

The analyses just shown were extended to a velocity range of V_e from 180 to 320 KEAS with trim at $V_e = 250$ KEAS and for c.g. locations at $0.30c$, $0.25c$, and $0.20c$. Figure 3.2 summarizes the effect of c.g. location on longitudinal control demands when changing airspeed from a trim position for the assumed aircraft in steady flight.

The airspeed changes from level-flight trim shown in Fig. 3.2 will result in the airframe either descending or ascending unless thrust changes are made. Example 3.1 assumed that thrust related derivatives were zero; i.e., $\partial C_L/\partial T = 0$ and $\partial C_m/\partial T = 0$. Also, the matrix $[A]$ in Example 3.1 varied only with respect to the influence of c.g. position on the C_{m_α} derivative in order to illustrate first principles. In practice, an altitude would be selected and then the influence of Mach number would be considered for each V_e with the corresponding stability derivatives used in the $[A]$ matrix. If static aeroelastic influences were of significance, these

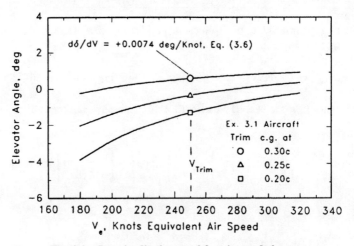

Fig. 3.2 **Longitudinal control for airspeed change.**

too would be included when estimating the longitudinal control variation with velocity.

Application of Eq. (3.2) to two C_L values gives the change in both α and δ to change the aircraft trim C_L from C_{L_1} to C_{L_2}. Here

$$\begin{Bmatrix} \alpha_2 - \alpha_1 \\ \delta_2 - \delta_1 \end{Bmatrix} = [A]^{-1} \begin{Bmatrix} 0 \\ C_{L_2} - C_{L_1} \end{Bmatrix}$$

or

$$\begin{Bmatrix} \Delta\alpha \\ \Delta\delta \end{Bmatrix} = [A]^{-1} \begin{Bmatrix} 0 \\ \Delta C_L \end{Bmatrix} \qquad (3.4)$$

Dividing all column elements in Eq. (3.4) by ΔC_L and taking the limit as ΔC_L tends to zero leads to a sensitivity analysis; i.e.,

$$\begin{Bmatrix} \frac{d\alpha}{dC_L} \\ \frac{d\delta}{dC_L} \end{Bmatrix} = [A]^{-1} \begin{Bmatrix} 0 \\ 1 \end{Bmatrix} = \frac{1}{\Delta} \begin{Bmatrix} -C_{m_\delta} \\ C_{m_\alpha} \end{Bmatrix} \qquad (3.5)$$

The controllability term of interest is the slope of the δ vs V_e curve shown in Fig. 3.2 at the trim velocity. Here

$$\frac{d\delta}{dV} = \frac{d\delta}{dC_L}\frac{dC_L}{dV} = -\frac{2C_L}{V}\frac{C_{m_\alpha}}{\Delta} \qquad (3.6)$$

When Eq. (3.6) is applied to Example 3.1 at the trim velocity of 250 KEAS with a 5% static margin, the $d\delta/dV$ derivative determination provides a slope of +0.0074 deg/kn as noted in Fig. 3.2. A low $d\delta/dV$ gradient may make holding a trim velocity difficult. For the example situation, the longitudinal control system hysteresis (i.e., a deadband typically due to friction) should be less than ±0.01 deg in order that the aircraft return to within 1 kn of the trim velocity following a disturbance induced either by the pilot or the atmosphere. Determination of control hysteresis depends on details of the control system (i.e., manual vs power- or servo-control) and can be estimated once the definition of the control surfaces has been established. Items such as control and tab hinge moments, as well as control cable friction, all play a role in the estimation.

The aircraft trim velocity, as shown in Fig. 3.2, corresponds to the pilot control force being zero. The pilot normally is provided with a trim control that implements trim velocity changes by methods such as movable trim tabs on an elevator control surface for a manual control or by a bias change in the position feedback for a power (hydraulic) control system that might be actuating a lead screw on a movable horizontal tail.

3.2.2 Maneuvering Flight

An important use of the longitudinal control is to provide an aircraft with a maneuvering capability. Both steady, vertical pull-ups and constant altitude banked turns will be considered in this section, subject to the constraint that the aircraft's velocity remains constant while in steady curvilinear flight (i.e., thrust may be added to compensate for the increased drag due to the α increase). Stability derivatives due to thrust changes also are assumed as zero. The assumption of steady flight implies that all transient dynamic aircraft responses, such as might be encountered by an abrupt entry into a pull-up, are neglected. Only the pitch damping term will have a bearing upon the analysis, whereas the influence of terms such as $d\alpha/dt$ and dq/dt will not be present. The constant altitude, turning flight will be presented to show the difference in the elevator per g from that obtained by steady pull-ups. Although either dimensional or dimensionless stability derivatives could be used in the analysis, the dimensionless form will be used initially followed by a conversion to the dimensional form.

Figure 3.3 depicts an idealized aircraft in a steady, vertical pull-up maneuver. Equilibrium of the z forces relative to the body axes may be expressed as

$$m(g + Vq) = QS[C_{L_\alpha}\alpha + C_{L_\delta}\delta] + QS[(C_{L_\alpha} + C_D)\Delta\alpha \\ + C_{L_q}(qc/2V) + C_{L_\delta}\Delta\delta] \tag{3.7}$$

It will be noted from Fig. 3.3 that the positive change of normal acceleration Δa_n, which acts in the positive z direction, is related to aircraft load factor n by the following statement:

$$\Delta a_n = -(n-1)g = -Vq \tag{3.8}$$

Because steady, unaccelerated level flight corresponds to

$$W = mg = QS[C_{L_\alpha}\alpha + C_{L_\delta}\delta]$$

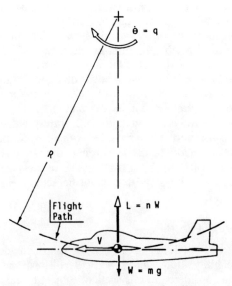

Fig. 3.3 Aircraft in a steady, vertical pull-up maneuver.

Eq. (3.7) simplifies to

$$0 = QS\left[(C_{L_\alpha} + C_D)\Delta\alpha + C_{L_q}\frac{c}{2V}q + C_{L_\delta}\Delta\delta\right] - mVq \quad (3.9)$$

Dividing through Eq. (3.9) by $(-m)$ converts the equation into dimensional stability derivative form, i.e.,

$$0 = Z_\alpha\Delta\alpha + (V + Z_q)q + Z_\delta\Delta\delta \quad (3.10)$$

where, as described in Chapter 2, the dimensional stability derivatives are

$$Z_\alpha = -\frac{QS}{m}(C_{L_\alpha} + C_D)$$

$$Z_\delta = -\frac{QS}{m}C_{L_\delta}$$

$$Z_q = -\frac{QS}{m}\frac{c}{2V}C_{L_q}$$

The change in pitching moment for the steady pull-up maneuver is given by

$$0 = C_{m_\alpha}\Delta\alpha + C_{m_q}\frac{qc}{2V} + C_{m_\delta}\Delta\delta \quad (3.11)$$

Multiplying by QSc/I_y provides the dimensional form with stability derivatives, as defined in Chapter 2, i.e,

$$0 = M_\alpha\Delta\alpha + M_qq + M_\delta\Delta\delta \quad (3.12)$$

where

$$M_\alpha = \frac{QSc}{I_y}C_{m_\alpha}$$

$$M_\delta = \frac{QSc}{I_y}C_{m_\delta}$$

$$M_q = \frac{QSc}{I_y}\frac{c}{2V}C_{m_q}$$

Equations (3.10) and (3.12) can be expressed in matrix form as

$$\begin{bmatrix} Z_\alpha & (V+Z_q) \\ M_\alpha & M_q \end{bmatrix}\begin{Bmatrix} \Delta\alpha \\ q \end{Bmatrix} = -\begin{Bmatrix} Z_\delta \\ M_\delta \end{Bmatrix}\Delta\delta \quad (3.13)$$

Because the pitch rate term q is related to normal acceleration by Eq. (3.8), Eq. (3.13) may be re-expressed as

$$\begin{bmatrix} Z_\alpha & -(V+Z_q)/V \\ M_\alpha & -M_q/V \end{bmatrix}\begin{Bmatrix} \Delta\alpha \\ \Delta a_n \end{Bmatrix} = -\begin{Bmatrix} Z_\delta \\ M_\delta \end{Bmatrix}\Delta\delta \quad (3.14)$$

When one divides both sides of Eq. (3.14) by $\Delta\delta$, takes $\Delta\delta$ to a zero limit, and premultiplies the resulting equation by the inverse of the 2×2 matrix, an expression for the acceleration sensitivity due to longitudinal control input is obtained, i.e.,

$$\left\{\begin{array}{c} \frac{d\alpha}{d\delta} \\ \frac{da_n}{d\delta} \end{array}\right\} = -\begin{bmatrix} Z_\alpha & -(V+Z_q)/V \\ M_\alpha & -M_q/V \end{bmatrix}^{-1} \left\{\begin{array}{c} Z_\delta \\ M_\delta \end{array}\right\}$$

which may be readily solved to yield

$$\left\{\begin{array}{c} \frac{d\alpha}{d\delta} \\ \frac{da_n}{d\delta} \end{array}\right\} = \frac{1}{\Delta}\begin{bmatrix} -M_q/V & (V+Z_q)/V \\ -M_\alpha & Z_\alpha \end{bmatrix} \left\{\begin{array}{c} Z_\delta \\ M_\delta \end{array}\right\} \quad (3.15)$$

where the determinant Δ is

$$\Delta = [Z_\alpha M_q - M_\alpha(V+Z_q)]/V$$

From Eq. (3.15), the acceleration in a steady, vertical pull-up will be recognized as

$$\frac{da_n}{d\delta} = \frac{V(Z_\alpha M_\delta - Z_\delta M_\alpha)}{[Z_\alpha M_q - M_\alpha(V+Z_q)]} \quad (3.16)$$

Equation (3.16) may alternatively be expressed in terms of elevator per g using dimensionless stability coefficients. Recognize that $da_n = -g\,dn$, and assume that $C_D \ll C_{L_\alpha}$ and $Z_q \ll V$; then one finds that

$$\frac{d\delta}{dn} = -\frac{C_L C_{m_\alpha} + (gc/2V^2)C_{L_\alpha}C_{m_q}}{C_{L_\alpha}C_{m_\delta} - C_{L_\delta}C_{m_\alpha}} \quad (3.17)$$

It will be noted in Eq. (3.17) that elevator per g in a pull-up maneuver at constant airspeed is a constant, and the constant will increase in value as the c.g. moves forward (i.e., due to the increase in C_{m_α}).

Example 3.2

Estimate the normal acceleration sensitivity for the DC-8 aircraft operating at a cruise flight condition, i.e., $M = 0.84$, $h = 33{,}000$ ft, and $(x/c)_{cg} = 0.15$.

The dimensional stability derivatives may be found in Appendix B.4, $V = 824.8$ ft-s^{-1}, $Z_\alpha = -664.3$ ft-s^{-2}, $M_\alpha = -9.149$ s^{-2}, $Z_q = 0.0$, $M_q = -0.924$ s^{-1}, $Z_\delta = -34.7$ ft-s^{-2}, and $M_\delta = -4.59$ s^{-2}. Substituting this information into Eq. (3.16) and solving gives

$$\frac{da_n}{d\delta} = 276.0 \text{ ft-s}^{-2}\text{-rad}^{-1} (= 0.150 \text{ g/deg})$$

Note that because plus a_n represents a negative load factor by the sign convention, then $dn/d\delta = -0.150$ g/deg. Physically this corresponds to a negative load factor being produced by a trailing-edge down (+) elevator control motion.

The M_α term in the determinant Δ of Eq. (3.15) is linearly related to the static margin $(\Delta x/c)_{np}$. When the c.g. is located at the airframe's neutral point, M_α becomes zero and the steady-state velocity can change without elevator control input (i.e., neutral speed stability). However, location of the c.g. at the neutral point does not result in the airframe's acceleration response from longitudinal control becoming infinite. The information contained in Δ allows one to define a maneuver point.

Definition: When the c.g. is located at the maneuver point, the airframe has infinite acceleration sensitivity to longitudinal control input commands.

On the assumption that $Z_q \ll V$, the denominator of Eq. (3.16) may be expressed as

$$Z_\alpha M_q - M_\alpha V = -\frac{QScV}{I_y} C_{L_\alpha} \left[\frac{\rho Sc}{4m} C_{m_q} + \left(\frac{\Delta x}{c}\right)_{np} \right] \quad (3.18)$$

The term in the brackets of Eq. (3.18) represents the location of the maneuver point relative to the c.g., i.e.,

$$\left(\frac{\Delta x}{c}\right)_{mp} = \left[\frac{\rho Sc}{4m} C_{m_q} + \left(\frac{\Delta x}{c}\right)_{np} \right]$$

and the dimensionless distance from the neutral point is given by

$$\left(\frac{\Delta x}{c}\right)_{mp} - \left(\frac{\Delta x}{c}\right)_{np} = \frac{\rho Sc}{4m} C_{m_q} \quad (3.19)$$

Because the pitch damping derivative is normally negative in sign, one finds that the maneuver point for an aircraft is located aft of the neutral point.

Example 3.3

Estimate the stability margin between the neutral and maneuver points for the DC-8 aircraft of Example 3.2 operating at a cruise flight condition according to Appendix B.4. Here, $\rho = 0.000797$ slug-ft^{-3}, $C_{m_q} = -14.60$, $S = 2600$ ft^2, $W = 230,000$ lb ($m = 7149$ slugs), and $c = 23.0$ ft.

Substitution of this data into Eq. (3.19) yields

$$\left(\frac{\Delta x}{c}\right)_{mp} - \left(\frac{\Delta x}{c}\right)_{np} = -0.0243$$

This result states that an estimate of the maneuver point location at the assumed flight condition for the DC-8 aircraft is 2.43% of the MAC aft of the neutral point.

A sketch of an aircraft in a constant altitude, steady banked flight with turn rate Ω is shown in Fig. 3.4. A typical turn rate for an aircraft under instrument flight

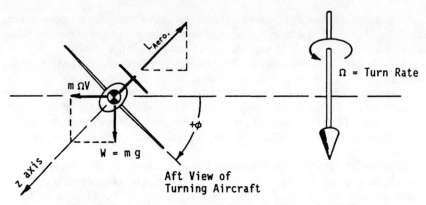

Fig. 3.4 **Aircraft in constant altitude, steady turning flight.**

conditions is 180 deg/min (i.e., 2-min turn). The turn rate is a vector quantity acting in a vertical direction and may be resolved into the aircraft's body-axis oriented pitch and yaw rates at a bank angle ϕ. In addition, the centrifugal force component experienced as side force by the aircraft during a coordinated turn is related to the lateral component of weight. These three relations may be stated as

$$r = \Omega \cos \phi \qquad (3.20)$$

$$q = \Omega \sin \phi \qquad (3.21)$$

$$mg \sin \phi = m\Omega V \cos \phi \qquad (3.22)$$

Applying Eq. (3.22) to Eq. (3.21) yields an expression for the aircraft pitch rate that is induced by the steady turn. Here

$$q = \frac{g}{V} \frac{(1 - \cos^2 \phi)}{\cos \phi} \qquad (3.23)$$

The aircraft's normal force (z direction) due to aerodynamic terms will be in equilibrium with the inertial components induced by the weight and centrifugal force vectors, i.e.,

$$Z + mg \cos \phi + m\Omega V \sin \phi = 0$$

which simplifies upon use of Eq. (3.22) to

$$Z = -(mg/\cos \phi)$$

Load factor n is considered positive acting upwards, which corresponds to a negative Z force. Therefore,

$$n = -(Z/mg) = 1/\cos \phi \qquad (3.24)$$

A steady turn at a 60-deg bank angle induces a load factor of $n = 2$, an effect that is quite apparent during flight when one extends an arm forward while executing a steeply banked turn. Application of Eq. (3.24) to Eq. (3.23) provides an alternate expression for the turn induced pitch rate in terms of load factor, i.e.,

$$q = \frac{g}{V}n\left(1 - \frac{1}{n^2}\right) = \frac{g}{V}\frac{(n-1)(n+1)}{n} \qquad (3.25)$$

An approach similar to that used in developing Eq. (3.9) can be followed to describe the aerodynamic forces required to maintain equilibrium, i.e.,

$$m\Delta a_n = -m(n-1)g = QS[(C_{L_\alpha} + C_D)\Delta\alpha + C_{L_q}(c/2V)q + C_{L_\delta}\Delta\delta]$$

Dividing through by $-m$ and simplifying yields

$$Z_\alpha \Delta\alpha + Z_\delta \Delta\delta = -\left[Z_q \frac{(n+1)}{Vn} + 1\right](n-1)g \qquad (3.26)$$

Substitution of Eq. (3.25) into Eq. (3.12) corresponds to maintaining aircraft pitching-moment equilibrium during the steady turn, i.e.,

$$M_\alpha \Delta\alpha + M_\delta \Delta\delta = -M_q[(n+1)/Vn](n-1)g \qquad (3.27)$$

Combining Eqs. (3.26) and (3.27) into matrix format and solving for $\Delta\alpha$ and $\Delta\delta$ provides the following relations:

$$\begin{Bmatrix}\Delta\alpha \\ \Delta\delta\end{Bmatrix} = -\frac{(n-1)}{Vn}\frac{g}{\Delta}\begin{bmatrix}M_\delta & -Z_\delta \\ -M_\alpha & Z_\alpha\end{bmatrix}\begin{Bmatrix}Z_q(n+1) + Vn \\ M_q(n+1)\end{Bmatrix} \qquad (3.28)$$

where

$$\Delta = Z_\alpha M_\delta - M_\alpha Z_\delta$$

Unlike a vertical pull-up maneuver, the elevator per g in a steady turn is not a linear function of load factor. The elevator required to develop load factor in the turn, from Eq. (3.28), is

$$\Delta\delta = -\frac{(n-1)}{Vn}\frac{g}{\Delta}\{Z_\alpha M_q(n+1) - M_\alpha[Z_q(n+1) + Vn]\} \qquad (3.29)$$

The preceding relation may be re-expressed in an approximate dimensionless stability coefficient form under the assumptions that both $C_D \ll C_{L_\alpha}$ and $Z_q \ll V$. Here

$$\Delta\delta = -\frac{(n-1)C_L}{\Delta}\left[C_{m_\alpha} + \frac{(n+1)}{2\mu n}C_{m_q}C_{L_\alpha}\right] \qquad (3.30)$$

where

C_L = lift coefficient when $\phi = 0$
μ = dimensionless mass coefficient, $2m/\rho Sc$
$\Delta = C_{L_\alpha}C_{m_\delta} - C_{m_\alpha}C_{L_\delta}$

Example 3.4

Estimate the elevator control required in steady banked turns for the DC-8 aircraft of Example 3.2 operating at a cruise-flight condition according to Appendix B.4. Here $\rho = 0.000797$ slug-ft^{-3}, $m = 7149$ slugs, $S = 2600$ ft^2, $c = 23.0$ ft, $C_{L_\alpha} = 6.744$ rad^{-1}, $C_{m_\alpha} = -2.017$ rad^{-1} for $(x/c)_{cg} = -0.15$, $C_{L_q} = 0.0$, $C_{m_q} = -14.60$, $C_{L_\delta} = 0.352$ rad^{-1}, $C_{m_\delta} = -1.008$ rad^{-1}, and $C_L = 0.326$. The dimensionless mass coefficient, which corresponds to the mass of the aircraft normalized with respect to the air contained in a volume defined by the product of $Sc/2$, is given by

$$\mu = \frac{2(7149 \text{ slugs})}{(0.000797 \text{ slug-ft}^{-3})(2600 \text{ ft}^2)(23.0 \text{ ft})} = 300.0$$

The determinant Δ for the aircraft when $(x/c)_{cg} = -0.15$ is

$$\Delta = (6.744)(-1.008) - (-2.017)(0.352) = -6.088 \text{ rad}^{-1}$$

The load factor for the aircraft at a 60-deg bank angle from Eq. (3.24) is

$$n = 1/\cos\phi = 2.0$$

The elevator control required to maintain $n = 2.0$ in a steady, banked turn may be estimated from Eq. (3.30) as

$$\Delta\delta = -0.121 \text{ rad} (= -6.95 \text{ deg})$$

Control angles for other bank angles may be found using the preceding two calculations as guidelines.

An estimate of control angle requirements for the aircraft with other c.g. locations requires a change in the C_{m_α} stability derivative, which can be determined once the neutral point is established.

The static margin for $(x/c)_{cg} = -0.15$ can be found from C_{m_α} and C_{L_α} using Eq. (2.14), i.e.,

$$\text{Static margin} = -\left(\frac{\Delta x}{c}\right)_{np} = -\frac{C_{m_\alpha}}{C_{L_\alpha}} = \frac{2.017 \text{ rad}^{-1}}{6.744 \text{ rad}^{-1}} = 0.299$$

It should be noted that the static margin result is relative to the origin of the x body axis, which corresponds to the c.g. location. It is frequently convenient to translate the neutral point location relative to the leading edge of the reference wing chord and express the number as a positive value when it is aft of the chord's leading edge. When that connotation is used, the location of the neutral point would be stated as being at 44.9% of the reference chord. An estimate of the C_{m_α} derivative when the c.g. is located at $0.20c$ may be made by subtracting -0.20 from -0.449 and then reapplying Eq. (2.14); i.e., for $(x/c)_{cg} = -0.20$, one finds that

$$C_{m_\alpha} = (6.744 \text{ rad}^{-1})(-0.249) = -1.679 \text{ rad}^{-1}$$

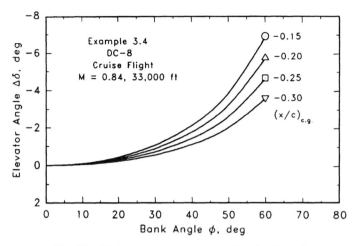

Fig. 3.5 Estimate of elevator control vs bank angle.

Application of Eq. (3.30) for the modified c.g. location when the aircraft is in a steady, banked turn of 60 deg yields

$$\Delta\delta = -0.082 \text{ rad } (= -4.69 \text{ deg})$$

The estimated variation of elevator angle as a function of bank angle is shown on Fig. 3.5 for various c.g. locations. Items to note include that 1) longitudinal control is not linear with bank angle and 2) moving the c.g. location forward places increased demands on the longitudinal control.

3.2.3 V–n Diagram

Longitudinal stability and control requirements for safe aircraft operation are typically determined by the maneuvering envelope, as shown in Fig. 3.6, which defines the boundary limits for velocity and normal load factor (i.e., V–n diagram). Figure 3.6 represents a typical, flaps-up maneuvering envelope for a transport category airplane as specified by the *U.S. Federal Aviation Regulations* (FAR) Part 25.[2] A similar maneuvering envelope is defined in FAR Part 23[3] for application to normal, utility, acrobatic, and commuter category airplanes. The maneuvering envelope for military aircraft is shown in the U.S. Military Specification, MIL-F-8785C.[4]

According to FAR Part 25, Sec. 25.337,[2] the positive and negative limiting load factors for transport category aircraft should not be less than

$$+n_{\text{limit}} = 2.1 + \left(\frac{24{,}000}{W + 10{,}000}\right)$$

$$-n_{\text{limit}} = 1$$

where W is the design maximum takeoff weight, in pounds.

The positive limiting load factors for other types of aircraft, such as normal and commuter, utility, and acrobatic categories, vary between $+n_{\text{limit}}$ of 3.8–6.0.[3]

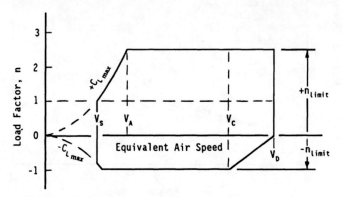

Fig. 3.6 Sketch of aircraft flaps-up maneuvering envelope.

The EASs are shown in Fig. 3.6 where

V_S = stall speed, flaps up, level flight, design weight, and c.g.
V_A = design maneuvering speed
V_C = cruise speed
V_D = design dive speed

High-performance aircraft will also have corresponding Mach number limits such as cruise Mach number M_C and design dive Mach number M_D and these Mach numbers will correspond to V_C and V_D at particular altitudes. The V_C/M_C velocity will typically not be greater than 80% of the V_D/M_D velocity. The stall speeds determined from flight tests establish the $+C_{L_{max}}$ limit at the lower velocity region of Fig. 3.6. Aircraft stall speed is defined as the minimum steady-flight speed at which the aircraft is controllable. Because the minimum flight speed is dependent on the deceleration rate at which stall is approached, the FARs specify that the airplane initially be trimmed between $1.2V_S$ and $1.4V_S$ and then longitudinal control be applied such that the deceleration does not exceed 1 kn/s until stall is reached (cf. Sec. 8.4, Stall Dynamics).

The physical structure must be able to withstand the limit loads imposed by operation both inside and on the boundaries of the $V-n$ diagram while the longitudinal controls are designed to ensure satisfactory flight within the envelope. Ultimate loads are typically 50% greater than the limit design loads, and the aircraft should be able to support ultimate loads without failure for at least 3 s (Ref. 2).

Another equally important $V-n$ diagram that is crucial to both structural design and aircraft control (for recovery from upsets) is the gust envelope, which defines the limit conditions for a set of specified gust velocity encounters. The gust envelope will be described in Chapter 9 when the gust response of aircraft is considered. The maneuvering envelope is primarily influenced by pilot input to the aircraft's longitudinal control system whereas the gust envelope is determined by the aircraft response to a set of vertical gust profiles, which experience has shown as representing a safe bound for all-weather operations.

3.3 Lateral-Directional Stability and Control

From an operational sense, lateral-directional stability traits are related to the aircraft's control requirements when demonstrating straight, steady sideslips. A stable sideslip behavior implies that sideslips to the right ($+\beta$) require the concurrent application of left-rudder ($+\delta_r$) and right-aileron ($+\delta_a$) control, where the control sign conventions are shown in Fig. 1.1. Furthermore, it is desirable that the variation of control angles should be approximately linear for sideslip angles between ± 15 deg or $\pm \beta_{max}$. Sideslip capabilities on the order of ± 10 deg are associated with maximum crosswind landings and takeoffs, which usually involve the aircraft in a deflected wing flap configuration.

Although β_{max} is limited as a function of airspeed for structural reasons, β information is not normally available to the pilot. Flight manuals for an aircraft usually include V–n diagram related information but seldom contain any β limits except by indirect references to steady-flight, crosswind limits. Limiting of β_{max} with flight dynamic pressures normally occurs by reducing the pilot's authority of rudder deflection as airspeed increases, an action that is transparent to the pilot. The pilot's control capability is directly related to the rudder hinge moments (HM), and their increase with dynamic pressure [i.e., $HM = f(\beta, \delta_r, Q, \ldots)$], ultimately establishes the control limits.

Multiengined aircraft impose another strong influence on the design authority of the lateral-directional control system. Consideration of the most critical engine failure condition during the takeoff phase of flight establishes the minimum-control speed criteria. The demonstration of the engine-out, minimum-control speed (V_{mc}) involves maintaining steady, straight flight with the critical engine out at a bank angle not exceeding 5 deg in the dead-engine high direction. Maximum yawing moment from rudder deflection during the V_{mc} demonstration sets the design objectives for the rudder control authority. During the V_{mc} demonstration, it is normally expected that the lateral control required to maintain equilibrium of the roll moments does not exceed about 75% of the maximum roll authority.

A rule of thumb for maximum roll control authority is that the use of maximum lateral control should correspond to the airplane developing a steady-state value of roll helix angle (i.e., $pb/2V$, cf. Sec. 2.3.3) of approximately 0.09. A roll control capability of this magnitude in the takeoff and landing configuration provides a lateral control margin in excess of that required for balancing the roll moments encountered during maximum sideslips under crosswind encounters. The lateral control margin is necessary to provide safe operation in atmospheric turbulence encounters, which are common occurrences during extreme crosswind situations. Civilian aircraft certification requirements[2,3] are not specific relative to lateral control capabilites; however, a former military specification[5] did provide firm guidelines.

In summary, lateral-directional control system design encompasses providing safe aircraft handling qualities by a competent pilot of engine-out thrust asymmetries, crosswind takeoffs and landings, trim of yaw and roll moment asymmetries, and correcting of aircraft transient dynamic responses during gusts, maneuvers, etc. These comments relate to static stability and control considerations for the lateral-directional axes of the aircraft. The following two sections will illustrate the principles involved in estimating the sideslip and engine-out capabilities for a representative aircraft.

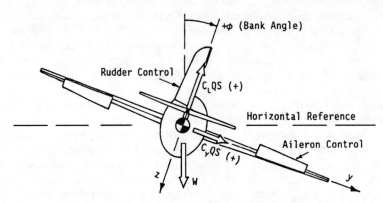

Fig. 3.7   Rear view of aircraft in a steady bank angle.

3.3.1   Aircraft in Sideslip

An aircraft in a steady sideslip requires concurrent equilibrium in side force and the roll and yaw moments. Linear equations will be considered to illustrate the principles, especially since experience has shown them as reasonable to describe modest sideslips during normal flight situations. The three equilibrium equations can be solved to find the values of lateral and directional control (δ_r and δ_a), and bank angle (ϕ) for a given sideslip angle (β).

Figure 3.7 shows the rear view of an aircraft in unaccelerated flight at a bank angle ϕ. Unlike the α and β angles, bank angle does not produce restorative moments. However, the aerodynamic forces in the y and z body-axis directions will be in balance with the corresponding dead-weight components.

In the y axis direction,

$$W \sin\phi + C_y QS = 0$$

and in the z axis direction,

$$W \cos\phi - C_L QS = 0$$

These two static force relations provide an equivalent statement that

$$C_y = -C_L \tan\phi$$

The three static equilibrium relations may be stated in a linear form as

$$-C_L \tan \Delta\phi = C_{y_\beta} \Delta\beta + C_{y_{\delta_r}} \Delta\delta_r + C_{y_{\delta_a}} \Delta\delta_a$$
$$0 = C_{\ell_\beta} \Delta\beta + C_{\ell_{\delta_r}} \Delta\delta_r + C_{\ell_{\delta_a}} \Delta\delta_a \qquad (3.31)$$
$$0 = C_{n_\beta} \Delta\beta + C_{n_{\delta_r}} \Delta\delta_r + C_{n_{\delta_a}} \Delta\delta_a$$

where the Δ notation has been used with all of the variables to indicate a change from the trimmed, wings-level flight condition.

STATIC STABILITY AND CONTROL

The terms in Eq. (3.31) may be rearranged, along with a small angle assumption on bank angle, to yield

$$\begin{bmatrix} +C_L & C_{y_{\delta_r}} & C_{y_{\delta_a}} \\ 0 & C_{\ell_{\delta_r}} & C_{\ell_{\delta_a}} \\ 0 & C_{n_{\delta_r}} & C_{n_{\delta_a}} \end{bmatrix} \begin{Bmatrix} \Delta\phi \\ \Delta\delta_r \\ \Delta\delta_a \end{Bmatrix} = - \begin{Bmatrix} C_{y_\beta} \\ C_{\ell_\beta} \\ C_{n_\beta} \end{Bmatrix} \Delta\beta \qquad (3.32)$$

Dividing both sides of Eq. (3.32) by $\Delta\beta$ and letting $\Delta\beta$ tend to a zero limit provides the following sideslip sensitivity derivatives:

$$\begin{Bmatrix} \frac{d\phi}{d\beta} \\ \frac{d\delta_r}{d\beta} \\ \frac{d\delta_a}{d\beta} \end{Bmatrix} = -[A]^{-1} \begin{Bmatrix} C_{y_\beta} \\ C_{\ell_\beta} \\ C_{n_\beta} \end{Bmatrix} \qquad (3.33)$$

where

$$[A] = \begin{bmatrix} +C_L & C_{y_{\delta_r}} & C_{y_{\delta_a}} \\ 0 & C_{\ell_{\delta_r}} & C_{\ell_{\delta_a}} \\ 0 & C_{n_{\delta_r}} & C_{n_{\delta_a}} \end{bmatrix}$$

The sideslip sensitivity derivatives of Eq. (3.33) may also be expressed in terms of dimensional stability derivatives, which should be evident when each row of Eq. (3.31) is multiplied by (QS/m), (QSb/I_x), and (QSb/I_z), respectively, i.e.,

$$\begin{Bmatrix} \frac{d\phi}{d\beta} \\ \frac{d\delta_r}{d\beta} \\ \frac{d\delta_a}{d\beta} \end{Bmatrix} = -[B]^{-1} \begin{Bmatrix} Y_\beta \\ L_\beta \\ N_\beta \end{Bmatrix} \qquad (3.34)$$

where

$$[B] = \begin{bmatrix} g & Y_{\delta_r} & Y_{\delta_a} \\ 0 & L_{\delta_r} & L_{\delta_a} \\ 0 & N_{\delta_r} & N_{\delta_a} \end{bmatrix}$$

Example 3.5

Estimate the sideslip sensitivity derivatives for the Lockheed Jetstar aircraft in the power approach condition described in Appendix B.6.

$$\begin{Bmatrix} Y_\beta \\ L_\beta \\ N_\beta \end{Bmatrix} = \begin{Bmatrix} -31.260 \\ -3.539 \\ 1.598 \end{Bmatrix}$$

$$[B] = \begin{bmatrix} 32.17 & 7.592 & 0.0 \\ 0.0 & 0.887 & 2.148 \\ 0.0 & -0.715 & -0.147 \end{bmatrix}$$

The sideslip sensitivity derivatives, according to Eq. (3.34), may be found by premultiplying the sideslip stability derivative column vector by the negative inverse of the $[B]$ matrix. The use of a matrix-oriented computer program is convenient for obtaining the following solution:

$$\begin{Bmatrix} \frac{d\phi}{d\beta} \\ \frac{d\delta_r}{d\beta} \\ \frac{d\delta_a}{d\beta} \end{Bmatrix} = \begin{Bmatrix} 0.483 \\ 2.072 \\ 0.792 \end{Bmatrix} \text{ rad/rad (or deg/deg)}$$

The results represent stable lateral-directional stability traits as described in Sec. 3.3. If one assumes that the lateral-directional controls behave in an approximate linear manner, the results for the example aircraft in a power approach flight condition indicate that a crosswind landing requiring $\beta = 10$ deg would be satisfied when bank angle $\phi = 4.8$ deg, rudder control $\delta_r = 21$ deg, and aileron control $\delta_a = 7.9$ deg.

It should be recognized that maximum lateral control deflection is approximately 20 deg and the analysis would indicate that there is an ample reserve of lateral control remaining to handle gust generated bank angle transients when encountering severe crosswind landing conditions.

3.3.2 Aircraft with Thrust Asymmetry

An important design consideration of the rudder control system for multiengined aircraft relates to establishing the minimum-control speed (V_{mc}) as described in Sec. 3.3. Takeoff performance (i.e., field length, climb gradients, etc.) is dependent on both stall speed and minimum-control speed with factors (e.g., $1.1V_{mc}$) applied to these speeds in order to ensure safe flight operation in case of an engine failure.

The yawing moment due to the indicated thrust asymmetry from an engine failure (Fig. 3.8) may be expressed in dimensionless coefficient form as

$$\Delta C_{n_{\text{eng}}} = \frac{T\eta_{\text{eng}}}{2QS} = \frac{1}{2}\left(\frac{T}{W}\right)\eta_{\text{eng}} C_L \qquad (3.35)$$

where

T = thrust force of a single propulsion unit
η_{eng} = dimensionless spanwise location of the failed engine, $2y_{\text{eng}}/b$

As in Sec. 3.3.1, linear equations will be considered to illustrate the principles. The thrust asymmetry may be included in the yaw moment equilibrium relation [Eq. (3.31)] when representing an engine-out condition. Propeller-driven aircraft would also introduce a rolling moment term in Eq. (3.31) due to propeller torque induced asymmetries. This roll moment term will be neglected in the subsequent material; consequently, the discussion will be more representative of jet-powered aircraft that are experiencing a thrust asymmetry due to engine failure. Static equilibrium in the lateral-directional equations using the stated assumptions can

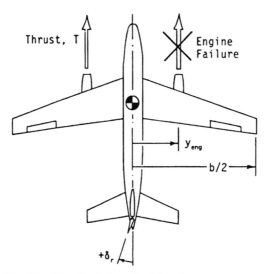

Fig. 3.8 Sketch of aircraft with a thrust asymmetry.

be represented as

$$-C_L \tan \Delta\phi = C_{y_\beta}\Delta\beta + C_{y_{\delta_r}}\Delta\delta_r + C_{y_{\delta_a}}\Delta\delta_a$$
$$0 = C_{\ell_\beta}\Delta\beta + C_{\ell_{\delta_r}}\Delta\delta_r + C_{\ell_{\delta_a}}\Delta\delta_a \quad (3.36)$$
$$-\tfrac{1}{2}(T/W)\eta_{\text{eng}} C_L = C_{n_\beta}\Delta\beta + C_{n_{\delta_r}}\Delta\delta_r + C_{n_{\delta_a}}\Delta\delta_a$$

or, alternatively, in a matrix format as

$$-\begin{Bmatrix} C_L \tan \Delta\phi \\ 0 \\ \tfrac{1}{2}(T/W)\eta_{\text{eng}} C_L \end{Bmatrix} = \begin{bmatrix} C_{y_\beta} & C_{y_{\delta_r}} & C_{y_{\delta_a}} \\ C_{\ell_\beta} & C_{\ell_{\delta_r}} & C_{\ell_{\delta_a}} \\ C_{n_\beta} & C_{n_{\delta_r}} & C_{n_{\delta_a}} \end{bmatrix} \begin{Bmatrix} \Delta\beta \\ \Delta\delta_r \\ \Delta\delta_a \end{Bmatrix} \quad (3.37)$$

The terms in Eq. (3.37) may be rearranged to solve for $\Delta\beta$, $\Delta\delta_r$, and $\Delta\delta_a$ as a function of C_L and $\Delta\phi$ as will be shown. The small-angle assumption will be used on bank angle such that $\tan \Delta\phi \cong \Delta\phi$. Here

$$\begin{Bmatrix} \Delta\beta \\ \Delta\delta_r \\ \Delta\delta_a \end{Bmatrix} = -[A]^{-1} \begin{Bmatrix} \Delta\phi \\ 0 \\ \tfrac{1}{2}(T/W)\eta_{\text{eng}} \end{Bmatrix} C_L \quad (3.38)$$

where the matrix $[A]$ corresponds to

$$[A] = \begin{bmatrix} C_{y_\beta} & C_{y_{\delta_r}} & C_{y_{\delta_a}} \\ C_{\ell_\beta} & C_{\ell_{\delta_r}} & C_{\ell_{\delta_a}} \\ C_{n_\beta} & C_{n_{\delta_r}} & C_{n_{\delta_a}} \end{bmatrix}$$

It will be noted from Eq. (3.38) that when an aircraft lift coefficient is selected, which corresponds to a given weight and airspeed, the solution for $\Delta\beta$, $\Delta\delta_r$, and

88   INTRODUCTION TO AIRCRAFT FLIGHT DYNAMICS

$\Delta\delta_a$ will depend on an assumed value for the bank angle change from trim, $\Delta\phi$. This observation implies that a pilot, during an engine-out simulation at a given weight and airspeed, finds $\Delta\beta$, $\Delta\delta_r$, and $\Delta\delta_a$ to be functions of the selected bank angle.

The minimum-control speed is established by decreasing the airspeed while maintaining the thrust asymmetry. V_{mc} is reached when a 5-deg bank angle occurs concurrently with the application of maximum rudder control while maintaining straight flight. If the selected aircraft weight is such that the stall speed (V_s) is equal to or greater than V_{mc}, unusual aircraft motions can occur due to the loss of lateral-directional control.

An example is given to illustrate the options available to the pilot to attain static equilibrium conditions in straight flight (i.e., $\dot{\psi} = 0$) with an engine-out condition. The analysis is based on stability derivative information available in the literature and cannot be expected to correspond to the manufacturer's actual flight-test results.

Example 3.6

Estimate the static equilibrium conditions for the Lockheed Jetstar aircraft with a critical engine out in the power approach flight condition. Use the dimensionless coefficient data of Appendix B.6 in making the estimates.

Assume that the outboard jet engine on the right-hand side (engine number 4) has failed and the remaining three engines are at an estimated takeoff thrust setting. From Appendix B.6, estimate $\eta_{eng} = 0.30$.

Because normal cruise weight is given as 38,200 lb, estimate the maximum takeoff weight as 40,000 lb. For a typical takeoff ground acceleration of $\frac{1}{3} g$, estimate each of the four jet engines develops a takeoff thrust of 3330 lb.

For the power approach condition ($W = 23,904$ lb)

$$\frac{1}{2}\left(\frac{T}{W}\right)\eta_{eng} = \frac{1}{2}\frac{(3330 \text{ lb})}{(23,904 \text{ lb})}(0.30) = 0.0209$$

The level-flight lift coefficient at $V = 223$ fps ($= 132.1$ KEAS) is

$$C_L = 0.743$$

The $[A]$ matrix in Eq. (3.38) is

$$[A] = \begin{bmatrix} C_{y_\beta} & C_{y_{\delta_r}} & C_{y_{\delta_a}} \\ C_{\ell_\beta} & C_{\ell_{\delta_r}} & C_{\ell_{\delta_a}} \\ C_{n_\beta} & C_{n_{\delta_r}} & C_{n_{\delta_a}} \end{bmatrix} = \begin{bmatrix} -0.722 & 0.175 & 0.0 \\ -0.087 & 0.022 & 0.053 \\ 0.148 & -0.066 & -0.014 \end{bmatrix}$$

Assume the bank angle $\Delta\phi$ in Eq. (3.38) is -3.0 deg (-0.0524 rad). The use of a matrix-oriented computer program is convenient for finding the solution in radians

$$\begin{Bmatrix} \Delta\beta \\ \Delta\delta_r \\ \Delta\delta_a \end{Bmatrix} = -[A]^{-1} \begin{Bmatrix} -0.0524 \\ 0.0 \\ 0.0209 \end{Bmatrix} 0.743 = \begin{Bmatrix} 0.0174 \\ 0.2942 \\ -0.0935 \end{Bmatrix}$$

or in degree notation

$$\begin{Bmatrix} \Delta\beta \\ \Delta\delta_r \\ \Delta\delta_a \end{Bmatrix} = \begin{Bmatrix} 1.00 \\ 16.86 \\ -5.36 \end{Bmatrix}$$

The result states that with the assumed aircraft configuration, weight, and airspeed, an outboard engine out, and the remaining three engines at an estimated takeoff thrust, static equilibrium in straight flight with $\Delta\phi = -3.0$ deg (dead-engine high) will occur with a sideslip of $\Delta\beta = +1.0$ deg (nose to the left), rudder control of $\Delta\delta_r = 16.9$ deg (left-rudder pedal input), and aileron control of $\Delta\delta_a = -5.4$ deg (left-control wheel input).

Increasing the negative bank angle from -3.0 to -4.0 deg alters the results by reducing the amount of rudder control needed while allowing the nose to turn slightly into the dead-engine side, i.e., when $\Delta\phi = -4.0$ deg (-0.0698 rad), Eq. (3.38) yields

$$\begin{Bmatrix} \Delta\beta \\ \Delta\delta_r \\ \Delta\delta_a \end{Bmatrix} = \begin{Bmatrix} -0.0187 \\ 0.2191 \\ -0.1217 \end{Bmatrix} \text{rad} = \begin{Bmatrix} -1.07 \\ 12.55 \\ -6.97 \end{Bmatrix} \text{deg}$$

Bank angles for the engine-out scenario were considered for an angular range of $-6.0 \le \Delta\phi \le 0$ deg, and results are shown in Fig. 3.9. In addition, the airspeed was reduced by 10 KEAS corresponding to $C_L = 0.873$, and similar static equilibrium results are also shown in Fig. 3.9.

All data are based on linearized stability derivative coefficients, which implies that control deflections greater than about ± 20 deg are not valid. When engine-out estimates are made using best available airframe stability and control data (e.g., wind-tunnel test data), iterative procedures are frequently required to solve the nonlinear control system traits.

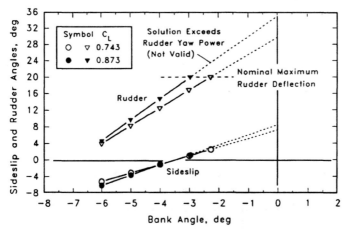

Fig. 3.9 Engine-out results for Example 3.6.

Figure 3.9 shows the tradeoff between rudder control and sideslip angle to establish static equilibrium for the assumed engine-out condition in Example 3.6. It is important to note that static equilibrium with an assumed nominal maximum rudder control can be maintained while speed is decreased by increasing the bank angle. A 10-kn airspeed decrease altered bank angle from $\Delta\phi = -2.3$ to -3.0 deg while applying full rudder control. At no time during the estimation process did the demands on the lateral control system near a limiting control situation.

The minimum-control speed determination for an aircraft is normally conducted in a takeoff configuration at the most critical weight and c.g. conditions.[2] Although Fig. 3.9 would indicate that increasing the bank angle during a demonstration reduces the demands on directional control, certification requirements define the limit bank angle as being 5 deg. The rationale for the selection of 5 deg as a bank angle limit during V_{mc} demonstrations appears arbitrary; however, its use does provide a reasonable ground clearance limit for low-wing, multiengined aircraft. The use of $1.1 V_{mc}$ as a takeoff speed at lighter airplane weights provides the flight operation with an added safety margin, which time has proven to be prudent.

References

[1]Perkins, C. D., and Hage, R. E., *Airplane Performance Stability and Control*, Wiley, New York, 1949, pp. 90–112.

[2]Anon., "Part 25, Airworthiness Standards: Transport Category Airplanes," *Federal Aviation Regulations*, U.S. Government Printing Office, Washington, DC, March 1993.

[3]Anon., "Part 23, Airworthiness Standards: Normal, Utility, Acrobatic and Commuter Category Airplanes," *Federal Aviation Regulations*, U.S. Government Printing Office, Washington, DC, Feb. 1991.

[4]Anon., "Flying Qualities of Piloted Airplanes," MIL SPEC, MIL-F-8785C, U.S. Government Printing Office, Washington, DC, Nov. 1980.

[5]Anon., "Flying Qualities of Piloted Airplanes," MIL SPEC, MIL-F-8785B, U.S. Government Printing Office, Washington, DC, Aug. 1969.

Problems

3.1. Consider a longitudinal control system, shown in Fig. P3.1, where the stick force exerted by the pilot is as given in the following nomenclature. Assume that the stick force F_s is zero at the trim velocity V when $\alpha = \alpha_1$ and elevator $\delta = \delta_1$ due to use of the trim tab set at δ_t. Then

$$F_{s_1} = QG(Sc)_e\left[C_{h_\alpha}(\alpha_1 - \alpha_0) + C_{h_\delta}\delta_1 + C_{h_{\delta t}}\delta_t\right]$$

where

F_s = pilot stick force (positive is pull), lb
G = control system gearing, rad/ft-lb
$(Sc)_e$ = product of elevator reference area and chord, ft^3
Q = dynamic pressure, lb-ft^{-2}
C_h = elevator hinge moment coefficient, nose up is positive

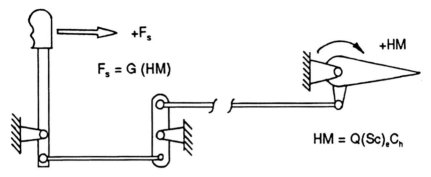

Fig. P3.1 Manual longitudinal control system.

$C_{h_\alpha} = \partial C_h / \partial \alpha$
$C_{h_\delta} = \partial C_h / \partial \delta$
$C_{h_{\delta t}} = $ trim tab effectiveness, $\partial C_h / \partial \delta_t$

a) With δ_t held constant at the trim position, show that the control force sensitivity with respect to C_L in unaccelerated flight is given by

$$\frac{d(F_s/Q)}{dC_L} = \frac{G(Sc)_e}{\Delta}(C_{h_\delta}C_{m_\alpha} - C_{h_\alpha}C_{m_\delta})$$

where the determinant is

$$\Delta = (C_{m_\alpha}C_{L_\delta} - C_{m_\delta}C_{L_\alpha})$$

b) Consider the aircraft of Example 3.1 with a manual longitudinal control system (Fig. P3.1) where the dimensionless elevator hinge moment derivatives are given by

$$C_{h_\delta} = -0.28 \text{ rad}^{-1} \quad \text{and} \quad C_{h_\alpha} = -0.06 \text{ rad}^{-1}$$

Find the stick-free neutral point.

Remarks: It was shown by Eq. (3.6) that the derivative of $d\delta/dV$ became zero when the c.g. was located at the stick-fixed neutral point. At this c.g. position, aircraft velocity could change without movement of the longitudinal control. From the pilot's standpoint, an equally important neutral point location is the stick-free neutral point because a c.g. position at this location would correspond to having no variation in stick force with changes in C_L (or V) about the reference flight condition. Because a manual longitudinal control system can take many forms including the presence of feel springs, bob weights, etc., only the basic aerodynamic terms were considered in order to illustrate the principles. Note that the stick-free neutral point is significantly influenced by the elevator floating derivative C_{h_α}.

3.2. Consider the Convair 880 jet transport as described in Appendix B.5 for condition 2. Assume that the jet engine takeoff thrust = 14,000 lb/engine and that the outboard jet engine is located at 70% of the wing semispan. If the starboard (R.H.)

number 4 engine failed while a go-around maneuver was initiated using takeoff thrust on the remaining three engines, determine the sideslip (β), rudder (δ_r), and spoiler (δ_s) angles for maintaining straight flight when the bank angle ϕ was -4.0 deg. Note that when the aircraft is in the equilibrium condition the dead engine is high and the pilot applies pedal force by stepping on the live engine(s) side.

3.3. Using the information from Problem 3.2:

a) Find the bank angle when rudder deflection (δ_r) is 20 deg while maintaining straight flight.

b) Decrease the aircraft velocity by 10 kn and redetermine the bank angle when rudder deflection is 20 deg. Assume that the stability derivatives used in Problem 3.2 remain unaltered for the increase in aircraft C_L. Note when using maximum rudder control during a most adverse engine-out flight condition, bank angle can be used to maintain straight flight while decreasing velocity, provided the aircraft is not near to a stall condition.

3.4. Using the stability information from Appendix B.5 for the Convair 880 jet transport in a landing configuration, condition 1, estimate (δ_r/β), (δ_s/β), and (ϕ/β). Do the answers make physical sense (i.e., left rudder and right lateral control for positive sideslip)?

4
Airframe Equations of Motion

4.1 Background

The purpose of this chapter is to establish the complete set of motion equations for an aircraft followed by a linearization process to put the time-dependent terms in a form that is tractable for both analysis and interpretation. Because there are three mutually perpendicular directions for both displacement and angular rotation, six differential equations will be needed to govern the complete motion of the vehicle.

The equations of motion will be a consequence of conservation laws in the three-dimensional space for linear and angular momentum. It is ingrained in our minds that force $[F = \mathrm{d}(mV)/\mathrm{d}t = ma]$ is related to a conservation of linear momentum in a fixed reference frame; however, it will be necessary in the case of airframe dynamics to consider other vector concepts in order to describe the system relative to body-fixed coordinates that can rotate in space. Early introductions to the mechanics of rotating systems are usually simplified to a problem such as a rotating disk (i.e., a phonograph record on a turntable) to introduce terms such as Coriolis forces. In our discussions, we will consider the general case where the rotation vector is defined by three components of angular displacement.

A linear algebra approach will be used for ease in understanding coordinate system transformations. A distinction will be made between inertial (i.e., fixed to an Earth orientation) and body-axis coordinates. For reasons to become clearer in the subsequent material, the use of body-axis coordinates will prove advantageous when considering both the linear and nonlinear equations.

The transformation from fixed, flat-Earth coordinates to the airframe body axes will follow a commonly accepted sequence of angle changes. For this transformation, the term Euler angles will be used, a term borrowed from gyroscope principles in engineering mechanics.

The externally applied forces and moments acting on the airframe will primarily be aerodynamic in nature (i.e., neglecting propulsion and thrust vectoring) due to airstream velocities, airframe attitudes, and control deflections. Descriptions of the aerodynamics will use linearized forms (stability derivatives) as described in Chapter 2. Tabulations of representative stability derivatives for several aircraft are given in Appendix B. The equations of motion will be linearized such that dynamic responses will represent perturbations from initial trimmed equilibrium conditions. Time response solutions will be in real time, which is a result of expressing the stability derivatives in a dimensional form according to accepted standards.

4.2 Euler Angle Transformations

The two coordinate systems considered in airframe dynamics are based on the inertial and body-axis references. Some vector quantities, such as airplane weight, remain fixed to a vertical direction relative to an inertial reference frame. This requires a coordinate transformation so that weight components will be in the airframe body-axis system. Because the airframe can go through a wide range of orientations during flight, it is desirable to establish an angular transformation about the airplane c.g. from inertial to body-axis coordinates that systematically expresses the body-axis components.

A multitude of strategies could be used to rotate the coordinate from inertial to body axes. In flight mechanics, it is customary to use a sequence of angular rotations following the Euler angle transformation rules to change the coordinate system. A theorem by Euler (see Ref. 1) states: "Any number of rotations about different axes through a point must, in the end, remain equivalent to a single rotation."

The Euler angle transformation consists of the following sequence of three rotations.

1) Starting with the inertial coordinate system where a vector is defined as X_1 with components x_1, y_1, and z_1, rotate about the inertial z_1 axis by angle Ψ to define a second coordinate system. In the second coordinate system, the vector is described as X_2 with components x_2, y_2, and z_2. Note that because the rotation was about the z_1 axis, the z_2 component value equals z_1.

2) Next, rotate about the y_2-axis direction by angle Θ to yield a third coordinate system where the vector is expressed as X_3 with components x_3, y_3, and z_3. For this case, rotation about the y_2 axis yields invariance relative to the y_2 component, i.e., $y_3 = y_2$.

3) Finally, rotate about the x_3 axis direction by angle Φ to yield the vector X_B with body-axis components of x_B, y_B, and z_B. In this case, the rotation about the x_3 axis results in the x_3 and x_B components being the same.

The sequence of Euler angle transformations is shown in Fig. 4.1. Each of the three individual coordinate rotations resemble the T_θ orthogonal transformation

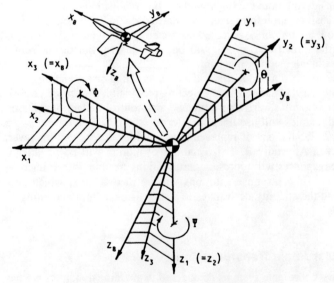

Fig. 4.1 Euler angle rotation sequence.

AIRFRAME EQUATIONS OF MOTION

described by Eq. (D.13) in Appendix D with only minor changes required due to differences in axis orientations. It is important to recognize that the transformations do not alter the vector in either magnitude or spatial orientation. The transformation sequence may be expressed symbolically as

$$X_1 \Longrightarrow T_\Psi \Longrightarrow X_2 \Longrightarrow T_\Theta \Longrightarrow X_3 \Longrightarrow T_\Phi \Longrightarrow X_B$$

The following symbology will be used for convenience in notation to describe the trigonometric functions used in the transformations:

$$C_\Theta = \cos \Theta; \quad C_\Phi = \cos \Phi; \quad C_\Psi = \cos \Psi$$
$$S_\Theta = \sin \Theta; \quad S_\Phi = \sin \Phi; \quad S_\Psi = \sin \Psi$$

1) Transformation of vector $X_1 \Longrightarrow T_\Psi \Longrightarrow X_2$ may be seen in Fig. 4.2a. In matrix notation, the transformation is

$$\begin{Bmatrix} x_2 \\ y_2 \\ z_2 \end{Bmatrix} = \begin{bmatrix} C_\Psi & S_\Psi & 0 \\ -S_\Psi & C_\Psi & 0 \\ 0 & 0 & 1 \end{bmatrix} \begin{Bmatrix} x_1 \\ y_1 \\ z_1 \end{Bmatrix} \quad z_1 = z_2$$

$$X_2 = T_\Psi X_1 \quad \text{or} \quad X_1 = T_\Psi^T X_2$$

2) Transformation of vector $X_2 \Longrightarrow T_\Theta \Longrightarrow X_3$ may be seen in Fig. 4.2b. In matrix notation, the transformation is

$$\begin{Bmatrix} x_3 \\ y_3 \\ z_3 \end{Bmatrix} = \begin{bmatrix} C_\Theta & 0 & -S_\Theta \\ 0 & 1 & 0 \\ S_\Theta & 0 & C_\Theta \end{bmatrix} \begin{Bmatrix} x_2 \\ y_2 \\ z_2 \end{Bmatrix} \quad y_2 = y_3$$

$$X_3 = T_\Theta X_2 \quad \text{or} \quad X_2 = T_\Theta^T X_3$$

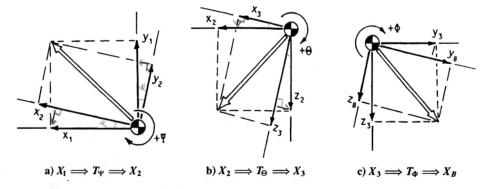

a) $X_1 \Longrightarrow T_\Psi \Longrightarrow X_2$ b) $X_2 \Longrightarrow T_\Theta \Longrightarrow X_3$ c) $X_3 \Longrightarrow T_\Phi \Longrightarrow X_B$

Fig. 4.2 Euler angle transformations.

3) Transformation of vector $X_3 \Longrightarrow T_\Phi \Longrightarrow X_B$ may be seen in Fig. 4.2c. In matrix notation, the transformation is

$$\begin{Bmatrix} x_B \\ y_B \\ z_B \end{Bmatrix} = \begin{bmatrix} 1 & 0 & 0 \\ 0 & C_\Phi & S_\Phi \\ 0 & -S_\Phi & C_\Phi \end{bmatrix} \begin{Bmatrix} x_3 \\ y_3 \\ z_3 \end{Bmatrix}$$

$$X_B = T_\Phi X_3 \quad \text{or} \quad X_3 = T_\Phi^T X_B$$

The final transformation from inertial coordinates is the ordered multiplication of the individual transformation matrices as shown by

$$\begin{Bmatrix} x_B \\ y_B \\ z_B \end{Bmatrix} = T_\Phi T_\Theta T_\Psi \begin{Bmatrix} x_1 \\ y_1 \\ z_1 \end{Bmatrix} \tag{4.1}$$

where each of the transformation matrices represents an orthogonal matrix; i.e., the transpose equals its inverse. It is straightforward to express a vector in inertial coordinates when body-axis coordinates are known by the matrix transpose relations for a product of matrices [cf. Eq. (D.8) in Appendix D]. Therefore,

$$\begin{Bmatrix} x_1 \\ y_1 \\ z_1 \end{Bmatrix} = [T_\Phi \quad T_\Theta \quad T_\Psi]^T \begin{Bmatrix} x_B \\ y_B \\ z_B \end{Bmatrix} \tag{4.2}$$

The total transformation expression of Eq. (4.1) can be obtained by a systematic multiplication process, i.e.,

$$T_\Theta T_\Psi = \begin{bmatrix} C_\Theta & 0 & -S_\Theta \\ 0 & 1 & 0 \\ S_\Theta & 0 & C_\Theta \end{bmatrix} \begin{bmatrix} C_\Psi & S_\Psi & 0 \\ -S_\Psi & C_\Psi & 0 \\ 0 & 0 & 1 \end{bmatrix} = \begin{bmatrix} C_\Theta C_\Psi & C_\Theta S_\Psi & -S_\Theta \\ -S_\Psi & C_\Psi & 0 \\ S_\Theta C_\Psi & S_\Theta S_\Psi & C_\Theta \end{bmatrix}$$

followed by

$$T_\Phi T_\Theta T_\Psi = \begin{bmatrix} 1 & 0 & 0 \\ 0 & C_\Phi & S_\Phi \\ 0 & -S_\Phi & C_\Phi \end{bmatrix} T_\Theta T_\Psi$$

For convenience in notation, let us define a (3 × 3) matrix $[S]$ where

$$S = T_\Phi T_\Theta T_\Psi$$

$$S = \begin{bmatrix} (C_\Psi C_\Theta) & (S_\Psi C_\Theta) & (-S_\Theta) \\ (C_\Psi S_\Theta S_\Phi - S_\Psi C_\Phi) & (S_\Psi S_\Theta S_\Phi + C_\Psi C_\Phi) & (C_\Theta S_\Phi) \\ (C_\Psi S_\Theta C_\Phi + S_\Psi S_\Phi) & (S_\Psi S_\Theta C_\Phi - C_\Psi S_\Phi) & (C_\Theta C_\Phi) \end{bmatrix} \tag{4.3}$$

Because three separate rotational transformations involving orthogonal matrices took place in expanding Eq. (4.3), the preceding seemingly complicated array involving products of trigonometric functions can be inverted just by transposition about the lead diagonal. The inverse of Eq. (4.3) is shown for sake of completeness as

$$S^T = \begin{bmatrix} (C_\psi C_\Theta) & (C_\psi S_\Theta S_\Phi - S_\psi C_\Phi) & (C_\psi S_\Theta C_\Phi + S_\psi S_\Phi) \\ (S_\psi C_\Theta) & (S_\psi S_\Theta S_\Phi + C_\psi C_\Phi) & (S_\psi S_\Theta C_\Phi - C_\psi S_\Phi) \\ (-S_\Theta) & (C_\Theta S_\Phi) & (C_\Theta C_\Phi) \end{bmatrix} \quad (4.4)$$

If the airframe velocity components were known in the fixed, body-axis coordinate system, then the absolute velocity components in an inertial reference frame would follow using the transformation relations of Eqs. (4.2) and (4.4), i.e., knowing V_B, find V_{inertial} by

$$\begin{Bmatrix} \frac{dx_1}{dt} \\ \frac{dy_1}{dt} \\ \frac{dz_1}{dt} \end{Bmatrix} = S^T \begin{Bmatrix} u \\ v \\ w \end{Bmatrix} \quad (4.5)$$

where $V_B = [u \; v \; w]^T$.

Equation (4.5) can be integrated with respect to time in order to obtain the position of the airframe c.g. in inertial space. This is a technique suitable for use in an aircraft simulator when it is desired to track a maneuvering aircraft. It is also necessary to know the time history of the Euler angles as well as the velocities in order to obtain a complete description for simulation purposes.

The airframe weight will enter into the equations of motion as a force component in the body-axis reference frame. First, consider the aircraft weight ($W = mg$) in an inertial reference frame. In inertial coordinates, the weight may be expressed as

$$W_{\text{inertial}} = [0 \; 0 \; mg]^T$$

The weight components in body axes are found using the transformation of Eq. (4.3), i.e.,

$$W_{\text{body}} = [S] \begin{Bmatrix} 0 \\ 0 \\ mg \end{Bmatrix} = (mg) \begin{Bmatrix} -S_\Theta \\ C_\Theta S_\Phi \\ C_\Theta C_\Phi \end{Bmatrix} = \begin{Bmatrix} F_x \\ F_y \\ F_z \end{Bmatrix}_{\text{gravity}} \quad (4.6)$$

It is also important to recognize that not all transformation matrices are orthogonal. Consider the situation of an aircraft experiencing an angular rotation rate vector Ω with body-axis components defined by

$$\Omega = \begin{Bmatrix} p \\ q \\ r \end{Bmatrix}$$

where

p = roll rate about x axis, $d\phi/dt$
q = pitch rate about y axis, $d\theta/dt$
r = yaw rate about z axis, $d\psi/dt$

The vector Ω can also be described by components aligned with the Euler angle rates, $d\Phi/dt$, $d\Theta/dt$, and $d\Psi/dt$ defined by $\Delta\Omega_\Phi$, $\Delta\Omega_\Theta$, and $\Delta\Omega_\Psi$, respectively. Here

$$\Omega = \Delta\Omega_\Phi + \Delta\Omega_\Theta + \Delta\Omega_\Psi$$

The $\{\Delta\Omega\}_\Phi$ term can be put into body-axis components by

$$\Delta\Omega_\Phi = \{\Delta\Omega\}_\Phi = T_\Phi \begin{Bmatrix} \frac{d\Phi}{dt} \\ 0 \\ 0 \end{Bmatrix}$$

Next, the $\{\Delta\Omega\}_\Theta$ term can be put into body-axis components by

$$\Delta\Omega_\Theta = \{\Delta\Omega\}_\Theta = T_\Phi T_\Theta \begin{Bmatrix} 0 \\ \frac{d\Theta}{dt} \\ 0 \end{Bmatrix}$$

Last, the $\{\Delta\Omega\}_\Psi$ term can be put into body-axis components by

$$\Delta\Omega_\Psi = \{\Delta\Omega\}_\Psi = T_\Phi T_\Theta T_\Psi \begin{Bmatrix} 0 \\ 0 \\ \frac{d\Psi}{dt} \end{Bmatrix}$$

These three matrix transformations combined with a slight amount of matrix partitioning yields a single matrix transformation relating the body-axis angular rates to the Euler angle rates, i.e.,

$$\begin{Bmatrix} p \\ q \\ r \end{Bmatrix} = \begin{bmatrix} 1 & 0 & -S_\Theta \\ 0 & C_\Phi & C_\Theta S_\Phi \\ 0 & -S_\Phi & C_\Theta C_\Phi \end{bmatrix} \begin{Bmatrix} \frac{d\Phi}{dt} \\ \frac{d\Theta}{dt} \\ \frac{d\Psi}{dt} \end{Bmatrix} \quad (4.7)$$

Because the Euler angle rate components of Ω are not normal to each other, the transformation of Eq. (4.7) is not an orthogonal matrix. However, because the transformation matrix is not singular, an inverse does exist. In this manner, the Euler angle rates in terms of the body-axis angle rates can be found by the following relationship that is valid subject to the constraint that $|\Theta| \neq \pi/2$:

$$\begin{Bmatrix} \frac{d\Phi}{dt} \\ \frac{d\Theta}{dt} \\ \frac{d\Psi}{dt} \end{Bmatrix} = \begin{bmatrix} 1 & S_\Phi \tan\Theta & C_\Phi \tan\Theta \\ 0 & C_\Phi & -S_\Phi \\ 0 & S_\Phi \sec\Theta & C_\Phi \sec\Theta \end{bmatrix} \begin{Bmatrix} p \\ q \\ r \end{Bmatrix} \quad (4.8)$$

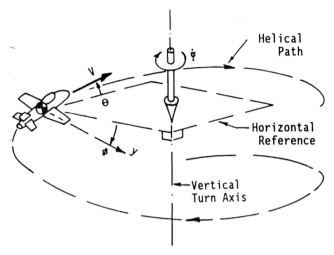

Fig. 4.3 Aircraft in a steady climbing turn.

Integration of the equations, which in general requires numerical techniques, can be performed to obtain the time history of the Euler angles $\Psi(t)$, $\Theta(t)$, and $\Phi(t)$. These comments have particular application when numerically establishing the airframe orientation in a flight simulator.

An application of Eq. (4.7) from a static stability and control viewpoint corresponds to an aircraft in a steady, banked climbing (or descending) turn, which is sketched on Fig. 4.3 for the situation of a positive (right-hand turn) turn rate of $\Omega = d\Psi/dt$. The aircraft body axes are oriented such that the velocity vector V is tangential to the helical path and aligned with the x axis passing through the c.g. Note the following two points.

1) The velocity vector is inclined at an Euler angle Θ relative to the horizontal plane.

2) The aircraft z axis is inclined at an Euler angle Φ with rotation being about the x axis.

The induced body-axis angular rates may be found by using Eq. (4.7).

$$\begin{Bmatrix} p \\ q \\ r \end{Bmatrix} = \begin{Bmatrix} -S_\Theta \\ C_\Theta S_\Phi \\ C_\Theta C_\Phi \end{Bmatrix} \frac{d\Psi}{dt} \qquad (4.9)$$

The limiting conditions to note from Eq. (4.9) include the following.

1) When the climb angle is vertical (i.e., $\Theta = \pi/2$), the aircraft's body (stability axis) roll rate is the negative of $d\Psi/dt$. The other two body-axis rates, q and r, are zero.

2) When the climb angle is zero (i.e., a steady, level banked turn), the turn rate translates into induced body-axis angular rates of

$$p = 0; \qquad q = S_\Phi \left(\frac{d\Psi}{dt} \right); \qquad \text{and} \qquad r = C_\Phi \left(\frac{d\Psi}{dt} \right)$$

A static stability and control analysis of an aircraft in a steady, climbing turn should include the influence of the p, q, and r induced angular velocities on the damping-type stability derivatives, as well as the influence of Θ and Ψ Euler angles on the lift and side-force coefficients. Example 3.4 illustrates the elevator control required during a constant altitude, steady turn using an induced pitch angle rate that is in accord with Eq. (4.9).

4.3 Rotation of a Rigid Body

The material of this section is relative to a rotating rigid body and includes a review of elementary vector operations, a description of the mass moment of inertia matrix, the principles of matrix diagonalization, development of the angular momentum matrix, and the momentum conservation relations in both fixed and rotating coordinate systems.

4.3.1 Kinetic Energy of a Rigid Body in Pure Rotation

Consider a fixed coordinate system having unit vectors e_1, e_2, and e_3 in three-dimensional space as shown in Fig. D.1 in Appendix D. The magnitudes in the direction of the unit vector, x_1, x_2, and x_3, form the components of a vector defined as R. A solid body with its c.g. located at the origin, as shown in Fig. 4.4, is rotating with an angular velocity of Ω with corresponding components in the $\{\Omega\}$ matrix of ω_1, ω_2, and ω_3.

The body will be considered as having N elements of mass, defined by Δm_i, for $i = 1, 2, \ldots, N$, where each element is located at $\{R_i\}$.

Each element has a velocity induced by the rotation, which is expressed by the vector cross product as

$$V_i = \Omega \times R_i \quad \text{or} \quad \{V_i\} = [\hat{\omega}]\{R_i\} \tag{4.10}$$

where the use of $[\hat{\omega}]$ as a matrix form of a vector cross product is described in Appendix D.

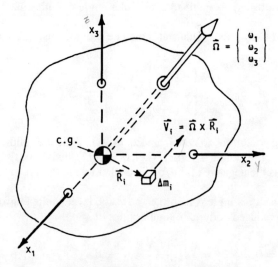

Fig. 4.4 Rigid body in rotation.

An expansion of the velocity components for the ith mass element yields

$$\{V_i\} = \begin{Bmatrix} \omega_2 x_3 - \omega_3 x_2 \\ \omega_3 x_1 - \omega_1 x_3 \\ \omega_1 x_2 - \omega_2 x_1 \end{Bmatrix} \quad (4.11)$$

The contribution to kinetic energy by the ith mass element is related to the dot or inner product of the velocity vector with itself, i.e.,

$$\Delta T_i = \tfrac{1}{2} \Delta m_i \{V_i\}^T \{V_i\}$$

The total kinetic energy, denoted by the scalar quantity T, is obtained by summing the individual elements for $i = 1, \ldots, N$, i.e.,

$$T = \frac{1}{2} \sum_{i=1}^{N} \Delta m_i \{V_i\}^T \{V_i\} \quad (4.12)$$

The inner product of the velocity, $\{V_i\}^T\{V_i\}$, is a scalar quantity applicable to the ith mass element. Expansion of the inner product yields

$$\{V_i\}^T\{V_i\} = \omega_1^2(x_2^2 + x_3^2) + \omega_2^2(x_1^2 + x_3^2) + \omega_3^2(x_1^2 + x_2^2)$$
$$- 2\omega_1\omega_2(x_1 x_2) - 2\omega_1\omega_3(x_1 x_3) - 2\omega_2\omega_3(x_2 x_3) \quad (4.13)$$

Finally, we proceed to an integral form of Eq. (4.12) in the limit by letting the index $N \to \infty$ and $\Delta m_i \to dm$. Then

$$T = \frac{1}{2}\omega_1^2 \int_{\text{Vol}} (x_2^2 + x_3^2)\,dm + \frac{1}{2}\omega_2^2 \int_{\text{Vol}} (x_1^2 + x_3^2)\,dm + \frac{1}{2}\omega_3^2 \int_{\text{Vol}} (x_1^2 + x_2^2)\,dm$$
$$- \omega_1\omega_2 \int_{\text{Vol}} (x_1 x_2)\,dm - \omega_1\omega_3 \int_{\text{Vol}} (x_1 x_3)\,dm - \omega_2\omega_3 \int_{\text{Vol}} (x_2 x_3)\,dm \quad (4.14)$$

The integral expressions are defined as mass moments of inertia, i.e.,

$$I_1 = \int_{\text{Vol}} (x_2^2 + x_3^2)\,dm$$

$$I_2 = \int_{\text{Vol}} (x_1^2 + x_3^2)\,dm$$

$$I_3 = \int_{\text{Vol}} (x_1^2 + x_2^2)\,dm$$

along with the cross products of inertia, i.e.,

$$I_{12} = I_{21} = \int_{\text{Vol}} (x_1 x_2)\,dm$$

$$I_{13} = I_{31} = \int_{\text{Vol}} (x_1 x_3)\,dm$$

$$I_{23} = I_{32} = \int_{\text{Vol}} (x_2 x_3)\,dm$$

Using these definitions for mass moments of inertia, the total rigid-body kinetic energy due to rotation becomes

$$T = \tfrac{1}{2}(I_1\omega_1^2 + I_2\omega_2^2 + I_3\omega_3^2 - 2I_{12}\omega_1\omega_2 - 2I_{13}\omega_1\omega_3 - 2I_{23}\omega_2\omega_3) \quad (4.15)$$

or, alternatively, as a quadratic in matrix form

$$T = \tfrac{1}{2}\{\Omega\}^T[I_m]\{\Omega\} \quad (4.16)$$

where the symmetric mass moment of inertia matrix is given by

$$[I_m] = \begin{bmatrix} I_1 & -I_{12} & -I_{13} \\ -I_{12} & I_2 & -I_{23} \\ -I_{13} & -I_{23} & I_3 \end{bmatrix}$$

4.3.2 Angular Momentum

Newton's laws dealing with the conservation of momentum apply to both the linear and angular forms of momentum. In a fixed inertial frame of reference (i.e., nonrotating coordinates) the linear conservation law may be expressed by

$$\sum_{\text{body}} F = \frac{d(mV)}{dt} \quad \text{or} \quad \sum_{\text{body}} \{F\} = \frac{d(m\{V\})}{dt}$$

where linear momentum considered at the c.g. is given by $m\{V\}$ for a body of mass m. This relation states that the external forces applied to the body result in a change in linear momentum.

Similarly, conservation of angular momentum states that the external moments applied to the body result in a change of angular momentum, which shall be defined using the vector quantity H or $\{H\}$. That is,

$$\sum_{\text{body}} M = \frac{d(H)}{dt} \quad \text{or} \quad \sum_{\text{body}} \{M\} = \frac{d(\{H\})}{dt}$$

The angular momentum vector quantity is the integrated sum of the moments of linear momentum from all of the infinitesimal elements (shown in Fig. 4.4 for the ith element). Here

$$\{H\} = \int_{\text{Vol}} \{R \times (\Omega \times R)\} dm \quad (4.17)$$

where the term $(\Omega \times R)$ is the local velocity contribution to the linear momentum from the infinitesimal element of mass dm whose position is given by R. The vector triple product in Eq. (4.17) can be shown to have the following equivalence in matrix notation:

$$\{A\} \times (\{B\} \times \{C\}) = \{B\}(\{A\}^T\{C\}) - \{C\}(\{A\}^T\{B\}) \quad (4.18)$$

For the situation of Eq. (4.17), substitute as follows. Let

$$\{R\} = \{A\}, \quad \{\Omega\} = \{B\}, \quad \text{and} \quad \{R\} = \{C\}$$

then Eq. (4.17) becomes

$$\{H\} = \int_{\text{Vol}} \{\{\Omega\}|R|^2 - \{R\}(\{R\}^T\{\Omega\})\}\,dm$$

where

$|R|^2$ = scalar quantity, $(x_1^2 + x_2^2 + x_3^2)$
$\{R\}^T\{\Omega\}$ = scalar quantity, $(\omega_1 x_1 + \omega_2 x_2 + \omega_3 x_3)$

Expansion of the preceding expression yields

$$\begin{Bmatrix} H_1 \\ H_2 \\ H_3 \end{Bmatrix} = \int_{\text{Vol}} \left\{ \begin{Bmatrix} (x_2^2 + x_3^2)\omega_1 \\ (x_1^2 + x_3^2)\omega_2 \\ (x_1^2 + x_2^2)\omega_3 \end{Bmatrix} - \begin{Bmatrix} x_1 x_2 \omega_2 + x_1 x_3 \omega_3 \\ x_1 x_2 \omega_1 + x_2 x_3 \omega_3 \\ x_1 x_3 \omega_1 + x_2 x_3 \omega_2 \end{Bmatrix} \right\} dm \quad (4.19)$$

Once the mass moment of inertia terms are recognized in Eq. (4.19), the angular momentum vector may be compactly stated as

$$\{H\} = [I_m]\{\Omega\} \quad (4.20)$$

In light of Eq. (4.20), it is worthwhile to reconsider the angular momentum conservation relation as previously stated, i.e.,

$$\sum_{\text{body}} \{M\} = \frac{d(\{H\})}{dt} = \frac{d[I_m]}{dt}\{\Omega\} + [I_m]\frac{d\{\Omega\}}{dt}$$

Unless there is symmetry in the mass moment of inertia matrix, the time derivative of $[I_m]$ in a fixed coordinate system will be difficult to evaluate because, when the body rotates under the influence of external moments, $[I_m]$ becomes a time-dependent function $[I_m(t)]$.

Note: The preceding observation provides the motivation to express the conservation laws in a rotating coordinate system fixed to the body because analysis becomes much simpler when the mass moment of inertia matrix is invariant with time.

4.3.3 Momentum Conservation in Rotating Coordinates

Based on the preceding comments, it is appropriate to consider a vector in a rotating vs a fixed coordinate system in order to keep the mass moment of inertia derivative invariant with time. In expressing Newton's momentum conservation laws, it is necessary to relate the time rate of change of the linear (and angular) momentum vectors to the applied forces (and moments). For sake of generality, let us consider a vector $\{A\}$, which is expressed as a function of time, i.e., $\{A(t)\}$. It should be recognized that the meaning of the vector is physically invariant with the choice of coordinate system. Let us assume that the vector is described in a rotating coordinate system.

In a vector context, A may be described as

$$A = ia_x + ja_y + ka_z \quad \text{or} \quad \{A\} = [a_x \quad a_y \quad a_z]^T$$

where the unit vectors $i, j,$ and k are attached to a coordinate system that is rotating with an angular velocity of Ω. The rotation vector is given by

$$\Omega = i\omega_x + j\omega_y + k\omega_z \quad \text{or} \quad \{\Omega\} = [\omega_x \quad \omega_y \quad \omega_z]^T$$

The time derivative, using the chain rule, can be expressed as

$$\frac{dA}{dt} = i\frac{\partial a_x}{\partial t} + j\frac{\partial a_y}{\partial t} + k\frac{\partial a_z}{\partial t} + a_x\frac{\partial i}{\partial t} + a_y\frac{\partial j}{\partial t} + a_z\frac{\partial k}{\partial t}$$

where

$\partial a_x/\partial t$ = partial derivative of the $a_x(t)$ term while considering the $i, j,$ and k unit vectors in the $x, y,$ and z coordinates as fixed with respect to time

$\partial i/\partial t$ = partial derivative of the i unit vector while considering the $a_x, a_y,$ and a_z components as fixed with respect to time

The time derivative of the i unit vector is indicated as an example in Fig. 4.5 by the velocity of the endpoint (P) of the unit vector as it rotates due to the angular velocity $\{\Omega\}$. The trajectory of (P) may be described as the hodograph of i due to $\{\Omega\}$. In addition, it will be noted in Fig. 4.5 that the velocity of the unit vector i at any instant will be normal to both the angular velocity and the unit vectors.

Therefore, the partial derivative of the rotating unit vector is

$$\frac{\partial i}{\partial t} = \Omega \times i = [\hat{\omega}][1 \quad 0 \quad 0]^T$$

and, similarly,

$$\frac{\partial j}{\partial t} = \Omega \times j = [\hat{\omega}][0 \quad 1 \quad 0]^T; \quad \frac{\partial k}{\partial t} = \Omega \times k = [\hat{\omega}][0 \quad 0 \quad 1]^T$$

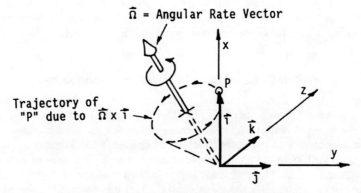

Fig. 4.5 Time derivative of a rotating unit vector.

By adding the terms, the total derivative is obtained as

$$\frac{d\{A\}}{dt} = \frac{\partial\{A\}}{\partial t} + [\hat{\omega}]\{A\} \qquad (4.21)$$

which represents the time derivative of a vector A in a rotating coordinate system. Applying Eq. (4.21) to Newton's momentum conservation laws, with the rotating coordinate system located at the c.g. of the body, yields

$$\sum_{\text{body}}\{F\} = \frac{d(m\{V\})}{dt} = m\frac{\partial\{V\}}{\partial t} + m[\hat{\omega}]\{V\} \qquad (4.22)$$

$$\sum_{\text{body}}\{M\} = \frac{d([I_m]\{\Omega\})}{dt} = [I_m]\frac{\partial\{\Omega\}}{\partial t} + [\hat{\omega}][I_m]\{\Omega\} \qquad (4.23)$$

where

m = mass of the body
$\{V\}$ = velocity, $[u \; v \; w]^T$
$\{H\}$ = angular momentum, $[I_m]\{\Omega\}$
$[I_m]$ = mass moment of inertia matrix fixed to body axes
$\{\Omega\}$ = rotation rate in body axes, $[\omega_x \; \omega_y \; \omega_z]^T = [p \; q \; r]^T$
$[\hat{\omega}]$ = skew symmetric matrix of rotation rates, cf. Appendix D

The mass moment of inertia matrix simplifies from the form of Eq. (4.16) for a body that is symmetric (e.g., an airplane) about the x–z plane because it is evident that

$$I_{12} = I_{21} = \int_{\text{Vol}} (x_1 x_2) dm = \int_{\text{Vol}} (xy) dm = I_{xy} = 0$$

$$I_{23} = I_{32} = \int_{\text{Vol}} (x_2 x_3) dm = \int_{\text{Vol}} (yz) dm = I_{yz} = 0$$

The mass moment of inertia matrix for an aircraft in body-axis notation, where the subscripts 1, 2, and 3 are associated with the x, y, and z axes, respectively, becomes

$$[I_m] = \begin{bmatrix} I_1 & 0 & -I_{13} \\ 0 & I_2 & 0 \\ -I_{13} & 0 & I_3 \end{bmatrix} = \begin{bmatrix} I_x & 0 & -I_{xz} \\ 0 & I_y & 0 \\ -I_{xz} & 0 & I_z \end{bmatrix} \qquad (4.24)$$

The mass moment of inertia matrix for an aircraft will change due to rotation of the reference body-axis coordinates about the lateral y axis. The transformation relations can be developed from the kinetic energy description for a rotating body [Eq. (4.16)], when the rotation vector is assumed to lie in the x–z plane.

Example 4.1: Principal Mass Moments of Inertia

Consider the kinetic energy for an aircraft that is rotating solely due to p (roll rate) and r (yaw rate) angular velocity components as shown in Fig. 4.6.

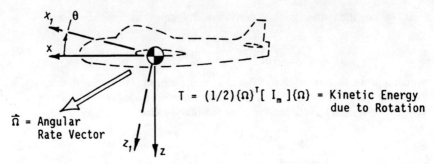

Fig. 4.6 Rotation of body axes by angle θ.

The kinetic energy associated with the assumed angular rate vector may be expressed by Eq. (4.16) as

$$T = \frac{1}{2}[p \ r]\begin{bmatrix} I_x & -I_{xz} \\ -I_{xz} & I_z \end{bmatrix}\begin{Bmatrix} p \\ r \end{Bmatrix} \quad (4.25)$$

which when expanded into the scalar, quadratic form becomes

$$T = \frac{1}{2}(I_x p^2 - 2I_{xz} pr + I_z r^2)$$

Next, consider a rotation of the x–z coordinate system by a nose-up rotation of $+\theta$ into the x_1–z_1 coordinate system as shown in Fig. 4.6. The coordinate rotation alters the angular rate vector (a first-order tensor quantity) by

$$\begin{Bmatrix} p \\ r \end{Bmatrix} = \begin{bmatrix} C_\theta & S_\theta \\ -S_\theta & C_\theta \end{bmatrix}\begin{Bmatrix} p_1 \\ r_1 \end{Bmatrix} \quad \text{or} \quad \{\Omega\} = [T_\theta]\{\Omega_1\}$$

Substituting this vector rotation relationship and using the orthogonal matrix property of the T_θ transformation matrix provide an alternate expression for the kinetic energy in the rotated coordinate system. Remember that the scalar value of kinetic energy is invariant with a coordinate rotation. The kinetic energy expression becomes

$$T = \frac{1}{2}[p_1 \ r_1][T_\theta]^T \begin{bmatrix} I_x & -I_{xz} \\ -I_{xz} & I_z \end{bmatrix}[T_\theta]\begin{Bmatrix} p_1 \\ r_1 \end{Bmatrix}$$

which when expanded into the scalar, quadratic form becomes

$$T = \frac{1}{2}(I'_x p_1^2 - 2I'_{xz} p_1 r_1 + I'_z r_1^2)$$

where the alternate form of $[I'_m]$ as defined in Eq. (4.26) is used in the quadratic expansion, i.e.,

$$[I'_m] = [T_\theta]^T [I_m][T_\theta] \quad (4.26)$$

This expression, showing the effect of a coordinate rotation on the mass moment of inertia matrix, is the property of a second-order tensor quantity. The reader will recognize other quantities with similar coordinate rotation relations as being the stress and strain matrices encountered in solid mechanics.

The $[I'_m]$ matrix of Eq. (4.26), when expanded, will include products of sines and cosines,

$$I'_x = I_x C_\theta^2 + I_z S_\theta^2 + I_{xz} S_{2\theta}$$
$$I'_z = I_x S_\theta^2 + I_z C_\theta^2 - I_{xz} S_{2\theta} \quad (4.27)$$
$$I'_{xz} = \tfrac{1}{2}(I_z - I_x) S_{2\theta} + I_{xz} C_{2\theta}$$

When the cross product of inertia term I'_{xz} is zero, then the x_1–z_1 coordinate system becomes the principal axes and the corresponding mass moments of inertia are defined as the principal values of the mass moment of inertia.

The rotation angle θ required to identify the principal axes is obtained from Eq. (4.27) by the relation

$$\theta = \tfrac{1}{2}\arctan\{2 I_{xz}/(I_x - I_z)\} \quad (4.28)$$

4.4 Airframe Equations of Motion

The full set of nonlinear equations describing the motion for an airframe having rotation in a body-fixed coordinate system can be obtained by performing the vector operations of Eqs. (4.22) and (4.23). The relations between external forces and moments and the corresponding momentum conservation laws are given by

$$\begin{Bmatrix} F_x \\ F_y \\ F_z \end{Bmatrix} = m \begin{Bmatrix} \frac{du}{dt} \\ \frac{dv}{dt} \\ \frac{dw}{dt} \end{Bmatrix} + m \begin{Bmatrix} qw - rv \\ ru - pw \\ pv - qu \end{Bmatrix} \quad (4.29)$$

$$\begin{Bmatrix} M_x \\ M_y \\ M_z \end{Bmatrix} = \begin{Bmatrix} I_x \dot{p} - I_{xz}\dot{r} \\ I_y \dot{q} \\ I_z \dot{r} - I_{xz}\dot{p} \end{Bmatrix} + \begin{Bmatrix} qr(I_z - I_y) - pq I_{xz} \\ pr(I_x - I_z) + (p^2 - r^2) I_{xz} \\ pq(I_y - I_x) + qr I_{xz} \end{Bmatrix} \quad (4.30)$$

It will be recognized in Eqs. (4.29) and (4.30) that a simplification would occur in the expressions if the body axes were aligned with the principal mass moment of inertia axes (i.e., $I_{xz} = 0$). Nevertheless, inertial related quantities such as $mq(t)w(t)$ and $mq(t)r(t)$ will lead to the set of six equations being nonlinearly coupled. In general, solutions to the full set of nonlinear equations require the use of numerical computing techniques. However, considerable insight into the airframe dynamics may be gained by simplifying the set of governing airframe equations of motion [Eqs. (4.29) and (4.30)] into a linearized form based on small perturbation motions about an initial trimmed flight condition.

4.5 Linearized Equations of Motion

Although it is recognized that an aircraft can change its flight configuration quite rapidly (i.e., transition from climb to cruise, decelerate from descent to approach, deflect wing flaps and extend landing gear), the linearization process will be considered relative to a fixed flight condition. The subsequent linearized equations of motion will allow a determination of an airframe's small amplitude dynamic behavior about a trim point. Important assumptions for the perturbation analysis include the following.

1) Small motion dynamics will be initiated from wings-level flight with the aircraft controls and thrust set for a trim, static equilibrium flight condition.

2) The body axes will be aligned with the initial velocity vector, and as such, will become known as the airframe's stability axes.

The forces acting on the airframe will be due both to the aerodynamic and gravitational terms and may be expressed as

$$F_x = X_0 + \Delta X(t) - mg(\sin\Theta)$$
$$F_y = Y_0 + \Delta Y(t) + mg(\cos\Theta)(\sin\Phi) \quad (4.31)$$
$$F_z = Z_0 + \Delta Z(t) + mg(\cos\Theta)(\cos\Phi)$$

where the mg related terms are due to body-weight components and the Euler angles Θ and Φ are as developed in Eq. (4.6). Terms such as X_0 represent the equilibrium value of the corresponding force component whereas $\Delta X(t)$ is the force term due to the time-varying aerodynamic forces caused by aircraft attitude, motion, and control deflections from trim.

The moments acting on the aircraft relative to the c.g. and oriented in stability axes will follow the L, M, and N convention as shown in Fig. 1.1. The moment components are defined as

$$M_x = L = L_0 + \Delta L(t)$$
$$M_y = M = M_0 + \Delta M(t) \quad (4.32)$$
$$M_z = N = N_0 + \Delta N(t)$$

where the 0 subscript refers to initial trim conditions whereas the typical $\Delta L(t)$ type of term will correspond to the time-varying aerodynamic moments that arise due to aircraft attitude, motion, and control deflections from trim.

Before linearization, the set of six coupled equations of motion as contained in Eqs. (4.29–4.32) will appear as

$$\begin{cases} X_0 + \Delta X(t) - mg(\sin\Theta) = m(\dot{u} + qw - rv) \\ Y_0 + \Delta Y(t) + mg(\cos\Theta)(\sin\Phi) = m(\dot{v} + ru - pw) \\ Z_0 + \Delta Z(t) + mg(\cos\Theta)(\cos\Phi) = m(\dot{w} + pv - qu) \\ L_0 + \Delta L(t) = I_x\dot{p} - I_{xz}\dot{r} + (I_z - I_y)qr - I_{xz}pq \\ M_0 + \Delta M(t) = I_y\dot{q} + (I_x - I_z)pr + I_{xz}(p^2 - r^2) \\ N_0 + \Delta N(t) = I_z\dot{r} - I_{xz}\dot{p} + (I_y - I_x)pq + I_{xz}qr \end{cases} \quad (4.33)$$

In a manner akin to the force and moment statements, linearization of the time-dependent variables will start with a definition of an initial equilibrium value plus a small perturbation term, i.e.,

$$\begin{aligned}
u &= u_0 + u(t); & p &= p_0 + p(t) = p(t) \\
v &= v_0 + v(t) = v(t); & q &= q_0 + q(t) = q(t) \\
w &= w_0 + w(t) = w(t); & r &= r_0 + r(t) = r(t) \\
\Theta &= \Theta_0 + \theta(t); & \Phi &= \Phi_0 + \phi(t) = \phi(t)
\end{aligned} \quad (4.34)$$

It should be noted in Eq. (4.34) that the initial condition assumption of the motion starting from wings-level flight with body axes corresponding to the airframe stability axes allows the elimination of terms such as v_0, w_0, p_0, q_0, r_0, and Φ_0. In addition, $u_0 = V_0$ by the assumptions.

At time $t = 0$, when the small perturbation motions are to begin, the relations of Eq. (4.33) simplify to the static equilibrium conditions on the airframe force and moments, i.e.,

$$\begin{aligned}
X_0 - mg(\sin \Theta_0) &= 0 \\
Y_0 &= 0 \quad (\text{since } \Phi_0 = 0) \\
Z_0 + mg(\cos \Theta_0) &= 0 \\
L_0 = M_0 = N_0 &= 0
\end{aligned} \quad (4.35)$$

The force relations involve trigonometric terms that can be linearized using the small perturbation angle assumptions, i.e.,

$$\begin{aligned}
\sin \Theta &= \sin \Theta_0 + (\cos \Theta_0)\theta(t) \quad \text{for } \theta(t) \ll 1 \\
\cos \Theta &= \cos \Theta_0 - (\sin \Theta_0)\theta(t) \\
\sin \Phi &= \phi(t) \quad \text{for } \Phi_0 = 0 \quad \text{and} \quad \phi(t) \ll 1 \\
\cos \Phi &= 1.0
\end{aligned} \quad (4.36)$$

As an example of the linearization process, consider the X force relation of Eq. (4.33). Subtracting the initial condition for X_0 from Eq. (4.35) along with use of the identity concerning $\cos[\Theta_0 + \theta(t)]$ leaves

$$\Delta X(t) - mg(\cos \Theta_0)\theta(t) = m[\dot{u} + q(t)w(t) - r(t)v(t)]$$

The higher-order nonlinear terms such as $q(t)w(t)$ are also dropped during the linearization process based on order of magnitude considerations. The end result for the three linearized force equations is

$$\begin{aligned}
\Delta X(t) - mg(\cos \Theta_0)\theta(t) &= m\dot{u} = mV_0(\dot{u}/V_0) \\
\Delta Y(t) + mg(\cos \Theta_0)\phi(t) &= m(\dot{v} + rV_0) = mV_0(\dot{\beta} + r) \\
\Delta Z(t) - mg(\sin \Theta_0)\theta(t) &= m(\dot{w} - qV_0) = mV_0(\dot{\alpha} - q)
\end{aligned} \quad (4.37)$$

where

$\alpha(t)$ = angle of attack perturbation, $w(t)/V_0$
$\beta(t)$ = sideslip angle perturbation, $v(t)/V_0$

The linearized roll, pitch, and yawing moment equations are obtained more readily for the following reasons.

1) The initial conditions on L_0, M_0, and N_0 are all zero due to the airframe being initially at trim.

2) Higher-order terms such as $p(t)r(t)$ and $p(t)^2$ are dropped by order of magnitude considerations.

In summary, the linearized, small perturbation equations of motion are

$$\Delta X(t) - mg(\cos\Theta_0)\theta(t) = mV_0(\dot{u}/V_0)$$
$$\Delta Y(t) + mg(\cos\theta_0)\phi(t) = mV_0(\dot{\beta} + r)$$
$$\Delta Z(t) - mg(\sin\Theta_0)\theta(t) = mV_0(\dot{\alpha} - q)$$
$$\Delta L(t) = I_x\dot{p} - I_{xz}\dot{r} \quad (4.38)$$
$$\Delta M(t) = I_y\dot{q}$$
$$\Delta N(t) = I_z\dot{r} - I_{xz}\dot{p}$$

The aerodynamic force and moment terms remaining in Eq. (4.38) will be established using Taylor series expansions (cf. Chapter 2) of the perturbation variables as typically indicated by

$$\Delta X(t) = \Delta X[(u/V_0), \alpha, \dot{\alpha}, q, \delta, \beta, \dot{\beta}, p, r, \ldots]$$

Contributions to the Taylor series expansion of the $\Delta X(t)$ force term due to the influence of $\beta(t)$, $\dot{\beta}(t)$, $p(t)$, and $r(t)$ will not normally be a factor for fixed-wing aircraft. However, coupling with the lateral-directional related motions by the longitudinally oriented force (or moment) term would be of import when considering a rotary-winged vehicle. Because the emphasis of this book is focused on fixed-wing aircraft, the cross coupling of aerodynamic derivatives will not be considered in subsequent Taylor series expansions.

4.6 Matrix Formulation of the Equations of Motion

The linearized equations of motion may be separated into longitudinal and lateral-directional equation sets on the assumption that a fixed-wing aircraft is being considered. It is at this point in the mathematical developments that a logic branch would occur by introducing rotary-winged aircraft considerations to yield relationships, such as found in the text by Prouty.[2]

The aerodynamic forces and moments in Eq. (4.38) will be grouped as longitudinal: axial force $\Delta X(t)$, normal force $\Delta Z(t)$, and pitching moment $\Delta M(t)$ and as lateral directional: side force $\Delta Y(t)$, rolling moment $\Delta L(t)$, and yawing moment $\Delta N(t)$.

The first-order Taylor series expansions of the preceding terms (cf. Chapter 2) will include partial derivatives relative to rate terms such as $\dot{\alpha}$ ($= d\alpha/dt$), p ($= d\phi/dt$), and r ($= d\psi/dt$). Aerodynamic interpretations of the rate terms will be based on quasi-steady concepts. This assumption implies that the frequencies involved in the motions are sufficiently low so that it is reasonable to expect an aerodynamic transfer function describing the derivative to have unit gain and zero phase shift. By contrast, this assumption would not be valid in an aeroelastic (flutter) study where the frequencies would involve the coupling of aerodynamic and structural modes at frequencies 1–2 orders of magnitude higher than those involved solely in airframe flight mechanics.

Another assumption made in developing the stability derivatives is that any terms due to acceleration of the fluid (in this case air) that might give rise to apparent mass considerations will be neglected. Again, the assumption would not be valid if the frequencies involved were in the neighborhood of the airframe structural frequencies. In contrast, apparent mass terms have import in discussions of surface ship dynamics because the fluid (water) has a density approaching that of the vehicle.

4.6.1 Longitudinal Dynamics

The partial derivatives used in the Taylor series expansions of $\Delta X(t)$, $\Delta Z(t)$, and $\Delta M(t)$ will include terms such as

$$\frac{u(t)}{V}, \quad \alpha(t) = \frac{w(t)}{V}, \quad \dot{\alpha} = \frac{d\alpha}{dt}, \quad q = \frac{d\theta}{dt}, \quad \text{and} \quad \delta(t)$$

The subscript 0 on the initial freestream velocity will be dropped in the following presentations because it should be clear that the use of stability axes for the onset of the motion implies

$$\text{velocity} = [u_0^2 + v_0^2 + w_0^2]^{1/2} = V_0 = V$$

The axial velocity perturbation $u(t)$ is normalized relative to the freestream velocity V so that the perturbation derivative has a similar order of magnitude to the $\alpha(t)$ derivative when making the Taylor series expansions. The $\alpha(t)$ perturbation in the following derivations is the same as the $\Delta\alpha(t)$ term described in Sec. 2.2.2 (i.e., the Δ prefix is being dropped for the sake of notational simplicity).

The expansion for the aerodynamic axial force may be expressed as

$$\Delta X(t) = \Delta X_u[u(t)/V] + \Delta X_\alpha \alpha(t) + \Delta X_{\dot{\alpha}} \dot{\alpha} + \Delta X_q q(t) + \Delta X_\delta \delta(t) \quad (4.39)$$

where

$\Delta X_u = \Delta X$ force change per unit (u/V), $\partial \Delta X / \partial (u/V)$, lb
$\Delta X_\alpha =$ change of ΔX force per unit α, $\partial \Delta X / \partial \alpha$, lb · rad^{-1}
$\Delta X_{\dot{\alpha}} =$ change of ΔX force per unit $\dot{\alpha}$, $\partial \Delta X / \partial \dot{\alpha}$, lb · s · rad^{-1}
$\Delta X_q =$ change of ΔX force per unit q, $\partial \Delta X / \partial q$, lb · s · rad^{-1}
$\Delta X_\delta =$ change of ΔX force per unit δ, $\partial \Delta X / \partial \delta$, lb · rad^{-1}

The $\Delta X_{\dot{\alpha}}$ and ΔX_q partial derivative terms are frequently neglected because of their small value relative to the other terms in the expansion. The ΔX_α term has a strong dependence on the aircraft lift coefficient C_L.

The expansion for the aerodynamic normal force may be expressed as

$$\Delta Z(t) = \Delta Z_u[u(t)/V] + \Delta Z_\alpha \alpha(t) + \Delta Z_{\dot{\alpha}} \dot{\alpha} + \Delta Z_q q(t) + \Delta Z_\delta \delta(t) \quad (4.40)$$

where, typically, $\Delta Z_\alpha = \partial \Delta Z / \partial \alpha$ is the change of ΔZ_α force per unit α, lb × rad^{-1}.

The ΔZ_α term has a strong dependence on the aircraft lift-curve slope. The $\Delta Z_{\dot\alpha}$ and ΔZ_q terms in Eq. (4.40) will occasionally be given in listings of aircraft derivatives, but will not be dominant in their influence on airframe dynamics.

Similarly, the expansion for the pitching moment may be expressed as

$$\Delta M(t) = \Delta M_u[u(t)/V] + \Delta M_\alpha \alpha(t) + \Delta M_{\dot\alpha} \dot\alpha$$
$$+ \Delta M_q q(t) + \Delta M_\delta \delta(t) \qquad (4.41)$$

where, typically, $\Delta M_\alpha = \partial \Delta M / \partial \alpha$ is the ΔM moment change per unit α, ft × lb × rad^{-1}.

The ΔM_α term is dependent on the aircraft c.g. location relative to the control-fixed neutral point. ΔM_q in Eq. (4.41) is a function of the aircraft pitch-damping derivative and will be found to have a dominant influence on the damping behavior of the aircraft's faster (short-period) natural flight dynamic mode.

The preceding force and moment perturbation terms relate to a change in a force or moment due to a small change in a physical quantity. It is convenient to simplify the expressions of Eqs. (4.39–4.41) by using the following definitions of dimensional stability derivatives. Typically, let us define for subscript $i = \alpha, \dot\alpha, q,$ and δ

$$X_i = \Delta X_i / m$$
$$Z_i = \Delta Z_i / m$$
$$M_i = \Delta M_i / I_y$$

for subscript $u = \alpha, \dot\alpha, q,$ and δ

$$X_u = \Delta X_u / (mV)$$
$$Z_u = \Delta Z_u / (mV)$$
$$M_u = \Delta M_u / (I_y V)$$

After making these substitutions, the longitudinal portions of Eq. (4.38) take the form of dimensional stability derivatives for the coefficients, i.e.,

$$V(\dot u/V) = V X_u(u/V) + X_\alpha \alpha - g \cos \Theta_0 \theta + X_\delta \delta$$
$$(V - Z_{\dot\alpha})\dot\alpha = V Z_u(u/V) + Z_\alpha \alpha + (V + Z_q)q - g \sin \Theta_0 \theta + Z_\delta \delta$$
$$- M_{\dot\alpha}\dot\alpha + \dot q = V M_u(u/V) + M_\alpha \alpha + M_q q + M_\delta \delta$$

To complete these expressions into a state vector format, also consider the identity relation

$$\dot\theta = q \quad \leftarrow \text{fourth equation}$$

If we define a longitudinal state vector by

$$\{x\} = [u/V \quad \alpha \quad q \quad \theta]^T$$

along with a single control term by δ, then the coupled set of four, first-order, ordinary differential equations that describe the linearized airframe longitudinal dynamics become

$$[I_n]\{\dot{x}\} = [A_n]\{x\} + \{B_n\}\delta \qquad (4.42)$$

where

$$[I_n] = \begin{bmatrix} V & 0 & 0 & 0 \\ 0 & (V - Z_{\dot{\alpha}}) & 0 & 0 \\ 0 & -M_{\dot{\alpha}} & 1 & 0 \\ 0 & 0 & 0 & 1 \end{bmatrix} = \text{inertial matrix}$$

← place the fourth equation in the row that makes the matrix orthogonal

$$[A_n] = \begin{bmatrix} VX_u & X_\alpha & 0 & -g\cos\Theta_0 \\ VZ_u & Z_\alpha & (V + Z_q) & -g\sin\Theta_0 \\ VM_u & M_\alpha & M_q & 0 \\ 0 & 0 & 1 & 0 \end{bmatrix}$$

and

$$\{B_n\} = [X_\delta \quad Z_\delta \quad M_\delta \quad 0]^T$$

Because the inertial matrix $[I_n]$ is nonsingular, premultiplication of both sides of Eq. (4.42) by $[I_n]^{-1}$ will yield the recognized form of the governing state equations as

$$\{\dot{x}\} = [A]\{x\} + \{B\}\delta \qquad (4.43)$$

where

$[A]$ = longitudinal airframe plant matrix, $[I_n]^{-1}[A_n]$
$\{B\}$ = longitudinal airframe control matrix, $[I_n]^{-1}\{B_n\}$

The use of dimensional stability derivatives in Eq. (4.43), as described in Chapter 2 and listed in Table 4.1, will lead to all time histories being expressed in real time (seconds) and frequencies (radians/second).

In Eq. (4.43) both the plant matrix and the control vector have been altered so that the time derivative of the state vector $\{\dot{x}\}$ is presented in an uncoupled form. It may be shown that the plant matrix $[A]$ and control vector $\{B\}$ become

$$A_{11} = X_u \qquad A_{12} = X_\alpha/V \qquad A_{13} = 0 \qquad A_{14} = -(g/V)\cos\Theta_0$$

$$A_{21} = VZ_u/(V - Z_{\dot{\alpha}}) \qquad A_{22} = Z_\alpha/(V - Z_{\dot{\alpha}})$$

$$A_{23} = (V + Z_q)/(V - Z_{\dot{\alpha}}) \qquad A_{24} = -g\sin\Theta_0/(V - Z_{\dot{\alpha}})$$

$$A_{31} = VM_u + M_{\dot{\alpha}}VZ_u/(V - Z_{\dot{\alpha}}) \qquad A_{32} = M_\alpha + M_{\dot{\alpha}}Z_\alpha/(V - Z_{\dot{\alpha}})$$

$$A_{33} = M_q + M_{\dot{\alpha}}(V + Z_q)/(V - Z_{\dot{\alpha}}) \qquad A_{34} = 0$$

$$A_{41} = A_{42} = A_{44} = 0 \qquad A_{43} = 1$$

and

$$B_1 = X_\delta/V \qquad B_2 = Z_\delta/(V - Z_{\dot{\alpha}})$$

$$B_3 = M_\delta + M_{\dot{\alpha}}Z_\delta/(V - Z_{\dot{\alpha}}) \qquad B_4 = 0$$

Table 4.1 Dimensional stability derivatives

Term	Description	Units
X_u	$-\frac{QS}{mV}\left(2C_D + M\frac{\partial C_D}{\partial M}\right)$	s^{-1}
X_α	$\frac{QS}{m}\left(C_L - \frac{\partial C_D}{\partial \alpha}\right)$	ft·s^{-2}
$X_{\dot\alpha}$	$-\frac{QS}{m}\left(\frac{c}{2V}\right)\frac{\partial C_D}{\partial(\dot\alpha c/2V)}$	ft·s^{-1}
X_q	$-\frac{QS}{m}\left(\frac{c}{2V}\right)\frac{\partial C_D}{\partial qc/2V}$	ft·s^{-1}
X_δ	$-\frac{QS}{m}\frac{\partial C_D}{\partial \delta}$	ft·s^{-2}
Z_u	$-\frac{QS}{mV}\left(2C_L + M\frac{\partial C_L}{\partial M}\right)$	s^{-1}
Z_α	$-\frac{QS}{m}\left(C_D + \frac{\partial C_L}{\partial \alpha}\right)$	ft·s^{-2}
$Z_{\dot\alpha}$	$-\frac{QS}{m}\left(\frac{c}{2V}\right)\frac{\partial C_L}{\partial(\dot\alpha c/2V)}$	ft·s^{-1}
Z_q	$-\frac{QS}{m}\left(\frac{c}{2V}\right)\frac{\partial C_L}{\partial(qc/2V)}$	ft·s^{-1}
Z_δ	$-\frac{QS}{m}\frac{\partial C_L}{\partial \delta}$	ft·s^{-2}
M_u	$\frac{QSc}{I_y V}M\frac{\partial C_m}{\partial M}$	ft^{-1}·s^{-1}
M_α	$\frac{QSc}{I_y}\frac{\partial C_m}{\partial \alpha}$	s^{-2}
$M_{\dot\alpha}$	$\frac{QSc}{I_y}\left(\frac{c}{2V}\right)\frac{\partial C_m}{\partial(\dot\alpha c/2V)}$	s^{-1}
M_q	$\frac{QSc}{I_y}\left(\frac{c}{2V}\right)\frac{\partial C_m}{\partial(qc/2V)}$	s^{-1}
M_δ	$\frac{QSc}{I_y}\frac{\partial C_m}{\partial \delta}$	s^{-2}

4.6.2 Lateral-Directional Dynamics

The partial derivatives used in the Taylor series expansions of $\Delta Y(t)$, $\Delta L(t)$, and $\Delta N(t)$ will include terms such as

$$\beta(t) = \frac{v(t)}{V}, \qquad p = \frac{d\phi}{dt}, \qquad \text{and} \qquad r = \frac{d\psi}{dt}$$

as well as the control deflection term δ, which could represent either a lateral control, such as provided by ailerons and/or spoilers, or a directional control, such as a rudder input. Unlike the case of longitudinal dynamics where both α and $\dot\alpha$ terms were considered, $\dot\beta = d\beta/dt$ influences will be neglected under the assumption of quasi-steady aerodynamics.

There will be no aerodynamic perturbation terms due to bank angle ϕ inasmuch as bank angle by itself does not result in an aerodynamic force or moment. In addition, there are no aerodynamic perturbation terms developed due to heading angle change ψ because the airframe dynamics do not make a distinction when on a north, south, east, or west heading. However, heading rate terms will develop damping moments.

The side-force aerodynamic terms may be expressed in a small perturbation form by

$$\Delta Y(t) = \Delta Y_\beta \beta(t) + \Delta Y_p p(t) + \Delta Y_r r(t) + \Delta Y_\delta \delta(t) \qquad (4.44)$$

where

$\Delta Y_\beta = \Delta Y$ force change per unit β, $\partial \Delta Y/\partial \beta$, lb · rad^{-1}
$\Delta Y_p = \Delta Y$ force change per unit p, $\partial \Delta Y/\partial p$, lb · s · rad^{-1}
$\Delta Y_r = \Delta Y$ force change per unit r, $\partial \Delta Y/\partial r$, lb · s · rad^{-1}
$\Delta Y_\delta = \Delta Y$ force change per unit δ, $\partial \Delta Y/\partial \delta$, lb · rad^{-1}

The roll moment aerodynamic terms may be stated as

$$\Delta L(t) = \Delta L_\beta \beta(t) + \Delta L_p p(t) + \Delta L_r r(t) + \Delta L_\delta \delta(t) \qquad (4.45)$$

where, typically, $\Delta L_\beta = \partial \Delta L/\partial \beta$ is the change of ΔL moment per unit β, ft · lb · rad^{-1}.

The ΔL_β term depends on an aerodynamic influence known as the dihedral effect (cf. Sec. 2.3.1) whereas ΔL_p is related to aircraft roll damping.

The yaw moment aerodynamic expansion is similarly stated as

$$\Delta N(t) = \Delta N_\beta \beta(t) + \Delta N_p p(t) + \Delta N_r r(t) + \Delta N_\delta \delta(t) \qquad (4.46)$$

where a typical derivative such as $\Delta N_\beta = \partial \Delta N/\partial \beta$ is the change of ΔN moment per unit β, ft · lb · rad^{-1}.

The ΔN_β term is due to the airframe's directional stability whereas the ΔN_r term is associated with the airframe's yaw damping. This latter term will be found to have a marked influence on the damping level of the aircraft's Dutch-roll mode, which will be described in Chapter 7.

As was done with the longitudinal stability derivatives, dimensional stability derivatives are defined for the lateral-directional aerodynamic perturbation terms by typical relations of

$$Y_i = \Delta Y_i/(m)$$

$$L_i = \Delta L_i/(I_x)$$

$$N_i = \Delta N_i/(I_z)$$

for subscript $i = \beta, p, r,$ or δ. In this manner, the lateral-directional portions of Eq. (4.38) become

$$V\dot\beta = Y_\beta \beta + Y_p p + g \cos \Theta_0 \phi + (Y_r - V)r + Y_\delta \delta$$

$$\dot p - (I_{xz}/I_x)\dot r = L_\beta \beta + L_p p + 0 + L_r r + L_\delta \delta$$

$$-(I_{xz}/I_z)\dot p + \dot r = N_\beta \beta + N_p p + 0 + N_r r + N_\delta \delta$$

And as in the longitudinal equation formulation, the state vector format requires the following identity statement:

$$\dot\phi = p$$

If we define a lateral-directional state vector by

$$\{x\} = [\beta \quad p \quad \phi \quad r]^T$$

and a control term δ, which could represent either aileron (δ_a) or rudder (δ_r) deflection, then the coupled set of four, first-order, ordinary differential equations becomes, in matrix notation,

$$[I_n]\{\dot{x}\} = [A_n]\{x\} + \{B_n\}\delta \tag{4.47}$$

where

$$[I_n] = \begin{bmatrix} V & 0 & 0 & 0 \\ 0 & 1 & 0 & -I_{xz}/I_x \\ 0 & 0 & 1 & 0 \\ 0 & -I_{xz}/I_z & 0 & 1 \end{bmatrix}$$

$$[A_n] = \begin{bmatrix} Y_\beta & Y_p & g\cos\Theta_0 & (Y_r - V) \\ L_\beta & L_p & 0 & L_r \\ 0 & 1 & 0 & 0 \\ N_\beta & N_p & 0 & N_r \end{bmatrix}$$

and

$$\{B_n\} = [Y_\delta \quad L_\delta \quad 0 \quad N_\delta]^T$$

The use of dimensional stability derivatives (cf. Chapter 2) will lead to all of the time histories being expressed in real time (seconds) and frequencies (radians/second). Definitions for the lateral-directional dimensional stability derivatives are listed in Table 4.2.

The matrix relation, Eq. (4.47), may be premultiplied by $[I_n]^{-1}$ to obtain a standard form for the governing state equation, i.e.,

$$\{\dot{x}\} = [A]\{x\} + \{B\}\delta \tag{4.48}$$

where

$[A]$ = lateral-directional airframe plant matrix, $[I_n]^{-1}[A_n]$
$\{B\}$ = lateral-directional airframe control vector, $[I_n]^{-1}\{B_n\}$

Equation (4.48) corresponds to a single-input/multiple-output (SIMO) system. However, if more than one control were under consideration during an analysis (e.g., both δ_a and δ_r), then the $4 \times 1\{B\}$ column vector would be replaced by a $4 \times 2[B]$ matrix to be acting in conjunction with a 2×1 control deflection column vector $\{\delta\} = [\delta_a \; \delta_r]^T$. In such a form, the dynamics would be considered as a SIMO system. The control matrix would be made up by partitioning the two separate control sensitivity vectors, i.e.,

$$[B] = [\{B_a\} \quad \{B_r\}]$$

The inertial matrix in Eq. (4.47) contains off-diagonal elements such as (I_{xz}/I_x) and (I_{xz}/I_z) due to the cross product of inertia terms. If the body axes were the principal axes for the aircraft, these terms would vanish. This situation is quite

AIRFRAME EQUATIONS OF MOTION

Table 4.2 Dimensional stability derivatives

Term	Description	Units
Y_β	$\dfrac{QS}{m}\dfrac{\partial C_y}{\partial \beta}$	ft · s^{-2}
Y_p	$\dfrac{QS}{m}\left(\dfrac{b}{2V}\right)\dfrac{\partial C_y}{\partial (pb/2V)}$	ft · s^{-1}
Y_r	$\dfrac{QS}{m}\left(\dfrac{b}{2V}\right)\dfrac{\partial C_y}{\partial (rb/2V)}$	ft · s^{-1}
Y_δ	$\dfrac{QS}{m}\dfrac{\partial C_y}{\partial \delta}$	ft · s^{-2}
L_β	$\dfrac{QSb}{I_x}\dfrac{\partial C_\ell}{\partial \beta}$	s^{-2}
L_p	$\dfrac{QSb}{I_x}\left(\dfrac{b}{2V}\right)\dfrac{\partial C_\ell}{\partial (pb/2V)}$	s^{-1}
L_r	$\dfrac{QSb}{I_x}\left(\dfrac{b}{2V}\right)\dfrac{\partial C_\ell}{\partial (rb/2V)}$	s^{-1}
L_δ	$\dfrac{QSb}{I_x}\dfrac{\partial C_\ell}{\partial \delta}$	s^{-2}
N_β	$\dfrac{QSb}{I_z}\dfrac{\partial C_n}{\partial \beta}$	s^{-2}
N_p	$\dfrac{QSb}{I_z}\left(\dfrac{b}{2V}\right)\dfrac{\partial C_n}{\partial (pb/2V)}$	s^{-1}
N_r	$\dfrac{QSb}{I_z}\left(\dfrac{b}{2V}\right)\dfrac{\partial C_n}{\partial (rb/2V)}$	s^{-1}
N_δ	$\dfrac{QSb}{I_z}\dfrac{\partial C_n}{\partial \delta}$	s^{-2}

remote in general. The alternate form for the governing state equation [Eq. (4.48)] shows the time derivative of the state vector as uncoupled, which corresponds to being premultiplied by a unit diagonal matrix. In the situation of Eq. (4.48), the elements of the plant matrix and the control vector become

$$A_{11} = Y_\beta/V \quad A_{12} = Y_p/V \quad A_{13} = (g/V)\cos\Theta_0 \quad A_{14} = (Y_r - V)/V$$

$$A_{21} = L'_\beta = G[L_\beta + N_\beta(I_{xz}/I_x)] \quad A_{22} = L'_p = G[L_p + N_p(I_{xz}/I_x)]$$

$$A_{23} = 0 \quad A_{24} = L'_r = G[L_r + N_r(I_{xz}/I_x)]$$

$$A_{31} = A_{33} = A_{34} = 0 \quad A_{32} = 1$$

$$A_{41} = N'_\beta = G[N_\beta + L_\beta(I_{xz}/I_z)] \quad A_{42} = N'_p = G[N_p + L_p(I_{xz}/I_z)]$$

$$A_{43} = 0 \quad A_{44} = N'_r = G[N_r + L_r(I_{xz}/I_z)]$$

whereas the representative control vector's components are

$$B_1 = Y_\delta/V \quad B_2 = L'_\delta = G[L_\delta + N_\delta(I_{xz}/I_x)]$$

$$B_3 = 0 \quad B_4 = N'_\delta = G[N_\delta + L_\delta(I_{xz}/I_z)]$$

and the constant G is equivalent to

$$G = 1/[1 - (I_{xz})^2/(I_x I_z)]$$

Both primed and unprimed roll and yaw moment dimensional stability derivatives are provided for a number of airframes in Appendix B.

References

[1]Halfman, R. L., *Dynamics: Particles, Rigid Bodies and Systems*, Vol. 1, Addison-Wesley, Reading, MA, 1962, p. 183.

[2]Prouty, R. W., *Helicopter Performance, Stability and Control*, PWS Publishers, Boston, MA, 1986, Chap. 9.

Problems

4.1. Assume that the Euler angles are $\Phi = 10$, $\Theta = 20$, and $\Psi = 15$ deg, respectively.

a) Evaluate Eq. (4.4) for the assumed Euler angle values.

b) Verify that the matrix is orthogonal by multiplication of the transformation matrix with its transpose.

4.2. Expand the vector triple product [Eq. (4.18)] to verify that

$$\{A\} \times (\{B\} \times \{C\}) = \{B\}(\{A\}^T\{C\}) - \{C\}(\{A\}^T\{B\})$$

4.3. Consider the A-4D aircraft at $M = 0.4$ and 15,000 ft, i.e., condition 2 in Appendix B.1. Assume that a coordinated, climbing turn has been established for that flight condition with $\Theta = 10$ deg and $\Phi = 45$ deg. Neglect the trim change due to thrust application for static equilibrium in the climbing turn. Find the δ_e, δ_a, and δ_r changes from wings-level flight to attain the climbing turn.

4.4. Find the orientation of the principal mass moments of inertia for the A-4D operating at $M = 0.4$ and 15,000 ft, i.e., condition 2 in Appendix B.1.

4.5. Verify the linearization steps used in establishing Eq. (4.37).

4.6. Verify the primed roll- and yaw-moment dimensional derivatives used in Eq. (4.48). Hint: Rearrange the equations of motion [Eq. (4.47)] such that the state vector becomes

$$\{x\} = [\beta \quad \phi \quad p \quad r]^T$$

The (4 × 4) inertial matrix can then be partitioned into four (2 × 2) matrices and will take on a diagonal form. The inversion of the partitioned diagonal matrix is straightforward.

5
Dynamic System Principles

5.1 Background

The governing momentum conservation relations were established in Chapter 4 in order to define the airframe equations of motion in both the nonlinear and linearized forms while using dimensional stability derivatives to describe the aerodynamic forces and moments associated with angle and velocity perturbations. Chapter 2 was devoted to describing the aerodynamic principles contained in the dimensional stability derivatives. The purpose of Chapter 5 is to both review and develop the basic principles for solving dynamic equations that will be needed when considering both longitudinal and lateral-directional flight dynamics in subsequent chapters.

As an introduction, the dynamics of first- and second-order linear, time-varying systems may be solved by many classical methods. For consistency with the remaining material in this chapter, Laplace transform concepts will be briefly reviewed with applications considered for solving both the homogeneous and the particular forms of the governing differential equations. Stability of the solutions will be considered initially using Routh's stability criteria. Consideration of these criteria will reinforce the understanding of where the real parts of the eigenvalues must be located in the complex Argand plane for satisfactory system behavior.

The remainder of the chapter will be devoted to an introduction of modern control theory principles using the concept of a state vector to describe the time-varying quantities in the equations of motion. A big advantage to modern control theory concepts is the ease of finding solutions in either the frequency or the time domain. Linear algebra concepts, as described in Appendix D, will prove useful when considering this portion of Chapter 5.

System stability in the sense of Lyapunov, which is based on state equation principles, will be introduced as a supplement to the Routh's stability criteria in order to broaden the understanding of stable systems. Lyapunov's equation will be described to assist in determining whether a dynamic system has asymptotic stability. This equation will prove valuable for other reasons when treating system response to random forcing functions, a subject to be considered in Chapter 9.

5.2 Laplace Transforms

The Laplace transform technique is very convenient for finding solutions to ordinary differential equations with constant coefficients. The transform converts the differential equation into an algebraic relation, which is readily solved in the transform domain. The solution of the original differential equation is then reduced to the calculation of the inverse of the transform variable of interest, which usually involves the ratio of two polynomials. The inverse transform frequently may be obtained by consulting a table of Laplace transforms; however, it may also be evaluated from the inverse integral using the calculus of residues.

The rules of this method were developed by the British electrical engineer, Oliver Heaviside,[1] late in the 19th century. Although the heuristic nature of the rules has proven invaluable to engineers, it remained for mathematicians later to rigorously prove them to be correct. Many reference books exist covering the subject.[2-5] In regard to the question of mathematical rigor, Heaviside has been quoted[6] as follows: "Shall I refuse my dinner because I do not fully understand the process of digestion?"

To describe the transform process, first consider a function $f(t)$, which vanishes for $t < 0$. The Laplace transform of the function is defined by

$$\mathcal{L}[f(t)] = F(s) = \int_0^\infty e^{-st} f(t) \, dt \quad \text{where Re } s > \alpha \qquad (5.1)$$

The inverse transform is given by the complex integral relation

$$\mathcal{L}^{-1}[F(s)] = f(t) = \frac{1}{2\pi i} \int_{c-i\infty}^{c+i\infty} e^{st} F(s) \, ds \quad \text{where } c > \alpha \qquad (5.2)$$

The constant α is a real number greater than the real parts of all of the singular points of $F(s)$. The validity of the Laplace transform requires the following.

1) The integral of Eq. (5.1) is absolutely convergent.
2) The function $f(t)$ is of bounded variation in any finite interval and $f(t) = 0$ for $t < 0$.
3) At a point of a finite discontinuity t_0, the value of $f(t_0)$ is equal to the mean value, i.e.,

$$f(t_0) = \lim_{\epsilon \to 0} \frac{f(t_0 + \epsilon) + f(t_0 - \epsilon)}{2}$$

Under these circumstances, the relationships between the transform Eq. (5.1), and its inverse Eq. (5.2), are essentially unique. A few Laplace transforms that are of particular import will be described by examples.

Impulse function. The Dirac delta impulse function $\delta(t)$ is defined to be zero if $t \neq 0$, to be infinite at $t = 0$, and to have the property that

$$\int_{-\infty}^{+\infty} \delta(t) \, dt = 1 \qquad (5.3)$$

This is a concise way of describing a function that is extremely large in a small region and zero outside the region while having a unit valued integral. An alternate description for $\delta(t)$ having the property of Eq. (5.3) is

$$\delta(t) = \begin{cases} 0 & \text{for } t \leq 0 \\ 1/\epsilon & \text{for } 0 < t < \epsilon \\ 0 & \text{for } t > 0 \end{cases} \qquad (5.4)$$

where ϵ is considered as a small quantity.

An interesting property of the impulse function can be seen by confirming that

$$\int_{-\infty}^{+\infty} f(t)\delta(t-a)\,dt = f(a) \tag{5.5}$$

where $f(t)$ is continuous in the neighborhood of $t = a$.
From Eq. (5.4), it follows that

$$\int_{-\infty}^{+\infty} f(t)\delta(t-a)\,dt = \frac{1}{\epsilon}\int_{a}^{a+\epsilon} f(t)\,dt = f(a + \eta\epsilon)$$

where $0 < \eta < 1$ by the mean-value theorem of differential calculus. Equation (5.5) becomes evident by noting that

$$\lim_{\epsilon \to 0} f(a + \eta\epsilon) = f(a)$$

Example 5.1

The Laplace transform for the impulse function may be obtained by applying Eq. (5.5) to the transform definition, Eq. (5.1). Consider $f(t) = \delta(t)$, the impulse function. Then,

$$F(s) = \int_{0}^{\infty} e^{-st}\delta(t)\,dt = 1 \tag{5.6}$$

Step function. The unit step function, $1(t)$ is defined as

$$1(t) = \begin{cases} 0 & \text{for } t < 0 \\ 1/2 & \text{for } t = 0 \\ 1 & \text{for } t > 0 \end{cases} \tag{5.7}$$

Example 5.2

The Laplace transform for the unit step function may be obtained from Eq. (5.1), i.e.,

$$F(s) = \int_{0}^{\infty} e^{-st} 1(t)\,dt = \int_{0}^{\infty} e^{-st}\,dt = -\frac{1}{s}e^{-st}\Big|_{0}^{\infty}$$

The requirement that the Re $s > 0$ for the existence of the transform and its inverse establishes the transform for a unit step function as

$$F(s) = 1/s \tag{5.8}$$

Exponential function. The exponential function e^{at} is important to consider for a number of reasons including, as will be shown later, its being a scalar form of the transition matrix, which is a matrix functional having an exponential heritage.

The infinite series form of e^{at} may be confirmed by application of a Taylor series expansion [Eq. (2.1)] to obtain

$$e^{at} = 1 + at + \frac{1}{2!}(at)^2 + \frac{1}{3!}(at)^3 + \cdots \qquad (5.9)$$

which can be recognized as having the following properties:

$$\frac{d(e^{at})}{dt} = a e^{at}$$

at $t = 0$,

$$e^{at} = 1$$

and

$$e^{a(t_1+t_2)} = e^{at_1} e^{at_2} = e^{at_2} e^{at_1}$$

Example 5.3

The Laplace transform for the exponential function subject to the constraints imposed on Eqs. (5.1) and (5.2) may be obtained by applying Eq. (5.1).

Consider $f(t) = e^{at} 1(t)$. Then,

$$F(s) = \int_0^\infty e^{-st} e^{at} \, dt = \frac{1}{s-a} \qquad (5.10)$$

where $\operatorname{Re} s > \operatorname{Re} a$ in order to guarantee that both the transform and its inverse exist.

Added insight to the power of the exponential function may be obtained by considering its argument as a pure complex number; i.e., let at in e^{at} be $i\theta$ where $i = \sqrt{-1}$. By constructing a Taylor series about $\theta = 0$, one finds from Eq. (5.9) that

$$e^{i\theta} = 1 + i\theta - \theta^2/(2!) - i\theta^3/(3!) + \theta^4/(4!) + \cdots$$

which may be grouped into its real and imaginary parts as

$$e^{i\theta} = [1 - \theta^2/(2!) + \theta^4/(4!) + \cdots] + i[\theta - \theta^3/(3!) + \theta^5/(5!) + \cdots]$$

From this expression, Euler's formula is evident, i.e.,

$$e^{i\theta} = \cos\theta + i \sin\theta \qquad (5.11)$$

It will be noted that the magnitude of $e^{i\theta}$ is one, i.e.,

$$|e^{i\theta}| = [\cos^2\theta + \sin^2\theta]^{1/2} = 1$$

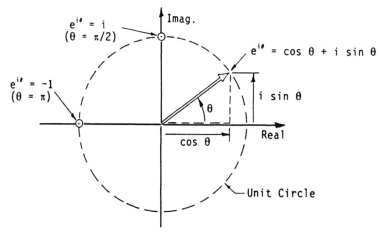

Fig. 5.1 Graphical interpretation of $e^{i\theta}$.

A graphical interpretation for $e^{i\theta}$ is shown in Fig. 5.1. From this figure, note that $e^{i\pi/2} = +i$, $e^{+\pi} = -1$, etc.

Illustrative examples using Euler's formula can be seen from Eq. (5.11) by considering the following relations:

$$e^{iA} = \cos A + i \sin A$$
$$e^{iB} = \cos B + i \sin B$$
$$e^{i(A+B)} = \cos(A+B) + i \sin(A+B)$$

from which expressions it may be seen that two familiar trigonometric relations evolve, i.e.,

$$\cos(A \pm B) = \cos A \cos B \mp \sin A \sin B$$
$$\sin(A \pm B) = \sin A \cos B \pm \cos A \sin B \quad (5.12)$$

A listing for the Laplace transforms of many functions that will prove useful in later discussions is provided by Table 5.1. It should be remembered that the function $f(t)$ applies for $t > 0$ and obeys the constraints of Eqs. (5.1) and (5.2). More complete tabulations may be found in many references, e.g., Erdelyi[5] and Ogata.[7] Verification of some more common transform relations in the tabulation are left as exercises for the reader.

The transform definition, Eqs. (5.1) and (5.2), and the relations of Table 5.1 may be extended to yield some useful properties for consideration when applying the Laplace transform technique.

Linearity. The linearity property of the Laplace transform may be seen from Eq. (5.1) by considering the following.

When $F(s)$ and $G(s)$ exist for the functions $f(t)$ and $g(t)$ respectively, then if

$$h(t) = A f(t) + B g(t)$$

Table 5.1 Laplace transforms

$f(t)$	$F(s)$
$\delta(t)$	1
$1(t)$	$1/s$
t	$1/s^2$
t^n	$n!/s^{n+1}$
e^{at}	$1/(s-a)$
$\cos \omega t$	$s/(s^2+\omega^2)$
$\sin \omega t$	$\omega/(s^2+\omega^2)$
df/dt	$sF(s) - f(0)$
$d^2 f/dt^2$	$s^2 F(s) - sf(0) - \dot{f}(0)$

it follows that

$$H(S) = AF(S) + BG(s)$$

Shift rule. A useful property is the shift rule. If $F(s)$ is the Laplace transform of $f(t)$, then $e^{-as}F(s)$ is the transform of the function $f(t-a)1(t-a)$ where $a > 0$ and $1(t-a)$ is the delayed unit step function.

The shift rule may be verified by considering Eq. (5.1)

$$\int_0^\infty e^{-st} f(t-a) 1(t-a) dt = \int_a^\infty e^{-st} f(t-a) dt = e^{-as} \int_0^\infty e^{-st} f(t) dt$$

where

$$f(t-a)1(t-a) = \begin{cases} 0 & \text{for } t < a \\ f(t-a) & \text{for } t > a \end{cases}$$

Example 5.4

Consider a unit square wave extending for time span T as shown in Fig. 5.2. The time function may be expressed as

$$f(t) = 1(t) - 1(t-T)$$

By the shift rule, the Laplace transform of the square wave becomes

$$F(s) = (1/s) - e^{-Ts}(1/s) = (1 - e^{-Ts})1/s$$

Attenuation rule. If $f(t)$ has a Laplace transform $F(s)$, then the Laplace transform of $e^{-at} f(t)$ is $F(s+a)$ providing that $a < \text{Re } s < \infty$. Again from Eq. (5.1) it follows that

$$\mathcal{L}[e^{-at}f(t)] = \int_0^\infty e^{-st} e^{-at} f(t) dt = F(s+a) \tag{5.13}$$

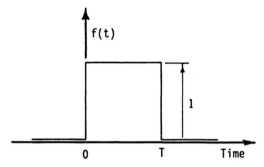

Fig. 5.2 Unit square wave.

Example 5.5

The attenuation rule is useful for recognizing attenuated trigonometric functions from their Laplace transforms. Consider from Table 5.1 that if

$$f(t) = \sin \omega t \quad \text{and} \quad g(t) = \cos \omega t$$

then

$$F(s) = \frac{\omega}{s^2 + \omega^2} \quad \text{and} \quad G(s) = \frac{s}{s^2 + \omega^2}$$

By the attenuation rule, it may be noted that

$$\mathcal{L}[e^{-at} \sin \omega t] = \frac{\omega}{(s+a)^2 + \omega^2}$$

$$\mathcal{L}[e^{-at} \cos \omega t] = \frac{(s+a)}{(s+a)^2 + \omega^2}$$

Final value theorem. The final value theorem allows one to determine the limit of $f(t)$ as $t \to \infty$ from the corresponding Laplace transform value. The theorem is stated as

$$\lim_{t \to \infty} f(t) = \lim_{s \to 0} s F(s) \qquad (5.14)$$

providing that 1) $f(t)$ and $df(t)/dt$ are Laplace transformable, 2) $F(s) = \mathcal{L}[f(t)]$, and 3) the limit of $f(t)$ as $t \to \infty$ exists.

The theorem may be proven by letting the transform variable s tend to zero in the Laplace transform of $df(t)/dt$,

$$\lim_{s \to 0} \int_0^\infty e^{-st} \left[\frac{d}{dt} f(t)\right] dt = \lim_{s \to 0} [s F(s) - f(0)]$$

but

$$\lim_{s \to 0} e^{-st} = 1$$

Hence,

$$f(t)|_0^\infty = f(\infty) - f(0) = \lim_{s \to 0} s F(s) - f(0)$$

Equation (5.14) is proven by noting that

$$\lim_{t\to\infty} f(t) = f(\infty)$$

Initial value theorem. The counterpart of the final value theorem is the initial value theorem, which is stated as

$$f(0+) = \lim_{s\to\infty} sF(s) \qquad (5.15)$$

providing that 1) $f(t)$ and $df(t)dt$ are Laplace transformable and 2) the limit of $F(s)$ exists as $s \to \infty$.

The proof of Eq. (5.15) is similar to that used for the final value theorem. The value of the function $f(t)$ at $t = 0$ is expressed as $f(0+)$ because the Laplace transform requires that the function be zero for time $t < 0$.

Convolution integral. A useful relationship is provided by the convolution integral. Consider the following integral, which involves the product of a function with a second delayed (convoluted) function:

$$\int_0^t f_1(\tau) f_2(t-\tau) d\tau$$

The convolution integral and its alternate form are frequently expressed in a symbolic form as

$$\int_0^t f_1(\tau) f_2(t-\tau) d\tau = \int_0^t f_1(t-\tau) f_2(\tau) d\tau = f_1(t) * f_2(t) \qquad (5.16)$$

If both $f_1(t)$ and $f_2(t)$ are Laplace transformable, then it can be shown that the Laplace transform of the convolution integral is

$$\mathcal{L}[f(t) * f_2(t)] = F_1(s) F_2(s) \qquad (5.17)$$

If $f_2(t - \tau)$ in Eq. (5.16) represented the impulse response of a physical system and $f_1(\tau)$ were the forcing function for $\tau > 0$, then the convolution integral could be interpreted as the Duhamel superposition integral[6] subject to the constraint that the system was initially at rest. Figure 5.3 illustrates this concept.

Inverse Laplace transform. Obtaining the inverse of the Laplace transform is an essential step in finding a solution. After a time-dependent linear system has been transformed into the complex s domain, the transform variable of interest may be solved by algebraic means. The reverse process of passing the solution from the complex domain to the real-time domain involves an inverse Laplace transformation. The inversion integral [Eq. (5.2)] involves the evaluation by a contour integration of a complex-natured integral having a finite number of singularities (poles). An alternate procedure for the integral inversion is based on the recognition that the variable of interest, $F(s)$, will frequently occur as the ratio of two polynomials, i.e.,

$$F(s) = \frac{B(s)}{A(s)}$$

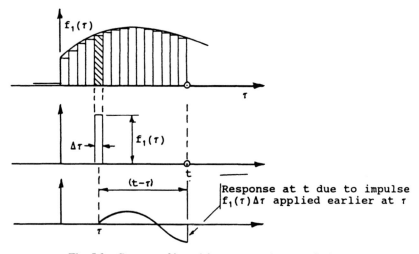

Fig. 5.3 Concept of impulsive response by convolution.

where $B(s)$ and $A(s)$ are polynomials in the complex variable s with the $B(s)$ polynomial's degree being less than that of the $A(s)$ polynomial. A partial fraction expansion method proves useful for finding the inverse Laplace transform in that situation based on the linear nature of the transform process.

The numerator and denominator will be assumed as consisting of m zeros and n poles, respectively, where $m < n$. In that case, the transform can be expressed as

$$F(s) = \frac{B(s)}{A(s)} = \frac{K(s+z_1)(s+z_2)\cdots(s+z_m)}{(s+p_1)(s+p_2)\cdots(s+p_n)}$$

with $m < n$, where the poles and zeros can be either real or complex quantities. Continuing on the assumption that $F(s)$ has distinct poles, it can be expanded into a sum of simple partial fractions, i.e.,

$$F(s) = \frac{a_1}{s+p_1} + \frac{a_2}{s+p_2} + \cdots + \frac{a_n}{s+p_n} \tag{5.18}$$

where the numerators are constants, which correspond to residues of the complex $F(s)$ function identified at each of the singular poles. The kth constant a_k is found by the evaluation of the residue at the pole located at $s = -p_k$, i.e.,

$$a_k = \left[(s+p_k)\frac{B(s)}{A(s)}\right]_{s=-p_k} \tag{5.19}$$

The inverse Laplace transform for the kth pole will be recognized as

$$\mathcal{L}^{-1}\left[\frac{a_k}{s+p_k}\right] = a_k e^{-p_k t}$$

Hence, the inverse of $F(s)$ becomes

$$f(t) = \sum_{k=1}^{n} a_k e^{-p_k t} \tag{5.20}$$

It should be noted that a complex pole will always be paired with a complex conjugate pole because $f(t)$ is a real quantity. Each of the associated constants also will be complex conjugates. If

$$p_j = a + ib \quad \text{and} \quad p_{j+1} = a - ib$$

then it follows that

$$a_j = c + id \quad \text{and} \quad a_{j+1} = c - id$$

In that case, the two complex time-domain exponential functions will combine into a pair of real trigonometric functions, i.e.,

$$\begin{aligned} f_j(t) + f_{j+1}(t) &= (c + id)e^{-(a+ib)t} + (c - id)e^{-(a-ib)t} \\ &= 2e^{-at}(c \cos bt + d \sin bt) \end{aligned} \tag{5.21}$$

When the partial fraction expansion [Eq. (5.18)] discloses the presence of repeated poles, the coefficients take a slightly modified form. Consider a situation where a double pole exists at $s = -p_1$, i.e.,

$$F(s) = \frac{B(s)}{(s + p_1)^2} \tag{5.22}$$

In this case, the expansion takes the form

$$F(s) = \frac{a_1}{s + p_1} + \frac{a_2}{(s + p_1)^2}$$

which may be altered to give

$$(s + p_1)^2 F(s) = a_1(s + p_1) + a_2$$

The constant a_2 is found using the relation of

$$a_2 = \left[(s + p_1)^2 F(s)\right]_{s=-p_1} \tag{5.23}$$

Similarly,

$$a_1 = \frac{d}{ds}\left[(s + p_1)^2 F(s)\right]_{s=-p_1} \tag{5.24}$$

The extension of this procedure to higher-order repeated poles is a straightforward matter.[7]

5.3 First-Order Linear System

A brief study of a first-order system with constant coefficients provides insights in related topics such as the 1) low-pass filter (electrical engineering), 2) roll response approximation (aircraft flight mechanics), and 3) scalar form of the state equation (control theory). A first-order linear system is given by

$$\frac{dx}{dt} = \dot{x} = ax(t) + bu(t) \tag{5.25}$$

where

$x(t)$ = output response variable
$u(t)$ = input control variable
t = time
a, b = system constants

Although there are many methods for solving Eq. (5.25) including the use of an integrating factor, the method used next will be consistent with Sec. 5.2.

Assume that the system described by Eq. (5.25) is at rest for $t < 0$ and that the input variable $u(t)$ is Laplace transformable. Application of the Laplace transform rules of Table 5.1 yields

$$sX(s) - x(0) = aX(s) + bU(s) \tag{5.26}$$

where

$$X(s) = \mathcal{L}[x(t)] \qquad U(s) = \mathcal{L}[u(t)]$$

$X(s)$ may be found from Eq. (5.26),

$$X(s) = \frac{x(0)}{(s-a)} + b\frac{U(s)}{(s-a)} \tag{5.27}$$

Application of Table 5.1 along with the convolution integral rule [Eq. (5.17)] provides the inverse Laplace transform of Eq. (5.27), which corresponds to an answer in the time domain, i.e.,

$$x(t) = \mathcal{L}^{-1}[X(s)] = x(0)e^{at} + b\int_0^t e^{a(t-\tau)}u(\tau)d\tau \tag{5.28}$$

The solution [Eq. (5.28)] contains the response to the initial condition, $x(0)$, which is known as the homogeneous solution. The integral relation contains the solution to the input control or forcing function. This portion of the equation is known as the nonhomogeneous or particular solution. Because the system is linear by assumption, the total system response is the sum of the homogeneous and particular solutions. Equally important, Eq. (5.28) will be seen in subsequent material as the scalar form of the state vector response solution to an arbitrary input function.

Homogeneous solution. The first consideration of Eq. (5.28) will be given to the system responding solely to an initial condition $x(0)$. From Eq. (5.28), it may be seen that the homogeneous response is

$$x_H(t) = x(0)e^{at} \quad \text{for } t > 0 \tag{5.29}$$

It should be noted that the homogeneous solution has a stable physical meaning only when the coefficient $a < 0$; otherwise, the solution would become unbounded as $t \to \infty$. Additionally, the use of the Laplace transform also imposes a similar constraint upon the coefficient a.

If the initial condition were replaced by an impulse forcing function described by

$$bu(\tau) = x(0)\delta(\tau)$$

the solution for the system response would be identical to the homogeneous response of Eq. (5.29). This observation is based on using the property of the Dirac delta function, as shown by Eq. (5.5).

Step response. Consider the situation of a zero initial condition while the input is a step function having the form

$$u(t) = u_0 1(t) = \begin{cases} 0 & \text{for } t < 0 \\ u_0 & \text{for } t > 0 \end{cases}$$

The solution to Eq. (5.28) is found by evaluating the integral term, i.e.,

$$x(t) = bu_0 \int_0^t e^{a(t-\tau)} d\tau = (-a)^{-1} bu_0 [1 - e^{at}] \tag{5.30}$$

The coefficient a must be negative, as already noted, for the solution to have a realizable physical meaning. It is customary to recognize that the step response solution for $x(t)$ tends to a static value when all startup transients have decayed or subsided. It may be seen from Eq. (5.30) that a final steady-state (static) value is

$$x_{\text{stat}} = (-a)^{-1} bu_0$$

which in turn allows the response to be expressed as

$$x(t) = x_{\text{stat}}[1 - e^{at}] 1(t)$$

It is also customary, upon the recognition that $a < 0$, to define a system time constant τ where $\tau = -1/a$. The constant τ will have units of time if the variable t also is in units of time. The modified results become

$$x(t) = x_{\text{stat}}\left[1 - e^{-t/\tau}\right] 1(t) \tag{5.31}$$

In the time history of the system response to a step input function, shown in Fig. 5.4, a specific time to note is when $t = \tau$, for at that time the output response is

$$x(t) = x_{\text{stat}}[1 - e^{-1}] = 0.63 x_{\text{stat}}$$

Harmonic response. Another particular solution of significance is obtained when the input function is harmonic (trigonometric). Consider the control input in a general, complex-natured harmonic form, i.e., let

$$u(t) = u_0 e^{i\omega t} 1(t) = u_0 (\cos \omega t + i \sin \omega t) 1(t)$$

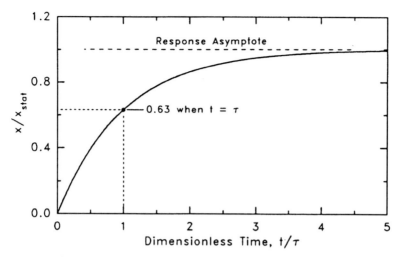

Fig. 5.4 First-order system response to a step function.

where

u_0 = amplitude of the harmonic input
ω = circular frequency of the input, rad · s^{-1}

The response will be complex natured, which is a mathematic approach that combines both trigonometric solutions in a single expression. The real and imaginary parts of the solution will correspond to the cosine and sine responses, respectively. In addition, the integral term in Eq. (5.28) will show the influence of a startup transient from the lower limit ($\tau = 0$) of the integral. Physically, the transient response is due to the system motion starting from an at-rest condition at $t = 0$. Substitution of the $u(t)$ input function into Eq. (5.11) yields

$$x(t) = e^{at} \int_0^t e^{-a\theta} b u_0 e^{i\omega\theta}\, d\theta = x_{\text{stat}}[1 + i\omega\tau]^{-1}\left[e^{i\omega t} - e^{-t/\tau}\right] \quad (5.32)$$

where the static response x_{stat} and the time constant τ have already been defined.

The long-term, steady-state harmonic response, after all startup transients have subsided, is described in a general form from Eq. (5.32) as

$$x(t) = x_{\text{stat}}[1 + i\omega\tau]^{-1} e^{i\omega t} \quad (5.33)$$

The complex term containing $\omega\tau$ may be simplified by Euler's formula to an equivalent complex form,

$$[1 + i\omega\tau]^{-1} = G(\omega)e^{i\phi} = G(\omega)(\cos\phi + i\sin\phi)$$

where

$G(\omega)$ = gain function, $[1 + (\omega\tau)^2]^{-1/2}$
ϕ = phase angle, $\tan^{-1}(-\omega\tau)$

The steady-state harmonic response may be summarized as

$$x(t) = x_{\text{stat}} G(\omega) e^{i(\omega t + \phi)}$$

or, alternatively, in terms of trigonometric functions due to an input of

$$u(t) = u_0 \begin{Bmatrix} \cos \omega t \\ \sin \omega t \end{Bmatrix}$$

the harmonic response is

$$x(t) = x_{\text{stat}} G(\omega) \begin{Bmatrix} \cos(\omega t + \phi) \\ \sin(\omega t + \phi) \end{Bmatrix} \tag{5.34}$$

The harmonic response for the linear, first-order system has several dominant traits including the following.
1) Gain function is maximum at $\omega = 0$ with a value of $G(\omega) = 1.0$.
2) When the applied frequency $\omega = 1/\tau$, the gain function $G(\omega = 1/\tau) = 0.707$. This frequency is described as the corner frequency when considering Bode plots.
3) Phase shift is zero at zero frequency.
4) The phase shift when $\omega = 1/\tau$ is $-\pi/4$ rad (-45 deg), which represents a phase lag between the output and the input.
5) Phase shift smoothly changes to $\phi = -\pi/2$ rad (-90 deg) as $\omega \to \infty$.

These harmonic response properties are shown in Fig. 5.5 for a first-order linear system having a time constant of $\tau = 0.2$ s. The plots of gain and phase angle have been presented using linear scales. An alternate format using logarithmic scaling (known as a Bode plot) also is convenient for interpreting frequency-response properties and is preferred by many engineers when performing feedback analysis to determine gain and phase margins.

5.4 Second-Order Linear System

Vehicle dynamics can frequently be modeled quite accurately by coupled, second-order systems that are dissipative in nature. These systems are described as being nonconservative, which is an alternate way of stating that the systems have an energy absorbing mechanism that may be represented by a rate-dependent or damping term. The current considerations will be restricted to linear systems with constant coefficients.

Before investigating the behavior of a multi-degree-of-freedom system (cf. Chapters 6 and 7), it is helpful to explore the traits of a single-degree-of-freedom second-order linear system. For simplicity, the system will be modeled by a spring/mass/damper arrangement as shown in Fig. 5.6 with the deformation variable being represented by a linear displacement. An angular displacement could be considered equally as well because the term deformation variable has a generic meaning that implies a perturbation from a position of static equilibrium.

If a free-body diagram were considered for the system of Fig. 5.6, Newton's law for the conservation of linear momentum would yield

$$m\ddot{y} = -c\dot{y} - ky + f(t) \tag{5.35}$$

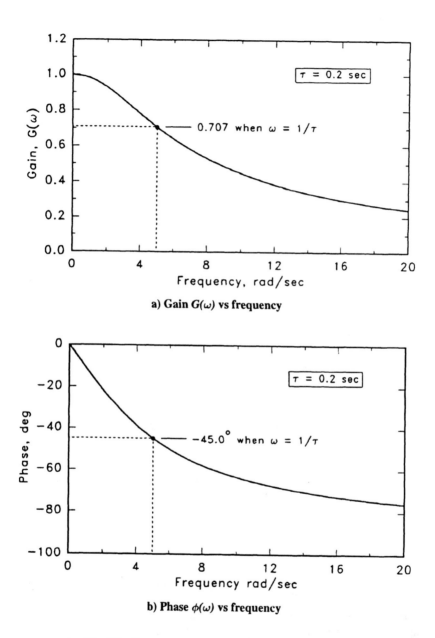

a) Gain $G(\omega)$ vs frequency

b) Phase $\phi(\omega)$ vs frequency

Fig. 5.5 Harmonic response of a first-order system.

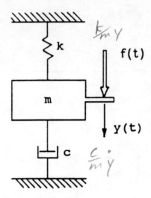

Fig. 5.6 Sketch of spring/mass/dashpot system.

where

y = displacement from the static equilibrium position
$\dot{y}$ = velocity, dy/dt
$\ddot{y}$ = acceleration, d^2y/dt^2
m = mass
k = spring constant
c = damping constant

The units are consistent such that (ky), $(c\dot{y})$, and $(m\ddot{y})$ represent forces. It is convenient to rewrite Eq. (5.35) into a more familiar form of an ordinary differential equation (ODE) with constant coefficients.

$$\ddot{y} + (c/m)\dot{y} + (k/m)y(t) = (1/m)f(t) \tag{5.36}$$

with the initial conditions of $y(0) = y_0$, which is initial displacement at $t = 0$, and $\dot{y}(0) = v_0$, which is initial velocity at $t = 0$.

Although there are many methods available for solving Eq. (5.36), it is appropriate to continue as in Sec. 5.3 and use Laplace transform techniques as outlined in Sec. 5.2. For reasons to become evident later, let us define the dimensionless damping ratio ζ and the undamped natural frequency ω_n (in rad · s^{-1}) where

$$\omega_n^2 = k/m, \qquad \zeta = (c/m)(1/2\omega_n)$$

Then Eq. (5.36) may be expressed as

$$\ddot{y} + 2\zeta\omega_n \dot{y} + \omega_n^2 y(t) = 1/m\, f(t) \tag{5.37}$$

On the assumption that both the forcing function $f(t)$ and the response $y(t)$ are Laplace transformable, Eq. (5.37) can be transformed into the complex s domain to provide $Y(s)$,

$$Y(s) = \frac{F(s)}{m(s^2 + 2\zeta\omega_n s + \omega_n^2)} + \frac{y_0(s + 2\zeta\omega_n) + v_0}{(s^2 + 2\zeta\omega_n s + \omega_n^2)} \tag{5.38}$$

The denominator in these fractional expansions is known as the characteristic equation and it has a pair of complex conjugate roots when the dimensionless damping coefficient is bounded by $0 < \zeta < 1$, i.e.,

$$s^2 + 2\zeta\omega_n s + \omega_n^2 = (s + \zeta\omega_n + i\sqrt{1-\zeta^2}\omega_n)(s + \zeta\omega_n - i\sqrt{1-\zeta^2}\omega_n)$$

When $0 < \zeta < 1$, it is helpful to recast the denominator as

$$s^2 + 2\zeta\omega_n s + \omega_n^2 = (s + \zeta\omega_n)^2 + \omega_d^2$$

where $\omega_d = \sqrt{1-\zeta^2}\omega_n$ (rad·s^{-1}) is the damped natural frequency.

The relations in the solution for $Y(s)$ [Eq. (5.38)] represent the Laplace transform of the particular and homogeneous responses, respectively. Because the transforms are linear, let us first consider the homogeneous response, i.e., the system response due to the initial conditions.

Response to y_0. The system response to an initial displacement y_0 may be expressed in a form such that the inverse transform is easily seen by application of Table 5.1 and the attenuation rule described in Sec. 5.2,

$$Y_1(s) = y_0 \frac{(s + \zeta\omega_n) + (\zeta\omega_n)}{(s + \zeta\omega_n)^2 + \omega_d^2}$$

which can be inverted to yield a time-domain solution as

$$y_1(t) = y_0 e^{-\zeta\omega_n t}\left(\cos\omega_d t + \frac{\zeta}{\sqrt{1-\zeta^2}}\sin\omega_d t\right) 1(t) \quad \text{for } 0 < \zeta < 1 \quad (5.39)$$

The use of the $1(t)$ symbol in the solution guarantees that the response is zero for $t < 0$. An alternate expression for Eq. (5.39) obtained by merging the trigonometric terms is

$$y_1(t) = y_0 \frac{e^{-\zeta\omega_n t}}{\sqrt{1-\zeta^2}}\cos(\omega_d t - \phi) 1(t) \quad \text{for } 0 < \zeta < 1$$

where

$$\phi = \tan^{-1}\frac{\zeta}{\sqrt{1-\zeta^2}}$$

The solution to $y_1(t)$ when $\zeta = 1$ is found by taking the limit of Eq. (5.39) as ζ tends to 1,

$$\lim_{\zeta \to 1} y_1(t) = y_0 e^{-\omega_n t} 1(t)$$

Verification of the preceding equation is left as an exercise, Problem 5.8.

When the damping coefficient $\zeta > 1$, the response due to initial conditions as well as due to both impulsive and step input functions will be nonoscillatory in nature. This can be recognized by considering the characteristic equation (the denominator in the transformed variable) where the roots become real for $\zeta > 1$,

$$s^2 + 2\zeta\omega_n s + \omega_n^2 = (s + \lambda_1)(s + \lambda_2) \quad \text{for } \zeta > 1$$

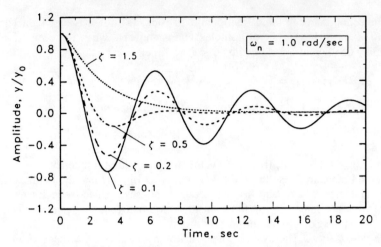

Fig. 5.7 Second-order system response to an initial displacement.

where

$$\lambda_{1,2} = -\zeta\omega_n \pm \sqrt{\zeta^2 - 1}\,\omega_n$$

The system response solutions will typically take the form of damped hyperbolic sines and cosines. The solution for $y_1(t)$ becomes

$$y_1(t) = y_0 e^{-\zeta\omega_n t}\left(\cosh\sqrt{\zeta^2 - 1}\,\omega_n t - \frac{\zeta}{\sqrt{\zeta^2 - 1}}\sinh\sqrt{\zeta^2 - 1}\,\omega_n t\right) \quad \text{for } \zeta > 1$$

Verification of this relation is left as an exercise, Problem 5.9.

The time-history response due to a unit displacement (i.e., $y_0 = 1$ at $t = 0$) is shown in Fig. 5.7 for several values of dimensionless damping ratio $\zeta = 0.1, 0.2, 0.5,$ and 1.5. Items to note in the response include the following.

1) The peak oscillation amplitudes show a gradual decay as time increases due to the presence of the $e^{-\zeta\omega_n t}$ term as a factor applied to the trigonometric response terms.

2) The period of the free oscillation increases as ζ increases because

$$\text{period } T = 2\pi/\omega_d$$

and ω_d decreases as ζ increases due to the $\sqrt{1 - \zeta^2}$ factor between ω_d and ω_n.

3) When dimensionless damping is above the critical value (i.e., $\zeta \geq 1$), the free response becomes nonoscillatory.

Log-decrement procedure. The log-decrement procedure is an experimental method for estimating the dimensionless damping ratio from a time history of the free-vibration response. The technique is especially suitable for systems with light damping, i.e., $0 < \zeta < 0.2$. Figure 5.8 is a time-history response for $\zeta = 0.1$, as seen in Fig. 5.7. In addition, the $\pm e^{-\zeta\omega_n t}$ envelope has been superimposed on the response to emphasize the exponential decay or subsidence of the free-vibration response.

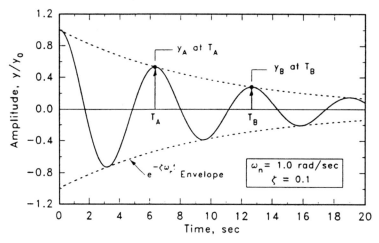

Fig. 5.8 Estimation of dimensionless damping ratio ζ.

It will be noted in Fig. 5.8 that the two consecutive peak amplitudes occur at times T_A and T_B (one period apart) with amplitudes y_A and y_B, respectively,

$$y_A = y_0 \exp(-\zeta\omega_n T_A) \cos(2\pi)$$

$$y_B = y_0 \exp[-\zeta\omega_n(T_A + 2\pi/\omega_d)] \cos(4\pi)$$

The ratio of the two consecutive amplitude peaks is

$$(y_A/y_B) = \exp[+\zeta\omega_n(2\pi/\omega_d)]$$

Taking the natural logarithm of the peak amplitude ratio yields

$$\ln(y_A/y_B) = +\zeta\omega_n(2\pi/\omega_d) = 2\pi\zeta/\sqrt{1-\zeta^2}$$

The light damping assumption provides a bound on the estimation accuracy when $\zeta = 0.2$, say, using the Maclaurin series approximation. Consider

$$\sqrt{1-\zeta^2} \cong 1 - 0.5(0.2)^2 = 0.98 \quad \text{for } \zeta = 0.2$$

This assumption permits a damping ratio estimate to be made using Eq. (5.40) that is within 2% of being accurate when $0 < \zeta < 0.2$. This is quite reasonable when performing experimental analyses. Hence, it may be concluded that

$$\zeta \cong (1/2\pi) \ln(y_A/y_B) \quad \text{for } 0 < \zeta < 0.2 \qquad (5.40)$$

Response to v_0. For the response to v_0,

$$Y_2(s) = \frac{v_0}{(s+\zeta\omega_n)^2 + \omega_d^2}$$

which inverts into the time domain as

$$y_2(t) = \frac{v_0}{\omega_d} e^{-\zeta\omega_n t} \sin\omega_d t \, 1(t) \quad \text{for } 0 < \zeta < 1 \qquad (5.41)$$

Impulse response. Assume that the initial conditions are zero and that the forcing function $f(t)$ is an impulse using the notation of the Dirac delta function, i.e., $f(t) = f_0 \delta(t)$. In this case, Eq. (5.38) becomes

$$Y_3(s) = \frac{f_0}{m} \frac{1}{(s + \zeta \omega_n)^2 + \omega_d^2}$$

and the inversion into the time domain is

$$y_3(t) = \frac{f_0}{m \omega_d} e^{-\zeta \omega_n t} \sin \omega_d t \, 1(t) \quad \text{for } 0 < \zeta < 1 \tag{5.42}$$

The similarity between the impulse response [Eq. (5.42)] and the response to an initial condition on velocity [Eq. (5.41)] allows one to compare solutions when $y_2(t) = y_3(t)$. In that case, it will be recognized that

$$f_0 = m v_0$$

which physically means that an impulse applied at $t = 0$ provides a jump in system momentum, i.e.,

$$f_0 \delta(t) = m[v(0+) - v(0-)]$$

Because the Laplace transform assumes the system to be at rest for $t < 0$, the $v(0-)$ term becomes zero and $v(0+)$ can be interpreted as the initial condition on velocity v_0.

Step response. Assume that the system is at rest for $t < 0$ and that a step input function is applied at $t = 0$, i.e., $f(t) = f_0 1(t)$. Then

$$Y_4(s) = \frac{f_0}{m} \frac{1}{s[(s + \zeta \omega_n)^2 + \omega_d^2]} = y_{\text{stat}} \frac{\omega_n^2}{s[(s + \zeta \omega_n)^2 + \omega_d^2]}$$

where $y_{\text{stat}} = f_0/k$ is the static response of the system due to f_0.

The $Y_4(s)$ transform can be expanded into a partial fraction form in order that the inverse transform can be easily found from Table 5.1,

$$Y_4(s) = y_{\text{stat}} \left[\frac{1}{s} - \frac{(s + 2\zeta \omega_n)}{(s + \zeta \omega_n)^2 + \omega_d^2} \right]$$

which can be inverted, when $0 < \zeta < 1$, into

$$y_4(t) = y_{\text{stat}} \left[1 - e^{-\zeta \omega_n t} \left(\cos \omega_d t + \frac{\zeta}{\sqrt{1 - \zeta^2}} \sin \omega_d t \right) \right] 1(t) \tag{5.43}$$

or alternately expressed as

$$y_4(t) = y_{\text{stat}} \left[1 - \frac{e^{-\zeta \omega_n t}}{\sqrt{1 - \zeta^2}} \cos(\omega_d t - \phi) \right] 1(t) \quad \text{for } 0 < \zeta < 1$$

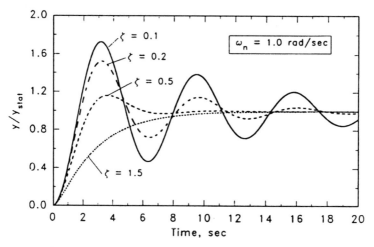

Fig. 5.9 Second-order system response to a step function.

where in this case

$$\phi = \tan^{-1} \frac{\zeta}{\sqrt{1-\zeta^2}}$$

A time history of the system's response is shown in Fig. 5.9 for several values of dimensionless damping ratio. Items of particular note include the following.

1) When $\zeta \to 0$, the peak dynamic overshoot tends to twice the static value, which can also be recognized in Eq. (5.43). This is a useful fact to keep in mind when making engineering design estimates due to abrupt changes in applied loading.
2) The response tends to the static value, y_{stat}, as $t \to \infty$.
3) The log-decrement procedure can be applied to a step function response curve for situations of light damping.

For the harmonic response, as was done when considering first-order systems, it will be assumed that the forcing function has the form of $f(t) = f_0 e^{i\omega t}$ and that $0 < \zeta < 1$. Then,

$$Y_5(s) = \frac{f_0}{m} \frac{1}{(s - i\omega)[(s + \zeta\omega_n)^2 + \omega_d^2]}$$

The partial fraction expansion takes the form of

$$Y_5(s) = \frac{f_0}{m} \left[\frac{a_1}{(s - i\omega)} + \frac{b_1}{(s + \zeta\omega_n - i\omega_d)} + \frac{b_2}{(s + \zeta\omega_n + i\omega_d)} \right]$$

The a_1 coefficient will yield the steady-state harmonic response, whereas the b_1 and b_2 coefficients will provide the response due to the startup of the forcing function at $t = 0$, i.e.,

$$y_5(t) = \frac{f_0}{m} \left[a_1 e^{i\omega t} + e^{-\zeta\omega_n t} \left(b_1 e^{i\omega_d t} + b_2 e^{-i\omega_d t} \right) \right] 1(t) \quad \text{for } 0 < \zeta < 1$$

Application of the Cauchy theorem for residues [Eq. (5.19)] gives

$$a_1 = \left. \frac{1}{(s+\zeta\omega_n)^2 + \omega_d^2} \right|_{s=+i\omega}$$

$$b_{1,2} = \left. \frac{1}{(s-i\omega)(s+\zeta\omega_n \pm i\omega_d)} \right|_{s=-\zeta\omega_n \mp \omega_d}$$

The transient response will vanish as t increases, due to the $e^{-\zeta\omega_n t}$ term, leaving the steady-state response, which is determined from the a_1 coefficient. Let the steady-state response be $y_6(t)$, then

$$y_6(t) = \lim_{t \to \infty} y_5(t) = \frac{f_0}{m} \frac{1}{[(\omega_n^2 - \omega^2) + 2i\zeta\omega_n\omega]} e^{i\omega t}$$

which simplifies to

$$y_6(t) = y_{\text{stat}} G(\omega/\omega_n) e^{i(\omega t + \phi)} \qquad (5.44)$$

where

$$y_{\text{stat}} = f_0/k$$

$$G\left(\frac{\omega}{\omega_n}\right) = \left\{ \left[1 - \left(\frac{\omega}{\omega_n}\right)^2\right]^2 + 4\zeta^2\left(\frac{\omega}{\omega_n}\right)^2 \right\}^{-1/2}$$

$$\phi = \tan^{-1} \frac{-2\zeta(\omega/\omega_n)}{[1 - (\omega/\omega_n)^2]}$$

or, alternatively, in terms of trigonometric functions due to an input of

$$f(t) = f_0 \begin{Bmatrix} \cos \omega t \\ \sin \omega t \end{Bmatrix}$$

the harmonic response is

$$y(t) = y_{\text{stat}} G\left(\frac{\omega}{\omega_n}\right) \begin{Bmatrix} \cos(\omega t + \phi) \\ \sin(\omega t + \phi) \end{Bmatrix} \qquad (5.45)$$

The harmonic response for the second-order system is shown in Fig. 5.10 by the gain and phase angle variations as functions of the dimensionless frequency ratio (ω/ω_n). The following will be noted in Fig. 5.10.
1) Peak response occurs near to $\omega/\omega_n = 1.0$ for a lightly damped system. The maximum gain is approximately $1/2\zeta$ when $0 < \zeta < 0.2$.
2) Peak response occurs at $\omega/\omega_n = 0$ when damping ratio $\zeta \geq 0.707$.
3) Phase lag is $\pi/2$ rad (90 deg) when $\omega/\omega_n = 1.0$ regardless of the damping ratio value.
Other frequency response properties of a second-order linear system are listed as problems for verification by the reader, cf. Problem 5.11.

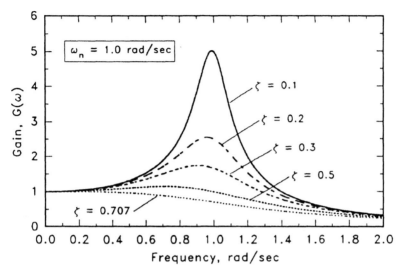

a) Gain $G(\omega/\omega_n)$ vs frequency ratio

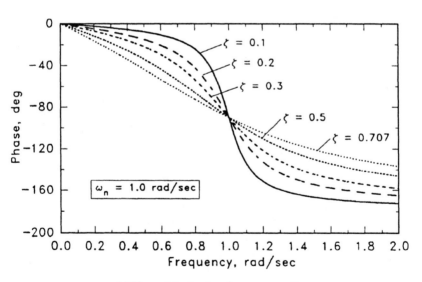

b) Phase $\phi(\omega/\omega_n)$ vs frequency ratio

Fig. 5.10 Harmonic response of a second-order system.

5.5 Stability

The stability of a physical system is usually associated with the response trajectory following the application of a disturbance to the system at its static equilibrium position. A stable system will return to its equilibrium state following the disturbance. Examples of stable responses for first- and second-order systems were shown in Secs. 5.3 and 5.4 by Figs. 5.4, 5.7, and 5.9. Linear system stability is assured if all of the poles of the transfer function $G(s)$ have real parts less than zero. Consider a system whose response in the complex s domain to a forcing function is given by

$$X(s) = G(s)F(s) \tag{5.46}$$

where

$X(s)$ = Laplace transform of the response variable, $\mathcal{L}[x(t)]$
$F(s)$ = Laplace transform of the input function, $\mathcal{L}[f(t)]$
$G(s)$ = transfer function between the input and the output

By principles described in Sec. 5.2, a partial fraction expansion can be made of $G(s)$ according to Eq. (5.18) to yield

$$G(s) = \frac{a_1}{s + p_1} + \frac{a_2}{s + p_2} + \cdots + \frac{a_n}{s + p_n}$$

The poles in the transfer function can be either real or complex. However, a stable linear system requires that

$$\text{Re}(p_m) < 0 \quad \text{for } m = 1, \ldots, n \tag{5.47}$$

The exponential decay behavior with the advent of time, a characteristic of a stable system, is apparent, as shown in Fig. 5.11, when Eq. (5.47) is satisfied for all of the system's poles. When the real part of a system pole is to the right of the imaginary axis, a divergent response will result.

Information concerning the system's stability can be obtained by solving for the roots of the characteristic equation in order to establish the pole locations or

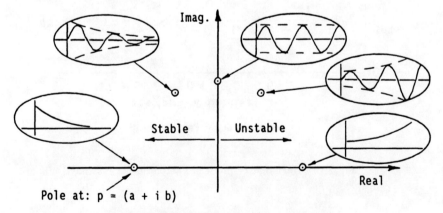

Fig. 5.11 Effect of pole location on response.

the system's eigenvalues as described in Appendix D. The question of whether the system is stable can also be determined without directly solving for the roots of the characteristic equation. This approach was frequently applied to stability issues at the turn of the 20th century when finding the roots of a characteristic equation having an order of four or greater was a tedious matter. A well-known method for determining whether a system is stable involves the evaluation of Routh's discriminants based on the coefficients of the system's characteristic equation. The method as described by Routh[8] is cited in many references including Ogata[7] and Etkin.[9]

In Routh's method, the coefficients of the characteristic equation are systematically arranged in a matrix array of the same order as the original equation. The number of sign changes for the terms in the first column determines the stability properties of the polynomial expression. The method does not actually locate the roots, but only determines whether there are roots present with positive real parts, and if there are, tells how many. Requirements for a stable system by Routh's method may be briefly stated without proof for third- and fourth-order polynomial equations as follows.

1) Consider a cubic characteristic equation with real-valued coefficients.

$$a_3 s^3 + a_2 s^2 + a_1 s + a_0 = 0$$

Stability is assured [i.e., Eq. (5.47) is satisfied] if and only if all of the coefficients $a_0, a_1, a_2,$ and a_3 along with the Routh discriminant

$$R = a_1 a_2 - a_0 a_3$$

are of the same sign.

2) Consider a quartic characteristic equation with real-valued coefficients,

$$a_4 s^4 + a_3 s^3 + a_2 s^2 + a_1 s + a_0 = 0$$

Stability is assured if and only if all of the coefficients $a_0, a_1, a_2, a_3,$ and a_4 along with the Routh discriminant

$$R = a_1 a_2 a_3 - a_1^2 a_4 - a_0 a_3^2$$

are of the same sign.

Routh's method is mentioned primarily to provide a historical perspective. A convenient method for stability analysis that was held in high favor by engineers during the mid-20th century involved the use of the analog computer. The analog computer, based on the properties of high-gain, direct current operational amplifiers, allowed electrical simulations to be made of physical systems. Stability traits were evident from the time-history traces of voltage output (by analogy) including estimates of system damping by use of the log-decrement procedure. These traces provided both an indication of the degree of system stability and the influence of physical parameter changes on the system dynamics. In retrospect, the analog computer served as a transition tool between the elegant mathematical principles embedded in Routh's stability criteria and the efficient numerical analysis methods available today using the desktop computer.

5.6 State-Space Fundamentals

It is a temptation to use the term modern control theory to describe a real-time oriented approach for solving control system problems. However, the term modern is only relative, and much of the mathematics involved has a long-time standing with a recent renewal of interest having developed as a consequence of digital computer improvements. Physical systems with multiple inputs and outputs are in abundance, and a need exists for establishing feedback laws for these systems to satisfy many practical constraints, e.g., optimal control. Consequently, there is motivation to consider systems in the context of their conditions of state by a state-space viewpoint. It is helpful to relate state-space techniques to other familiar methods, such as those involved in frequency-domain analyses. Concepts such as the use of transfer functions in classical control theory become reinforced from alternate outlooks, e.g., state-space principles.

A physical system during its dynamic life has a status that may be evaluated or described by a measure of its state. The physical system is called a plant and could represent an airplane, ship, automobile, chemical processing factory, or an electrical circuit.

The system, as a plant, produces an output(s) under the impetus of an input(s). The description of the plant operation, status, or state is provided by the following terms.

1) *State variables* are the smallest set of variables such that knowledge at time $t = t_0$ together with the inputs for time $t \geq 0$ completely determine the system behavior for all times $t \geq t_0$.

2) The n state variables used to describe the behavior of a given system that can be considered as the n components of the *state vector*.

3) *State space* comprises the n-dimensional space spanned by the state vector.

In two dimensions, the state space reduces to a state plane, sometimes called a phase plane. A familiar example of a two-dimensional system is the harmonic oscillator where the second-order differential equation representing the system is completely described by knowledge of its displacement and velocity. Therefore, displacement and velocity serve in the role of state variables. When the dimensionality is greater than three, we refer to the space as a hyperspace, a nomenclature designed to cover up for human inadequacies in physical interpretation.

In three dimensions, the trajectory of the state vector in state space appears like that shown in Fig. 5.12. It should be noted that, although the trajectory is determined by the vector components, their values are time dependent because time is still the independent parameter.

Control theory using state-space formulations may seem somewhat abstract on initial encounter. It is the purpose of this section to show some sample formulations, some properties of linear, time-invariant systems, solution techniques, and a few equivalences in the frequency domain. The material will be limited initially to continuous systems, although discrete systems have become quite important in aeronautics because of fly-by-wire applications to recent high-performance aircraft designs.

5.6.1 State Equation Formulation

The discussion of state variable concepts will begin by stating two fairly simple rules dealing with ODEs having constant coefficients.

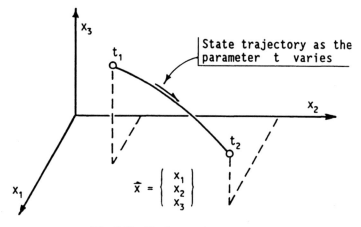

Fig. 5.12 Trajectory in state space.

1) A single, nth-order ODE can be re-expressed as n first-order coupled ODEs.
2) Two second-order ODEs (as one would find in a multi-degree-of-freedom system) can be re-expressed as four first-order coupled ODEs.

To illustrate the principle, consider a third-order equation having the form

$$y^{III} + a_2 y^{II} + a_1 y^{I} + a_0 y(t) = b_0 u(t) \tag{5.48}$$

To make the transformation to state-variable format, redefine the variables of interest as

$$x_1 = y(t)$$

$$x_2 = \frac{dy}{dt} = y^{I} = \dot{x}_1 \tag{5.49}$$

$$x_3 = \frac{d^2 y}{dt^2} = y^{II} = \dot{x}_2$$

Equations (5.48) and (5.49) can be combined into

$$\dot{x}_1 = x_2$$

$$\dot{x}_2 = x_3$$

$$\dot{x}_3 = -a_0 x_1 - a_1 x_2 - a_2 x_3 + b_0 u$$

or in matrix form as

$$\{\dot{x}\} = [A]\{x\} + \{B\}u \tag{5.50}$$

where the state vector is

$$\{x\} = [x_1 \quad x_2 \quad x_3]^T$$

the plant matrix is

$$[A] = \begin{bmatrix} 0 & 1 & 0 \\ 0 & 0 & 1 \\ -a_0 & -a_1 & -a_2 \end{bmatrix}$$

and the control vector (matrix) is

$$\{B\} = [0 \quad 0 \quad b_0]^T$$

It is helpful to consider the homogeneous form of Eq. (5.50) in order to show that the system's transient response dynamics are related to the eigenvalue problem described in Appendix D.3. Consider the preceding equation, where the response is determined by the initial conditions,

$$\{\dot{x}\} = [A]\{x\} \qquad (5.51)$$

Assume that a solution to Eq. (5.51) takes the form

$$\{x(t)\} = \{X\}e^{\lambda t}$$

then

$$\{\dot{x}(t)\} = \lambda\{X\}e^{\lambda t}$$

After canceling out the $e^{\lambda t}$ term, Eq. (5.51) becomes

$$\lambda\{X\} = [A]\{X\} \qquad (5.52)$$

which is identical in meaning to Eq. (D.15) found in Appendix D.3 when describing the eigenvalue problem.

Example 5.6

Consider the damped harmonic oscillator as described in Sec. 5.4. The governing differential equation was described as

$$\ddot{y} + 2\zeta\omega_n\dot{y} + \omega_n^2 y(t) = 1/m\, f(t) \qquad (5.37)$$

The equation can be converted to state variable format by letting the state vector be

$$\{x\} = [x_1 \quad x_2]^T$$

the plant matrix be

$$[A] = \begin{bmatrix} 0 & 1 \\ -\omega_n^2 & -2\zeta\omega_n \end{bmatrix}$$

and the control vector be

$$\{B\} = [0 \quad 1/m]^T$$

The eigenvalue problem, as described in Appendix D.3, starts with identifying the characteristic (modal) frequencies from the roots of the characteristic equation.

Example 5.7

To find the eigenvalues of the damped harmonic oscillator described in Example 5.6, consider the determinant of the characteristic equation as being equal to zero for a nontrivial solution,

$$|\lambda[I] - [A]| = \begin{vmatrix} \lambda & -1 \\ \omega_n^2 & (\lambda + 2\zeta\omega_n) \end{vmatrix} = \lambda^2 + 2\zeta\omega_n + \omega_n^2 = 0$$

which yields

$$\lambda_{1,2} = -\zeta\omega_n \pm i\sqrt{1-\zeta^2}\,\omega_n = -\zeta\omega_n \pm i\omega_d \quad \text{for } 0 < \zeta < 1$$

If the damping ratio $\zeta > 1$, then the roots would be real valued.

The next step in the eigenvalue problem is to identify the mode shapes (eigenvectors) that satisfy the matrix equation according to the eigenvalue relationship in accord with Eq. (5.52). The index n is used in order to distinguish the individual mode shapes. The harmonic oscillator example has two eigenvalues, with the second being the complex conjugate of the first one. A similar relationship will also hold true for the eigenvectors. Therefore, only the first eigenvector need be identified,

$$\lambda_n \{x^{(n)}\} = [A]\{x^{(n)}\} \tag{5.53}$$

Example 5.8

Using λ_1 from Example 5.7, the first eigenvector is found by using Eq. (5.53), i.e.,

$$(-\zeta\omega_n + i\omega_d)\{x^{(1)}\} = [A]\{x^{(1)}\}$$

which translates, when expanded, into two, coupled linear equations relating $x_1^{(1)}$ to $x_2^{(1)}$, i.e.,

$$(-\zeta\omega_n + i\omega_d)x_1^{(1)} = x_2^{(1)}$$
$$(-\zeta\omega_n + i\omega_d)x_2^{(1)} = -\omega_n^2 x_1^{(1)} - 2\zeta\omega_n x_2^{(1)}$$

Both relations contain the same information describing the two components of the eigenvector with the magnitudes arbitrarily scaled. From the first, simpler relationship one finds the eigenvector as

$$\{x^{(1)}\} = \begin{Bmatrix} x_1^{(1)} \\ x_2^{(1)} \end{Bmatrix} = \begin{Bmatrix} 1.0 \\ -\zeta\omega_n + i\omega_d \end{Bmatrix} = \begin{Bmatrix} 1.0 \\ \omega_n e^{i\phi} \end{Bmatrix} \tag{5.54}$$

where

$$\phi = \tan^{-1}\left(\sqrt{1-\zeta^2}\right)\big/(-\zeta)$$

with the phase angle being in the second quadrant because the second component $x_2^{(1)}$ represents a time derivative (velocity) of the first component $x_1^{(1)}$ (displacement). The phase angle represents a lead and varies from $+90$ to $+180$ deg as ζ varies from 0 to 1.

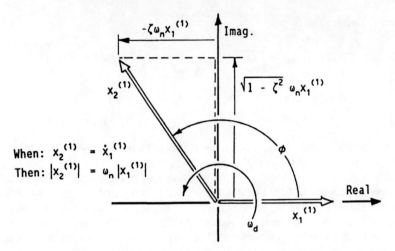

Fig. 5.13 Phasor representation for an eigenvector.

Example 5.8 provided the phase relationship between an eigenvector component and its time derivative using the harmonic oscillator as a sample system. This result will prove useful in Chapters 6 and 7 when estimating the phasing between certain components in several aircraft dynamic mode shape evaluations.

The eigenvector result in Eq. (5.54) represents quantities having an exponential decay ($e^{-\zeta\omega_n t}$) applied to a time-varying trigonometric function with a frequency ω_d. One way of describing the eigenvector components uses the concept of a phasor relation, shown in Fig. 5.13 at time $t = 0$ with the appropriate phase angle and lengths. The two phasor quantities will have a rotation velocity of ω_d rad-s^{-1} and can be visualized as rotating about the origin with their lengths decaying as an exponential spiral while keeping the phase angle relationship constant.

The phasor presentation of Fig. 5.13 can alternatively be visualized as a time-history plot by considering the component of each eigenvector component on the vertical axis as the vector components rotate. For the situation of an impulsive response, as indicated in Sec. 5.4 by Eq. (5.42), the shadow on the vertical axis would appear as shown in Fig. 5.14.

5.6.2 Transition Matrix

A natural question concerns finding a solution to the set of coupled, first-order equations as represented by Eq. (5.50). It would not be surprising to find that the solution had an appearance similar to the scalar counterpart, Eq. (5.28) in Sec. 5.3. Laplace transform methods will prove valuable for finding the solution.

It is valid to make a Laplace transform to the s domain for a function $y(t)$ subject to the constraints stated in Sec. 5.2,

$$\mathcal{L}[y(t)] = Y(s)$$

where

$$Y(s) = \int_0^\infty y(t) e^{-st}\, dt$$

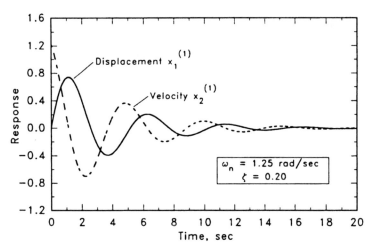

Fig. 5.14 Time-history equivalence from a phasor diagram.

and s is a complex quantity,

$$s = \sigma + i\omega$$

Likewise, a column vector where each element is subject to similar constraints may be transformed because all operations are linear, i.e.,

$$\mathcal{L}[\{x(t)\}] = \begin{Bmatrix} \mathcal{L}[x_1(t)] \\ \vdots \\ \mathcal{L}[x_n(t)] \end{Bmatrix} = \begin{Bmatrix} X_1(s) \\ \vdots \\ X_n(s) \end{Bmatrix} = \{X(s)\}$$

Application of the Laplace transform technique to the time-domain matrix terms of the state equation [Eq. (5.50)] yields

$$s\{X(s)\} - \{x(0)\} = [A]\{X(s)\} + [B]\{U(s)\}$$

from which one obtains

$$[sI - A]\{X(s)\} = \{x(0)\} + [B]\{U(s)\}$$

with the final result

$$\{X(s)\} = [sI - A]^{-1}\{x(0)\} + [sI - A]^{-1}[B]\{U(s)\} \qquad (5.55)$$

It should noted that the terms in Eq. (5.55) are grouped in accord with the rules of matrix algebra. The matrix term $[sI - A]^{-1}$ is denoted as the resolvent matrix in the literature.[10] Its Laplace inverse is known as the transition matrix $\phi(t)$. The Laplace inverse of Eq. (5.55), which includes using the convolution property of Eqs. (5.16) and (5.17), provides the system's state vector response in the time domain,

$$\{x(t)\} = \phi(t)\{x(0)\} + \int_0^t \phi(t - \tau)[B]\{u(\tau)\}d\tau \qquad (5.56)$$

As mentioned in Sec. 5.3, the total solution of Eq. (5.56) contains both the homogeneous response due to an initial condition of the state vector and the particular response due to the applied forcing function. A physical interpretation may be made to the transition matrix by considering only the homogeneous portion of Eq. (5.56),

$$\{x(t)\}_H = \phi(t)\{x(0)\} \tag{5.57}$$

The matrix $\phi(t)$ in Eq. (5.57) transitions the system from an initial state to a state at a later time t. Consequently, the definition is given for the transition matrix

$$\phi(t) = \mathcal{L}^{-1}\{[sI - A]^{-1}\} \tag{5.58}$$

An alternate expression for the transition matrix as an exponential matrix function will assist in establishing the properties of a transition matrix. In the scalar situation, it can be seen from Table 5.1 that

$$\mathcal{L}^{-1}\{(s-a)^{-1}\} = e^{at} \quad \text{for } t > 0$$

Alternatively, $(s - a)^{-1}$ can be expanded in the form of an infinite series to give

$$(s - a)^{-1} = (1/s) + (a/s^2) + (a^2/s^3) + \cdots$$

which can be inverted using properties of the Laplace transform given in Table 5.1 into an infinite time series,

$$\mathcal{L}^{-1}\{(s-a)^{-1}\} = 1 + at + (at)^2/2! + (at)^3/3! + \cdots$$

The infinite time series represents the exponential function e^{at} where a is a scalar constant. In a matrix sense, it would appear that the resolvent matrix could be expressed by a similar infinite series using the matrix A in place of the scalar a. Therefore, assume that

$$[sI - A]^{-1} = (I/s) + (A/s^2) + (A^2/s^3) + \cdots$$

Verification is done by multiplying both sides of the matrix expression by $[sI - A]$,

$$[sI - A][sI - A]^{-1} = [I - A/s] + [A/s - (A/s)^2] + [(A/s)^2 - (A/s)^3] + \cdots$$

$$= I - (A/s)^n + \cdots$$

Pre- and postmultiplication by the modal matrices $[P]^{-1}$ and $[P]$ [cf. similarity transformation, Eq. (D.30)] alters the matrix terms in the series to a diagonal form $[D]$ where the lead diagonal contains the eigenvalues,

$$[P]^{-1}[sI - A][sI - A]^{-1}[P] = I - (D/s)^n + \cdots = I \quad \text{as } n \to \infty$$

Diagonal terms in (D/s), such as (λ_i/s), will converge to zero in the limit as n tends to infinity in a manner like the nth power of (a/s) converged to zero in the scalar form. Demonstration of the resolvent matrix as a matrix series in powers of $(1/s)$ supports using an inverse Laplace transform to get an infinite time series for the transition matrix $\phi(t)$,

$$\mathcal{L}^{-1}\{[sI - A]^{-1}\} = \phi(t) = I + At + (At)^2/2! + (At)^3/3! + \cdots \quad (5.59)$$

where $\phi(t)$ has the form of an exponential function in a matrix sense and may be expressed in its alternate form as

$$\phi(t) = e^{At} \quad (5.60)$$

Useful properties of the transition matrix include the following.
1) When $t = 0$, $e^{At} = I$ (the unit diagonal matrix). This may be verified from Eq. (5.59) by setting $t = 0$.
2) Exponential functions,

$$e^{A(t_1 + t_2)} = e^{At_1} e^{At_2} = e^{At_2} e^{At_1}$$

These properties are deduced by considering the well-known scalar equivalents. Replace the scalar terms involving a by the matrix counterparts, A, in the product of the two infinite series.
3) The relation

$$e^{-At} = [e^{At}]^{-1}$$

may be seen by setting $t_2 = -t_1$ in statement 2. Because $(t_1 - t_1) = 0$, statement 1 completes the verification.
4) Here

$$Ae^{At} = e^{At} A$$

The unique property of the transition matrix being commutative with its A matrix may be seen by considering the series representation of the exponential matrix for At,

$$A[I + At + (At)^2/2! + (At)^3/3! + \cdots]$$
$$= [I + At + (At)^2/2! + (At)^3/3! + \cdots]A$$

It should be recognized that e^{At} also is commutative with A^{-1}. This, too, can be a useful relation.
5) The derivative of the transition matrix, $(d/dt)[e^{At}] = Ae^{At}$. Again, apply the time-derivative operation to each term in the power series of e^{At} to obtain

$$\left(\frac{d}{dt}\right)[e^{At}] = \left(\frac{d}{dt}\right)[I + At + (At)^2/2! + (At)^3/3! + \cdots] = Ae^{At}$$

The state equation solution [Eq. (5.56)] may be stated in an alternate form using the properties of e^{At} by

$$\{x(t)\} = e^{At}\{x(0)\} + e^{At} \int_0^t e^{-A\tau}[B]\{u(\tau)\}d\tau \quad (5.61)$$

The mathematical properties of the transition matrix are useful when developing principles of control theory, but the use of e^{At} when solving dynamic problems requires its being evaluated numerically. Before considering a numerical evaluation, a closed-form solution for the transition matrix will be obtained for a harmonic oscillator system in order to increase the understanding of the information contained in e^{At}.

Example 5.9

Find the transition matrix for the harmonic oscillator system. The solution method will involve the determination of the resolvent matrix followed by taking its inverse Laplace transform to find the transition matrix in accord with Eq. (5.58).
1) The second-order plant is defined by

$$A = \begin{bmatrix} 0 & 1 \\ -\omega_n^2 & -2\zeta\omega_n \end{bmatrix}$$

2) The $[sI - A]$ matrix is

$$[sI - A] = \begin{bmatrix} s & -1 \\ \omega_n^2 & (s + 2\zeta\omega_n) \end{bmatrix}$$

3) The inverse of $[sI - A]$, the resolvent matrix, may be obtained for the 2×2 matrix by inspection,

$$[sI - A]^{-1} = \frac{1}{\Delta(s)} \begin{bmatrix} (s + 2\zeta\omega_n) & +1 \\ -\omega_n^2 & s \end{bmatrix} = R(s)$$

where the characteristic determinant $\Delta(s)$ is

$$\Delta(s) = s^2 + 2\zeta\omega_n s + \omega_n^2 = (s + \zeta\omega_n)^2 + \omega_d^2$$

4) The inverse Laplace transform of the resolvent matrix $R(s)$ can be done term by term using familiar rules from Table 5.1,

$$R(1, 1) = (s + \zeta\omega_n + \zeta\omega_n)/\Delta(s)$$

$$\mathcal{L}^{-1}\{R(1, 1)\} = e^{-\zeta\omega_n t}\left[\cos \omega_d t + \left(\zeta/\sqrt{1 - \zeta^2}\right) \sin \omega_d t\right]$$

$$R(1, 2) = 1/\Delta(s)$$

$$\mathcal{L}^{-1}\{R(1, 2)\} = e^{-\zeta\omega_n t}\left[\left(1/\sqrt{1 - \zeta^2}\omega_n\right) \sin \omega_d t\right]$$

$$R(2, 1) = -\omega_n^2/\Delta(s)$$

$$\mathcal{L}^{-1}\{R(2, 1)\} = e^{-\zeta\omega_n t}\left[\left(-\omega_n/\sqrt{1 - \zeta^2}\right) \sin \omega_d t\right]$$

$$R(2, 2) = (s + \zeta\omega_n - \zeta\omega_n)/\Delta(s)$$

$$\mathcal{L}^{-1}\{R(2, 2)\} = e^{-\zeta\omega_n t}\left[\cos \omega_d t - \left(\zeta/\sqrt{1 - \zeta^2}\right) \sin \omega_d t\right]$$

5) The transition matrix is obtained by assembling terms from step 4,

$$e^{At} = e^{-\zeta\omega_n t}\left[\begin{bmatrix} 1 & 0 \\ 0 & 1 \end{bmatrix}\cos\omega_d t + \begin{bmatrix} (\zeta/\sqrt{1-\zeta^2}) & (1/\sqrt{1-\zeta^2}\omega_n) \\ (-\omega_n/\sqrt{1-\zeta^2}) & (-\zeta/\sqrt{1-\zeta^2}) \end{bmatrix}\sin\omega_d t\right]$$

(5.62)

The transition matrix for the harmonic oscillator [Eq. (5.62)] can be used to find the time histories of both displacement and velocity due to initial conditions by application of the homogeneous solution relation [Eq. (5.57)],

$$\{x(t)\}_H = e^{At}\begin{Bmatrix} y_0 \\ v_0 \end{Bmatrix}$$

If one assumed that $y_0 \neq 0$ while $v_0 = 0$, expansion of the matrix expression for the homogeneous response would simplify to

$$\begin{Bmatrix} x_1(t) \\ x_2(t) \end{Bmatrix} = y_0 e^{-\zeta\omega_n t}\left\{\begin{Bmatrix} 1 \\ 0 \end{Bmatrix}\cos\omega_d t + \begin{Bmatrix} (\zeta/\sqrt{1-\zeta^2}) \\ (-\omega_n/\sqrt{1-\zeta^2}) \end{Bmatrix}\sin\omega_d t\right\}$$

It will be noted that the first row of the homogeneous response solution agrees with the earlier result in Sec. 5.4, Eq. (5.39). In addition, information concerning the second component of the state vector (the velocity) is contained in the second row of the preceding solution. Similarly, if one assumed $y_0 = 0$ while $v_0 \neq 0$, accord would be found with the earlier result for the response to a velocity initial condition [Eq. (5.41)].

5.6.3 System Time Response

An output response other than the state vector solution can be obtained by a linear combination of the state vector with the input vector. This arrangement adds versatility by providing response solutions for variables other than the state vector (i.e., aircraft normal acceleration response due to elevator control input, Chapter 6). The generalized expression becomes

$$\dot{x} = Ax + Bu \qquad (5.50a)$$

$$y = Cx + Du \qquad (5.63)$$

where y is the system output variable.

A block diagram illustrating the information flow for Eqs. (5.50) and (5.63) is shown in Fig. 5.15. The double-line information flow corresponds to a vector whereas the single line represents a scalar variable. The figure reflects the concept of an open-loop system because none of the state variables are being introduced as feedback to modify the system's behavior. The airframe analyses in Chapters 6 and 7 will reflect open-loop conditions.

The system response will be described for three situations including the homogeneous case, along with the particular solutions due to both the unit impulse and the unit step function as control inputs.

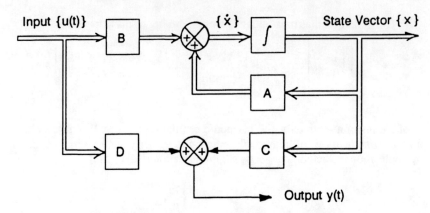

Fig. 5.15   Block diagram of an open-loop system.

Homogeneous solution. The state vector solution for the system's homogeneous response has been stated earlier as

$$\{x(t)\}_H = e^{At}\{x(0)\} \tag{5.64}$$

The output response, from Eq. (5.63), is

$$y_H(t) = Ce^{At}\{x(0)\}$$

where, for a scalar output, the C matrix would be a $1 \times n$ row vector for an nth order system.

For the impulse response, the unit impulse will be applied to a system having a single control input. For an aircraft situation, this control input could correspond to a rudder kick by the pilot. Let

$$\{u(\tau)\} = \delta(\tau)$$

be the Dirac delta function.

The general solution [Eq. (5.61)] becomes

$$\{x(t)\} = e^{At}\int_0^t e^{-A\tau}\{B\}\delta(\tau)\,d\tau = e^{At}\int_0^t e^{-A\tau}\delta(\tau)\,d\tau\{B\}$$

The preceding rearrangement groups the time-related scalar terms into a compact integral that can be readily solved using known properties of the Dirac delta function described by Eq. (5.5) and the transition matrix (i.e., $e^{A\tau} = I$ when $\tau = 0$). Hence, due to $u(\tau) = \delta(\tau)$,

$$\{x(t)\} = e^{At}\{B\} \tag{5.65}$$

The output response becomes

$$y(t) = Ce^{At}\{B\} \quad \text{for } t > 0 \tag{5.66}$$

For the step response, consider a unit-step function control input applied at $t = 0$ to a single control. Let

$$\{u(\tau)\} = 1(\tau)$$

be the unit step function.

The general solution, Eq. (5.61), becomes

$$\{x(t)\} = e^{At} \int_0^t e^{-A\tau}\{B\}1(\tau)d\tau = e^{At} \int_0^t e^{-A\tau}d\tau\{B\}$$

The integral expression can be evaluated by using the properties of the transition matrix,

$$\int_0^t e^{-A\tau} d\tau = -A^{-1}[e^{-At} - I] = [I - e^{-At}]A^{-1}$$

Therefore, due to $u(\tau) = 1(\tau)$,

$$\{x(t)\} = [e^{At} - I]A^{-1}\{B\} \tag{5.67}$$

and the output response for this case becomes

$$y = C[e^{At} - I]A^{-1}\{B\} + D \quad \text{for } t > 0 \tag{5.68}$$

5.6.4 System Frequency Response

A harmonic control input will be assumed having the form

$$u(t) = f_0 e^{i\omega t} 1(t) = f_0(\cos \omega t + i \sin \omega t)1(t)$$

where f_0 is the scalar amplitude of a single control input. This assumption would correspond to the situation of a pilot sinusoidally exciting an aircraft control at a frequency of ω rad-s^{-1}. Also, it will be assumed that the physical system (or airframe) was initially at an equilibrium state. The system response will include both the startup transient plus the long-term steady-state harmonic response. The steady-state harmonic system response to a harmonic control input represents the system motion after all startup transients have subsided.

The state equation for harmonic excitation is

$$\{\dot{x}\} = [A]\{x\} + f_0\{B\}e^{i\omega t} \tag{5.69}$$

and the Laplace transform is

$$\{X(s)\} = f_0[sI - A]^{-1}\{B\}\frac{1}{s - i\omega} = f_0\{G(s)\}\frac{1}{s - i\omega} \tag{5.70}$$

In Eq. (5.70) the $\{G(s)\}$ transfer function corresponds to

$$\{G(s)\} = [sI - A]^{-1}\{B\} = \frac{\{\text{Num}(s)\}}{\Delta(s)}$$

where

$\Delta(s)$ = characteristic polynomial, $(s + \lambda_1)(s + \lambda_2)\ldots(s + \lambda_n)$
λ_i = ith root of the nth-order polynomial
Num(s) = numerator

The use of the Laplace transform implies that Re $\lambda_i < 0$; i.e., the physical system is stable. The inverse Laplace transform of the transfer function matrix involves using partial fraction expansions as described in Sec. 5.2 where the coefficients are obtained by the method of residues evaluated at each of the poles,

$$\{X(s)\} = f_0 \sum_{j=1}^{n} \frac{(s+\lambda_j)}{(s-i\omega)} \{G(s)\}\bigg|_{s=-\lambda_j} \frac{1}{(s+\lambda_j)} + f_0 \{G(s)\}\bigg|_{s=i\omega} \frac{1}{(s-i\omega)}$$

$$= f_0 \sum_{j=1}^{n} \{H_j(i\omega)\} \frac{1}{(s+\lambda_j)} + f_0 \{G(i\omega)\} \frac{1}{(s-i\omega)}$$

where

$$\{H_j(i\omega)\} = \frac{(s+\lambda_j)}{(s-i\omega)} \{G(s)\}\bigg|_{s=-\lambda_j}$$

The response in the time domain becomes

$$\{x(i\omega t)\} = f_0 \sum_{j=1}^{n} \{H_j(i\omega)\} e^{i\lambda_j t} + f_0 \{G(i\omega)\} e^{i\omega t} \qquad (5.71)$$

The term in Eq. (5.71) involving the summation represents the system's transient response to the startup of the harmonic control input with frequency ω at $t = 0$. Each element in $\{H_j(i\omega)\}$ corresponds to the transient behavior of an eigenvector component for the jth mode. For a stable system, the startup transient will subside leaving only the steady-state frequency response.

The steady-state harmonic response can be due to either a cosine or a sine input, and both responses are expressed in the complex-natured result as a consequence of Euler's theorem,

$$\{x(i\omega t)\}_{ss} = f_0 \{G(i\omega)\} e^{i\omega t} \qquad (5.72)$$

Alternatively in terms of trigonometric functions for an input term described as

$$\{u(\omega t)\} = f_0 \begin{Bmatrix} \sin \omega t \\ \cos \omega t \end{Bmatrix}$$

the steady-state harmonic response becomes

$$\{x(\omega t)\}_{ss} = f_0 \begin{Bmatrix} \{G(\omega) \sin(\omega t + \phi)\} \\ \{G(\omega) \cos(\omega t + \phi)\} \end{Bmatrix}$$

where the jth term in the $\{G(\omega)\}$ column vector is obtained from the complex-natured frequency-response function by the following relationships.

Consider

$$G_j(i\omega) = G_j(\omega)(\cos\phi_j + i\sin\phi_j)$$

then

$$G_j(\omega) = [G_j(i\omega)G_j(-i\omega)]^{1/2}$$

and

$$\tan\phi_j = \frac{\text{Im } G_j(i\omega)}{\text{Re } G_j(i\omega)}$$

When the output of interest is a quantity other than a component of the state vector, Eq. (5.63) would apply using appropriate choices of the C and D matrices, i.e.,

$$y(\omega t)_{ss} = C\{x(\omega t)\}_{ss} + Du(\omega t) \tag{5.73}$$

5.7 Stability in the Sense of Lyapunov

The earlier discussion on system stability (Sec. 5.5) briefly described Routh's method (and discriminant) for determining the stability of a linear, time-invariant system without solving the governing differential equations. An alternate approach that applies to determining the stability of both linear and nonlinear systems without solving the differential equations is the second method of Lyapunov.[11] Discussions on Lyapunov's method appear in many texts (e.g., Ogata[7] and Shinners[12]).

Stability, as expressed earlier in Sec. 5.5, may be classified as asymptotic stability in the sense of Lyapunov. A limit cycle, such as aircraft wing rock, is considered as stable motion when stability is described in a more general context.

1) An equilibrium state $\{x_e\}$ of a physical system is said to be *stable* in the sense of Lyapunov when every solution starting within domain $S(\sigma)$ at $\{x_0\}$ remains within domain $S(\epsilon)$ as time t increases indefinitely. The trajectory is illustrated in two-dimensional state space in Fig. 5.16a.

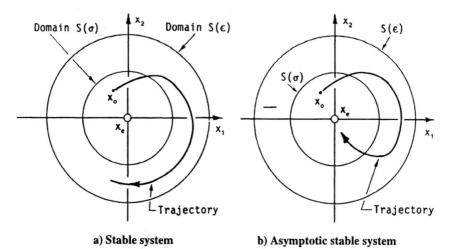

a) Stable system   b) Asymptotic stable system

Fig. 5.16   Sketch of stable trajectories.

2) A stable system has *asymptotic stability* when every solution starting within domain $S(\sigma)$ at $\{x_0\}$ reaches the static-equilibrium state $\{x_e\}$ in domain $S(\epsilon)$ as time t increases. This trajectory is shown in Fig. 5.16b.

3) *Asymptotic stability in the large* corresponds to asymptotic stability with the understanding that the initial condition $\{x_0\}$ can occur anywhere in state space; i.e., domain $S(\sigma)$ extends to the outer reaches (infinity).

From a practical standpoint, an aircraft can be stable and not have asymptotic stability. An example situation is the limit-cycle motion in flight known as wing rock. The motion is a finite amplitude roll angle oscillation that occurs on many high-performance jet aircraft at angles of attack near stall. Aircraft with satisfactory flying qualities will either be stable or have asymptotic stability in the sense of Lyapunov.

In Lyapunov's second method, stability can be determined without solving the differential equation using an approach that differs from Routh's stability criteria. Classical mechanics tells us that a vibratory system is stable if its total energy is continually decreasing following release from an initial condition until the system returns to its static equilibrium state. In this situation, total energy is a positive definite scalar function and the time derivative of total energy must be a negative definite function for the system to return to static equilibrium. In Lyapunov's method, a functional based on the state variables and resembling a fictitious energy function is considered. If a functional can be found that is positive definite and its time derivative is negative definite along the system's trajectories in state space, it can be concluded that the system is asymptotically stable.[7] The function satisfying these requirements is not unique; however, identifying one appropriate functional guarantees the system's asymptotic stability. The applications to be covered in this section will relate to linear, time-invariant systems.

A few comments are timely as to the meaning of positive definite and semidefinite functions, respectively, using examples having a quadratic form.

$V(\{x\}, t)$ is positive definite when

$$V(\{x\}, t) \begin{cases} > 0 & \text{for } \{x\} \neq 0 \quad \text{and} \quad t \geq t_0 \\ = 0 & \text{for } \{x\} = 0 \quad \text{and} \quad t \geq t_0 \end{cases}$$

An example would be

$$V = \{x\}^T \begin{bmatrix} 1 & 0 \\ 0 & 1 \end{bmatrix} \{x\} = x_1^2 + x_2^2$$

where $V = 0$ only when $x_1 = x_2 = 0$.

$V(\{x\}, t)$ is positive semidefinite when

$$V(\{x\}, t) \begin{cases} \geq 0 & \text{for } \{x\} \neq 0 \quad \text{and} \quad t \geq t_0 \\ = 0 & \text{for } \{x\} = 0 \quad \text{and} \quad t \geq t_0 \end{cases}$$

An example would be

$$V = \{x\}^T \begin{bmatrix} 1 & 1 \\ 1 & 1 \end{bmatrix} \{x\} = (x_1 + x_2)^2$$

where $V = 0$ when $x_1 = x_2 = 0$ and $x_1 = -x_2$.

Negative definite and semidefinite functions are similar to the preceding examples except for a sign change.

Because the stability definitions are given for the system in its homogeneous response situation, let us consider the linear, time-invariant system in the form given earlier,

$$\{\dot{x}\} = [A]\{x\} \tag{5.51}$$

where

$\{x\}$ = state vector of order n
$[A]$ = nonsingular plant matrix of order $n \times n$

Next, select a positive definite real symmetric matrix $[Q]$ of order $n \times n$ to form a possible Lyapunov function,

$$V(\{x\}) = \{x\}^T [Q]\{x\} \tag{5.74}$$

The time derivative of the Lyapunov function along any trajectory is

$$\begin{aligned}\dot{V} &= \{\dot{x}\}^T [Q]\{x\} + \{x\}^T [Q]\{\dot{x}\} \\ &= \{[A]\{x\}\}^T [Q]\{x\} + \{x\}^T [Q][A]\{x\} \\ &= \{x\}^T [[A]^T [Q] + [Q][A]]\{x\}\end{aligned}$$

Because $V(\{x\})$ was chosen as being positive definite, asymptotic stability requires that $\dot{V}$ be negative definite. Therefore,

$$\dot{V} = -\{x\}^T [R]\{x\}$$

where

$$[R] = -[[A]^T [Q] + [Q][A]] \tag{5.75}$$

is the positive definite matrix.

The matrix relations involving $[Q]$ in Eq. (5.75) may be solved by assuming a positive definite form for the matrix $[R]$ and then verifying that the matrix $[Q]$ is positive definite. If both $[R]$ in Lyapunov's equation [Eq. (5.75)] and $[Q]$ in Lyapunov's function [Eq. (5.74)] are positive definite, then the system has been shown to be asymptotic stable.

A quadratic form is positive definite if all of the leading principle minor determinants are positive using Sylvester's criterion (cf. any elementary matrix algebra text, such as Hohn[13]). The matrix $[Q]$ in Lyapunov's function [Eq. (5.74)] can be expressed as

$$[Q] = \begin{bmatrix} q_{11} & q_{12} & \cdots & q_{1n} \\ q_{21} & q_{22} & \cdots & q_{2n} \\ \vdots & \vdots & \ddots & \vdots \\ q_{n1} & q_{n2} & \cdots & q_{nn} \end{bmatrix}$$

The quadratic function is positive definite when the principle minor determinants satisfy

$$|q_{11}| > 0, \qquad \begin{vmatrix} q_{11} & q_{12} \\ q_{21} & q_{22} \end{vmatrix} > 0, \qquad |Q| > 0$$

Example 5.10

Use the second method of Lyapunov to verify that a second-order linear system, where $\omega_n = 1$ rad-s^{-1} and $\zeta = 0.1$, is asymptotically stable.

The plant matrix is

$$[A] = \begin{bmatrix} 0 & 1 \\ -1 & -0.2 \end{bmatrix}$$

Select an $[R]$ matrix that is positive definite. For a trial attempt, assume that it is a unit diagonal matrix. Lyapunov's equation becomes

$$-\begin{bmatrix} 1 & 0 \\ 0 & 1 \end{bmatrix} = \begin{bmatrix} 0 & -1 \\ 1 & -0.2 \end{bmatrix} \begin{bmatrix} q_{11} & q_{12} \\ q_{12} & q_{22} \end{bmatrix} + \begin{bmatrix} q_{11} & q_{12} \\ q_{12} & q_{22} \end{bmatrix} \begin{bmatrix} 0 & 1 \\ -1 & -0.2 \end{bmatrix}$$

Three simultaneous equations are obtained for finding the elements of the $[Q]$ matrix,

$$-1 = -2q_{12}$$
$$0 = q_{11} - 0.2q_{12} - q_{22}$$
$$-1 = 2q_{12} - 0.4q_{22}$$

The $[Q]$ matrix becomes

$$[Q] = \begin{bmatrix} 5.1 & 0.5 \\ 0.5 & 5.0 \end{bmatrix}$$

which is positive definite because

$$|5.1| > 0 \quad \text{and} \quad |Q| = 25.25 > 0$$

The damped harmonic oscillator has been confirmed as having asymptotic stability using the second method of Lyapunov.

In summary, the second method of Lyapunov has provided Lyapunov's equation [Eq. (5.75)] for determining whether a linear, time-invariant system has asymptotic stability subject to finding suitable positive definite functions. Example 5.10 illustrated the technique for the situation of a second-order plant. When the plant's order is greater than two, numerical techniques using available software (e.g., MATLAB™) will prove time effective.

In addition to indicating system stability without solving the differential equations, Lyapunov's equation has an important role when considering the response of a system subject to a white-noise type of random forcing function. This application will appear when considering the problem of aircraft response in atmospheric turbulence (cf. Sec. 9.5).

References

[1] Heaviside, O., *Electromagnetic Theory*, Dover, New York, 1950.

[2] Churchill, R. V., *Modern Operational Mathematics in Engineering*, McGraw–Hill, New York, 1954.

[3] Sokolnikoff, I. S., and Redheffer, R. M., *Mathematics of Physics and Modern Engineering*, McGraw–Hill, New York, 1958, pp. 754–768.

[4] Pipes, L. A., *Applied Mathematics for Engineers and Physicists*, 2nd ed., McGraw–Hill, New York, 1958, pp. 583–655.

[5] Erdelyi, A., *Tables of Integral Transforms*, Vol. 1, McGraw–Hill, New York, 1954, Chaps. 4–5.

[6] von Kármán, T., and Biot, M. A., *Mathematical Methods in Engineering*, McGraw–Hill, New York, 1940, p. 388.

[7] Ogata, K., *Modern Control Engineering*, 2nd ed., Prentice–Hall, Englewood Cliffs, NJ, 1990, pp. 16–38, 283–288, 722–736.

[8] Routh, E. J., *Advanced Rigid Dynamics*, Vol. II, Macmillan, London, 1930.

[9] Etkin, B., *Dynamics of Flight, Stability and Control*, 2nd ed., Wiley, New York, 1982, pp. 171–173.

[10] Friedland, B., *Control System Design*, McGraw–Hill, New York, 1986, pp. 68–70.

[11] Lyapunov, A. M., "On the General Problem of Stability of Motion," Ph.D. Thesis, Khorkov, Russia, 1892; reprinted (in French) *Annals of Mathematical Studies*, Vol. 17, Princeton Univ. Press, Princeton, NJ, 1949.

[12] Shinners, S. M., *Modern Control System Theory and Design*, Wiley, New York, 1992, pp. 589–598.

[13] Hohn, F. E., *Elementary Matrix Algebra*, Macmillan, New York, 1959, pp. 255–260.

Problems

5.1. Verify the Laplace transform for $f(t) = t$ when $t > 0$ as shown in Table 5.1. Hint: Integrate the transform integral by parts.

5.2. Table 5.1 shows $F(s) = 1/(s - a)$ when $f(t) = e^{at}$ for $t > 0$. By expanding the $F(s)$ function into an infinite series expression, deduce the Taylor series expansion for $f(t)$, Eq. (5.9).

5.3. Assume that the constant in e^{at} is a complex quantity (i.e., $a = i\omega$ where $i = \sqrt{-1}$) for $t > 0$. It will be found that the Laplace transform also is a complex quantity. By application of Euler's formula, deduce the Laplace transforms for the trigonometric functions $\cos \omega t$ and $\sin \omega t$.

5.4. Verify the Laplace transform for $\cos \omega t$ by applying the Laplace transform for a derivative to the $\sin \omega t$ transform.

5.5. Find $\sin \omega t$ and $\cos \omega t$ at $t = 0+$ using the initial value theorem. Note that the final value theorem will not apply to a sinusoidal function because the function is of bounded variation but a limit does not exist as $t \to \infty$. However, the initial value theorem will apply for finding $f(0+)$.

5.6. Apply the final value theorem for Laplace transforms to obtain the steady-state response of a first-order system due to a step input function.

5.7. Verify the alternate expression for Eq. (5.39).

5.8. When the damping coefficient tends to 1, show that:
a) The response $y_1(t)$ for Eq. (5.38) becomes

$$\lim_{\zeta \to 1} y_1(t) = y_0 e^{-\omega_n t} 1(t)$$

b) The response $y_2(t)$ for Eq. (5.40) becomes

$$\lim_{\zeta \to 1} y_2(t) = \frac{v_0}{\omega_n} e^{-\omega_n t} 1(t)$$

c) The response $y_4(t)$ for Eq. (5.42) becomes

$$\lim_{\zeta \to 1} y_4(t) = y_{\text{stat}}[1 - e^{-\omega_n t}] 1(t)$$

5.9. Show that when the damping coefficient $\zeta > 1$, the response of a second-order system to an initial condition on displacement y_0 is

$$y_1(t) = y_0 e^{-\zeta \omega_n t} \left(\cosh \sqrt{\zeta^2 - 1} \, \omega_n t - \frac{\zeta}{\sqrt{\zeta^2 - 1}} \sinh \sqrt{\zeta^2 - 1} \, \omega_n t \right)$$

Hint: The Laplace transforms for the hyperbolic functions are readily obtained by recognizing that

$$\cosh(at) = \tfrac{1}{2}(e^{at} + e^{-at})$$
$$\sinh(at) = \tfrac{1}{2}(e^{at} - e^{-at})$$

where a is a real quantity.

5.10. An impulsive type yaw moment input was applied to an instrumented aircraft by a pilot's rudder kick during a flight-test experiment. The pilot's control input excited the Dutch-roll mode of the aircraft. Telemetry records showed that the first positive sideslip peak was 2.60 deg and the third positive sideslip peak that occurred two oscillation periods later was 2.44 deg. Estimate the damping coefficient ζ for the aircraft's Dutch-roll mode at the test flight condition. Hint: The time difference between T_B and T_A was $4\pi/\omega_d$ (two oscillation periods).

5.11. Show that for a linear, second-order physical system, the steady-state harmonic response has the following properties:

a) Peak value of gain $G(\omega/\omega_n)$, G_{max}, occurs when

$$\omega/\omega_n = \begin{cases} \sqrt{1 - 2\zeta^2} & \text{for } 0 < \zeta < 1/\sqrt{2} \\ 0 & \text{for } \zeta > 1/\sqrt{2} \end{cases}$$

b) Peak value of gain $G(\omega/\omega_n)$, G_{max}, is

$$G_{max} = \begin{cases} \frac{1}{2\zeta\sqrt{1-\zeta^2}} & \text{for } 0 < \zeta < 1/\sqrt{2} \\ 1 & \text{for } \zeta > 1/\sqrt{2} \end{cases}$$

c) Bandwidth for the situation of light damping is

$$\Delta\omega/\omega_n = 2\zeta$$

Bandwidth on the second-order system is defined as the frequency difference for the two gain values of $G = (1/\sqrt{2})G_{max}$, which corresponds to the system's half-power points. The half-power concept comes from the electrical engineering community where power is proportional to voltage squared. In the case of a physical system, displacement is analagous to voltage.

5.12. Consider the harmonic oscillator, Example 5.9, for the case of $\zeta = 0.1$ and $\omega_n = 2.0$ rad-s^{-1}.
 a) Numerically evaluate the transition matrix at time $t = 1$ s and 2 s.
 b) Verify that $e^{2A} = e^A e^A$ numerically.
Comment: Although there is software available that can evaluate the transition matrix, it is suggested that the reader use a desktop computer only to validate the results.

5.13. Verify that the transition matrix $\phi(t)$ is the exponential matrix e^{At}, Eq. (5.60). Although there are many ways to show this relation, the following approach is suggested:
 a) Apply the similarity transformation to the resolvent matrix and the assumed matrix series.
 b) Apply the inverse-Laplace transform to the diagonalized matrix relations.
 c) Reapply the similarity transformation to show that

$$\phi(t) = e^{At} = Pe^{\lambda t}P^{-1}$$

where

$e^{\lambda t}$ = diagonal exponential matrix
λ_n = eigenvalue of the nth mode
P = modal matrix, cf. Appendix D.5

6
Longitudinal Dynamics

6.1 Background

This chapter considers the aircraft's longitudinal behavior about the pitch-axis reference frame. The system's dynamics will be developed by viewing the airframe as a multi-degree-of-freedom eigenvalue problem using control theory concepts described in Chapter 5. Descriptions of the mode shapes will be obtained and approximations will be made to the equations of motion in order to increase the understanding of the dynamics. It will be found that the fore and aft c.g. location (i.e., the distance forward from the neutral point) will play a large role in the dynamic behavior. In contrast, the airframe's lateral-directional dynamics are not significantly influenced by the c.g. position.

The linearized governing equations describing the longitudinal motion were cast into a set of four, first-order, coupled ODEs with constant coefficients, cf. Sec. 4.6.1. These equations used dimensional coefficients and described the time-varying perturbations of the state vector in real time. The equations were

$$V(\dot{u}/V) = VX_u(u/V) + X_\alpha \alpha - g\cos\Theta_0 \theta + X_\delta \delta$$

$$(V - Z_{\dot{\alpha}})\dot{\alpha} = VZ_u(u/V) + Z_\alpha \alpha + (V + Z_q)q - g\sin\Theta_0 \theta + Z_\delta \delta$$

$$-M_{\dot{\alpha}}\dot{\alpha} + \dot{q} = VM_u(u/V) + M_\alpha \alpha + M_q q + M_\delta \delta$$

$$\dot{\theta} = q$$

or, alternatively, in a linear algebra form

$$[I_n]\{\dot{x}\} = [A_n]\{x\} + \{B_n\}\delta \qquad (4.42)$$

where the longitudinal state vector was

$$\{x\} = [u/V \; \alpha \; q \; \theta]^T$$

and there was a single control term, δ, which normally would correspond to the deflection of either an elevator or a movable horizontal stabilizer.

The presence of the $\dot{\alpha}$ term on the left-hand side of the pitch moment equation is worthy of note. It will be found that when making approximations for the short-period mode, algebraic manipulations can be made to simplify the presence of the $\dot{\alpha}$ term. When investigating the complete linear system, which involves a state vector with four components, linear algebra solution techniques will be employed. The solution techniques for the state equations are exactly the same as would be used to describe the motion dynamics of an automobile, ship, submarine, and even a bridge structure. The choice of the state vector components and the plant matrix is problem specific, but the mathematical techniques are fundamental.

166 INTRODUCTION TO AIRCRAFT FLIGHT DYNAMICS

The aircraft's linearized longitudinal dynamics normally will consist of two pairs of complex conjugate roots, which will correspond to a fast (short-period) mode and a slow (long-period) mode. The long-period mode is known as the phugoid, having been given this name by Lanchester[1] using the Greek root phugos, which means flight as in flee; he mistook the meaning as a word for fly.

6.2 Aircraft Longitudinal Dynamics

The homogeneous form of the linearized equations of motion from Eq. (4.42) is given by

$$[I_n]\{\dot{x}\} = [A_n]\{x\} \qquad (6.1)$$

This set of first-order linear differential equations will be solved by example using computer tools available to many engineers and students, i.e., MATLAB.

Example 6.1

Consider the A-4D jet attack aircraft in level flight at $M = 0.6$, $h = 15{,}000$ ft, and c.g. at $0.25\bar{c}$, cf. Appendix B.1. Solve the eigenvalue problem and identify the mode shapes. The MATLAB listing is as follows:

```
disp(In)
            634.0      0.0       0.0       0.0
              0.0    634.0       0.0       0.0
              0.0      0.353    1.0000     0.0
              0.0      0.0       0.0       1.0000
disp(An)
           -8.179    -3.721      0.0     -32.174
          -65.94   -518.9      634.0       0.0
            0.25    -12.97     -1.071      0.0
             0.0      0.0       1.0000     0.0
% The plant matrix [A] in x-dot = A x is:
A=inv(In)*An;
disp(A)
           -0.0129   -0.0059     0.0      -0.0507
           -0.1040   -0.8185    1.0000     0.0
            0.2867  -12.6811   -1.4240     0.0
             0.0      0.0       1.0000     0.0
% Find the characteristic polynomial
P=poly(A); disp(P)
    1.0000    2.2554   13.8749    0.1939    0.0788
%   (s^4)     (s^3)    (s^2)     (s^1)     (s^0)
% Find the roots of the characteristic polynomial
R=roots(P); disp(R)
```

LONGITUDINAL DYNAMICS

```
      -1.1211±3.5472i; % Short-period mode (eigenvalue)
      -0.0065±0.0752i; % Long-period mode
% Find modal damping and undamped natural frequencies
[Wn,Z]=damp(R);
disp(Z)
       0.3014; % Short-period modal damping
       0.0867; % Long-period modal damping
disp(Wn);
       3.7202; % Short-period natural frequency, rad/sec
       0.0755; % Long-period natural frequency, rad/sec
% Find the eigenvectors
[V, D]=eig(A);
disp(V); % Display the conjugate paired eigenvectors
          -0.0016 ± 0.0035i    -0.5136 ∓ 0.2153i
           0.1539 ± 0.2117i    -0.0053 ∓ 0.0018i
          -0.7977 ± 0.4822i    -0.0576 ∓ 0.0243i
           0.1882 ± 0.1654i    -0.2541 ± 0.7882i
% Find magnitude and phasing for the short-period mode
MAG1=abs(V(:,1)); PHASE1=(180./pi)*angle(V(:,1));
disp(MAG1'/MAG1(2)); % Normalized to α component
       0.0146    1.0000    3.5614    0.9573
%      (u/V)       α         q         θ
disp(PHASE1' - PHASE1(2)); % Phase relative to α, deg
      61.3326     0.0     94.8636   -12.6760
%      (u/V)       α         q         θ
% Find magnitude and phasing for the long-period mode
MAG2=abs(V(:,3)); PHASE2=(180./pi)*angle(V(:,3));
disp(MAG2'/MAG2(1)); % Normalized to (u/V) component
       1.0000    0.0101    0.1122    1.4870
%      (u/V)       α         q         θ
disp(PHASE2' - PHASE2(1)); % Phase relative to (u/V), deg
       0.0      -3.9081    0.1014   265.1269 (=-94.8731)
%      (u/V)       α         q         θ
```

The MATLAB commands used to solve the eigenvalue problem of Example 6.1 included:

disp(A)   = display matrix $[A]$
inv(In)   = invert nonsingular matrix $[In]$
poly(A)   = find polynomial coefficients from $|\lambda I - A|$
roots(P)   = find roots of polynomial from row vector P
damp(R)   = find natural frequencies and modal damping from R
eig(A)   = eigenvectors and eigenvalues from square matrix A
abs(V(:,1))   = absolute number value in first column of matrix V
angle(V(:,1)) = phase angle of complex numbers in first column of V

The preceding commands are described in Appendix E.

6.2.1 Short-Period Mode

The short-period mode from Example 6.1 for the A-4D aircraft with an assumed flight condition and c.g. position may be summarized as

$$\lambda_{sp} = -1.1211 \pm 3.5472i$$

$$\zeta_{sp} = 0.3014 \text{ (a well-damped mode)}$$

$$\omega_n = 3.7202 \text{ rad/s}$$

$$\begin{Bmatrix} (u/V) \\ \alpha \\ q \\ \theta \end{Bmatrix} = \begin{Bmatrix} 0.0146 \angle 61.33 \text{ deg} \\ 1.000 \angle 0.0 \text{ deg} \\ 3.5614 \angle 94.86 \text{ deg} \\ 0.9573 \angle -12.68 \text{ deg} \end{Bmatrix}$$

The short-period eigenvector indicates a very small velocity perturbation (u/V) term relative to the α component, which lends credence to the assumption that $(u/V) = 0$ when making the short-period approximation. The pitch attitude term appears to be nearly the same in magnitude as the angle-of-attack component and lags α by a small phase angle. This implies that the aircraft c.g. trajectory will approach a horizontal path for the short-period mode. If the two components (α and θ) had been equal both in magnitude and phase, then the c.g. trajectory would have been a straight line.

It will be observed that the relationship between the pitch attitude and the corresponding rate component in the eigenvector is in accord with Eq. (5.54) as described in Sec. 5.6; i.e., magnitude scaling is

$$|q| = \omega_n |\theta| = 3.7202 * 0.9573 = 3.5614$$

and phase angle difference is

$$\Delta\phi = \tan^{-1}\left[\sqrt{1-\zeta^2}/(-\zeta)\right] = 107.54 \text{ deg}$$

The phasor representation of the short-period eigenvector is shown in Fig. 6.1.

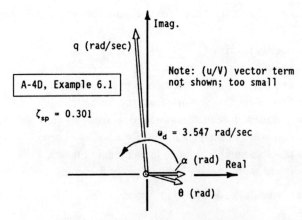

Fig. 6.1 Short-period phasor representation.

LONGITUDINAL DYNAMICS

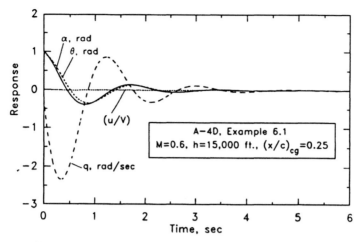

Fig. 6.2 Short-period response to phasor initial condition.

A time-history trace of the short-period response due to a unit initial condition of the eigenvector is shown in Fig. 6.2. The plot represents the projection of the exponentially decaying rotating phasors (shadows) on the real axis. The plot was obtained by using the initial condition of

$$\{x_0\}_{sp} = \begin{Bmatrix} 0.0070 + 0.0128i \\ 1.0000 \\ -0.3017 + 3.5486i \\ 0.9340 - 0.2101i \end{Bmatrix}$$

in conjunction with the MATLAB *initial* command and a time (row) vector that extended from 0 to 6 s by 0.05-s intervals, cf. Appendix E. The mode is well damped as can be noted by the rapid decay of the oscillatory response.

Although the aircraft's plant matrix contains information concerning all of the modes, the use of an initial condition corresponding to the short-period eigenvector provides assurance that only that mode will respond. This statement can be verified in the following manner.

Because the choice of coordinates used to span state space when describing the aircraft motion is not unique, it is equally valid to express the motion as a linear combination of the modal coordinates. The $\{x(t)\}$ coordinates are related to the modal form by the following transformation:

$$\{x(t)\} = [P]\{v(t)\} \tag{6.2}$$

where

$[P]$ = modal matrix, cf. Appendix D
$\{v(t)\}$ = modal response

As shown in Sec. 5.6, the homogeneous solution to the state equation is

$$\{x(t)\} = [e^{At}]\{x_0\} \tag{5.64}$$

The transformation to modal coordinates is made by applying Eq. (6.2),

$$[P]\{v(t)\} = [e^{At}][P]\{v_0\}$$

Next, premultiply both sides of the preceding expression by $[P]^{-1}$ to obtain

$$\{v(t)\} = [P]^{-1}[e^{At}][P]\{v_0\} = [e^{\lambda t}]\{v_0\} \qquad (6.3)$$

where

$[e^{\lambda t}]$ = diagonal exponential matrix
$\{v_0\}$ = initial condition for the modes

It can be seen that Eq. (6.3) represents the airframe's homogeneous response in uncoupled modal coordinates because the exponential matrix is in a diagonal form. When the modal initial condition is

$$\{v_0\} = [1\ 0\ 0\ 0]^T$$

as was chosen to illustrate the time history of the short-period phasor, the response

$$\{x(t)\} = [P]\{v(t)\}$$

will not contain any contributions from either of the long-period complex conjugate modes or the complex conjugate of the selected short-period mode.

6.2.2 Long-Period Mode

The long-period (phugoid) mode from Example 6.1 for the A-4D aircraft may be summarized as

$$\lambda_p = -0.0065 \pm 0.0752i$$

$$\zeta_p = 0.0867 \text{ (a lightly damped mode)}$$

$$\omega_n = 0.0755 \text{ rad/s}$$

The mode shape is

$$\begin{Bmatrix} (u/V) \\ \alpha \\ q \\ \theta \end{Bmatrix} = \begin{Bmatrix} 1.0000 \angle 0.0 \text{ deg} \\ 0.0101 \angle -3.91 \text{ deg} \\ 0.1122 \angle 0.10 \text{ deg} \\ 1.4870 \angle -94.87 \text{ deg} \end{Bmatrix}$$

The period of the phugoid mode is

$$T_p = \frac{2\pi}{\omega_d} = \frac{2\pi}{0.0752} = 83.55 \text{ s}$$

hence, the name of long period. Comparison with the short-period mode indicates a 47.2:1 ratio in the period lengths.

LONGITUDINAL DYNAMICS

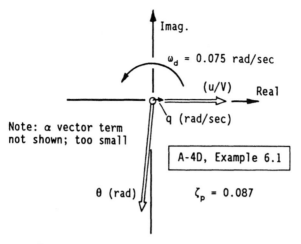

Fig. 6.3 Long-period phasor representation.

The phasor representation of the long-period mode is shown in Fig. 6.3. The phase lead between q and θ will be found to be in accord with the level of modal damping. Also the magnitude scaling between q and θ corresponds to the undamped natural frequency, i.e.,

$$|q| = \omega_n |\theta| = 0.0755 * 1.4870 = 0.1122$$

The time-history response of the phugoid mode is shown in Fig. 6.4 in a manner similar to the preceding illustration for the short-period mode. The influence of the light damping on the oscillation amplitude decay is clearly evident.

The long-period eigenvector illustrates a typical feature of the mode; namely, the α component is much smaller than the (u/V) component. This trait is exactly opposite from the behavior found in the short-period mode. Physically, the

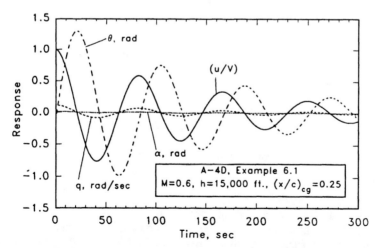

Fig. 6.4 Long-period response to phasor initial condition.

phugoid mode can be described as a damped vertical oscillation of the aircraft's c.g. trajectory about the horizontal flight path with the velocity decreasing (or increasing) during an altitude increase (or decrease). An interpretation for the oscillatory path of the c.g. is that the motion involves an exchange of aircraft potential energy (altitude) with kinetic energy (velocity), and a damping mechanism is provided by the energy loss associated with the aerodynamic drag term. Because the α component is orders of magnitude less than the (u/V) component, a modal approximation would typically be based on the assumption that the oscillation occurs with the aircraft remaining at a constant lift coefficient C_L. The phugoid oscillation appears approximately as shown in Fig. 2.4b.

Example 6.2

Estimate the altitude excursion for the lightly damped phugoid mode of the A-4D aircraft considered in Example 6.1. Assume that a 1% axial velocity perturbation is given by

$$(u/V) = 0.01 e^{-0.0065t} \cos(\omega_d t)$$

where

$$\omega_d = 0.0752 \text{ rad/s}; \quad \zeta\omega_n = 0.0065$$
$$V = 634.0 \text{ ft/s}; \quad u(0) = 6.34 \text{ ft/s}$$

Because rate of climb r/c is related approximately to the pitch attitude perturbation $\theta(t)$ relative to the horizontal by

$$r/c = \frac{dh}{dt} = V\theta(t) \tag{6.4}$$

the r/c variation may be expressed approximately as

$$\dot{h} = e^{-0.0065t} V(0.0149) \cos(\omega_d t - 94.87 \text{ deg})$$

The peak altitude excursion will occur every 1/2 period of the oscillation. An estimate for the peak altitude change due to a -6.34 ft/s phugoid mode velocity amplitude during an oscillation cycle is

$$\Delta h_{\max} \cong \frac{(634.0 \text{ ft/s})(0.0149 \text{ rad})}{(0.0755 \text{ rad/s})} = 125.1 \text{ ft}$$

The altitude excursions would be attenuated by the exponential factor

$$e^{-\zeta\omega_n t} = e^{-0.0065t}$$

In one period of the phugoid oscillation (83.6 s), the damping factor would be 0.58; i.e., the amplitudes of the velocity and pitch attitude perturbations would decrease by 42% each oscillation period. An initial ±125-ft altitude oscillation amplitude would be approximately ±73 ft on the second oscillatory cycle, etc.

6.2.3 Influence of Aircraft Center of Gravity Location

The longitudinal equations of motion are influenced by the aircraft's c.g. location primarily by the M_α dimensional stability derivative. The other dimensional derivatives in the plant are not significantly affected by the c.g. location for a stable aircraft. It will be recalled from Sec. 2.2.2 that the M_α derivative is related to C_{m_α} by the following equation:

$$M_\alpha = \frac{QSc}{I_y}\left(C_{m_\alpha}\right)_{cg} \quad (2.9a)$$

The dimensionless C_{m_α} term depends on the c.g. location relative to the aircraft's neutral point by

$$\left(C_{m_\alpha}\right)_{cg} = \left(\frac{\Delta x}{c}\right)_{np} C_{L_\alpha} \quad (2.14a)$$

For a statically stable aircraft, $(\Delta x/c)_{np}$ will be a negative number reflecting the fact that the (control-fixed) neutral point is aft of the c.g. Consequently, when one knows the location of the aircraft's neutral point at a given flight condition, the M_α derivative may be estimated for another c.g. position by a simple ratio expression, i.e.,

$$\frac{(M_\alpha)_{cg(1)}}{(M_\alpha)_{cg(2)}} = \frac{[(\Delta x/c)_{np}]_{cg(1)}}{[(\Delta x/c)_{np}]_{cg(2)}} \quad (6.5)$$

Example 6.3

For the A-4D aircraft of Example 6.1, estimate 1) the neutral and maneuver point locations and 2) the M_α dimensional derivative when $(x/c)_{cg} = -0.16$ and -0.29 (the forward and aft c.g. limits for the aircraft[2]). From $M_\alpha = -12.97 \text{ s}^{-2}$ for $(x/c)_{cg} = -0.25$, apply Eq. (2.9a) to find

$$\left(C_{m_\alpha}\right)_{cg} = \frac{(25{,}900 \text{ slug} \cdot \text{ft}^2)(-12.97 \text{ s}^{-2})}{(301 \text{ psf})(260 \text{ ft}^2)(10.8 \text{ ft})} = -0.3974 \text{ rad}^{-1}$$

From $Z_\alpha = -518.9 \text{ ft/s}^2$ (cf. Appendix B.1) and applying Eq. (2.12), find

$$\left(C_D + C_{L_\alpha}\right) = -\frac{(546 \text{ slugs})(-518.9 \text{ ft/s}^2)}{(301 \text{ psf})(260 \text{ ft}^2)} = 3.620$$

An estimate for the lift-curve slope follows after assuming that the aircraft $C_D = 0.020$ (i.e., 200 drag counts in flaps-up flight). Then

$$C_{L_\alpha} = 3.60 \text{ rad}^{-1}$$

The neutral point may be estimated from Eq. (2.14a), i.e.,

$$(\Delta x/c)_{np} = (-0.3974 \text{ rad}^{-1})/(3.60 \text{ rad}^{-1}) = -0.110$$

which corresponds to a neutral point location at $0.36\bar{c}$.

The maneuver point location relative to the neutral point may be found by using Eq. (3.19). From this calculation, the maneuver point location is estimated at $0.367\bar{c}$.

Apply Eq. (6.5) to estimate M_α at the other c.g. locations.

1) For $(x/c)_{cg} = -0.16$, find

$$M_\alpha = (-12.97\ \text{s}^{-2})\frac{(0.20)}{(0.11)} = -23.58\ \text{s}^{-2}$$

2) For $(x/c)_{cg} = -0.29$, find

$$M_\alpha = (-12.97\ \text{s}^{-2})\frac{(0.07)}{(0.11)} = -8.25\ \text{s}^{-2}$$

The effect of varying the c.g. location upon the longitudinal axis eigenvalues for the example aircraft was found by altering the M_α term in the plant matrix using available software (e.g., MATLAB). Table 6.1 summarizes the results. The real part of the short-period mode did not change appreciably for the $\pm 0.13c$ shift between the forward and aft c.g. limits. The influence of the M_α change is evident by the change in the complex part of the root (representing ω_d), which increased with a forward c.g. shift. The long-period eigenvalues did not change significantly for the c.g. in the normal range due to their being distant from the short-period roots.

An alternate way of visualizing the migration of the short-period roots as $(x/c)_{cg}$ varies in the normal range is shown in Fig. 6.5 by the o symbols. The complex region where the phugoid roots would be (cf. Table 6.1) is indicated but the roots are not shown for sake of clarity. The dashed-line curve for the short-period root locus, which represents a c.g. range beyond normal usage, shows both of the complex curves meeting on the real axis when $(x/c)_{cg} = -0.357$. The subsequent bifurcation of the roots along the real axis (e.g., the Δ symbols) shows one root going through the phugoid root region into an unstable root area, i.e., the right-hand side of the imaginary axis. The root locus curve, Fig. 6.5, is aircraft specific and is presented only to show general trends.

Those with an electrical engineering background are familiar with the construction of root-locus plots when introducing feedback gain to a system. The short-period root trajectory differs from the classical construction rules because of the way the c.g. dependent related variable enters into the plant.

Table 6.1 A-4D longitudinal eigenvalue summary

	Short period		Long period	
$(x/c)_{cg}$	Root, rad/s	ζ	Root, rad/s	ζ
-0.16	$-1.121 \pm 4.816i$	0.223	$-0.007 \pm 0.074i$	0.087
-0.25	$-1.121 \pm 3.547i$	0.301	$-0.007 \pm 0.075i$	0.087
-0.29	$-1.121 \pm 2.805i$	0.371	$-0.007 \pm 0.077i$	0.085
-0.357	$-1.124/-1.124$	1.000	$-0.004 \pm 0.099i$	0.038
-0.36	$-1.743/-0.512$	n/a	$-0.002 \pm 0.108i$	0.019
-0.38	$-2.778 + 0.527$	n/a	$-0.002 \pm 0.037i$	0.063

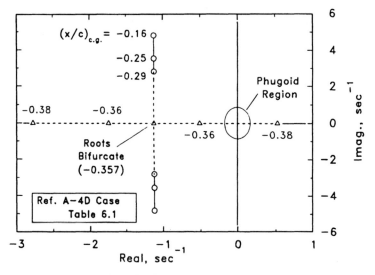

Fig. 6.5 Short-period root locus.

6.2.4 Eigenvalue Guidelines

Although the preceding sections have identified the properties of the two modes applicable to airframe longitudinal dynamics, no comments were made relative to their suitability for flight operation. Appropriate guidelines for airframe design purposes to yield satisfactory flying qualities are embedded in the certification requirements for U.S. civilian aircraft, *Federal Aviation Regulations* (FAR), Parts 23[3] and 25,[4] and the U.S. military specifications[5] dealing with aircraft flying qualities, MIL-F-8785C.

The civilian requirements[3,4] do not address the long-period dynamics directly but imply that the mode be stable by the return speed requirement when the control force is slowly released while demonstrating static longitudinal stability. The short-period mode is addressed[3,4] by the requirement that any short-period oscillation must be heavily damped with longitudinal controls both free and fixed.

The military specifications[5] are much more specific with regard to airframe dynamics and will yield acceptable aircraft flying qualities when used as a guideline during airframe design. Flying qualities (goodness factors) are defined by levels as shown in Table 6.2.

Table 6.2 Levels of flying qualities[5]

Level	Description
1	Flying qualities clearly adequate for the mission flight phase
2	Flying qualities adequate to accomplish the mission flight phase, but some increase in pilot workload or degradation in mission effectiveness, or both, exists
3	Flying qualities such that the airplane can be controlled safely, but pilot workload is excessive or mission effectiveness is inadequate (compromised), or both

The phugoid mode specifications[5] do not place a constraint on the oscillation frequency because the long period provides ample time for a pilot to make corrective control inputs to that mode following a flight disturbance. However, the level of modal damping is more specific; i.e., from Ref. 5:

> The long-period oscillations which occur when the airplane seeks a stabilized airspeed following a disturbance shall meet the following requirements:
> Level 1... ζ_p at least 0.04
> Level 2... ζ_p at least 0.0
> Level 3... T_2 (time to double amplitude) at least 55 seconds
> These requirements apply with the pitch control free and also with it fixed.

The short-period modal damping for the situations requiring accurate flight control provide damping ratio guideline limits as follows.

Level 1:
$$0.35 \leq \zeta_{sp} \leq 1.30$$

Level 2:
$$0.25 \leq \zeta_{sp} \leq 2.00$$

Level 3:
$$0.15 \leq \zeta_{sp}$$

The short-period frequency requirements[5] applicable to situations requiring accurate flight control are shown in Fig. 6.6 as a function of the $n/\alpha(g/\text{rad})$ ratio. Regions are defined by upper and lower boundaries for various flying quality levels. The (n/α) ratio corresponds to a derivative developed at the c.g. for an airframe

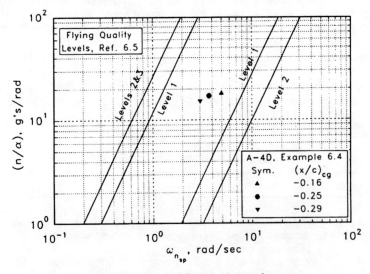

Fig. 6.6 **Short-period frequency requirements.**

LONGITUDINAL DYNAMICS

Table 6.3 A-4D short-period results

$(x/c)_{cg}$	(n/α), g/rad	ω_n, rad/s
−0.16 (forward)	13.91	4.945
−0.25 (normal)	14.87	3.720
−0.29 (aft)	15.30	3.021

during a steady pull-up maneuver as described in Sec. 3.2.2 by Eq. (3.15), i.e.,

$$\left(\frac{n}{\alpha}\right) \Rightarrow \frac{dn}{d\alpha} = -\frac{1}{g}\left(\frac{da_n}{d\delta}\right) \Big/ \left(\frac{d\alpha}{d\delta}\right) \tag{6.6}$$

The (n/α) ratio is a measure of the airframe's load factor sensitivity to an α change and is dependent on the airframe's c.g. location, flight configuration, and flight condition (velocity and altitude).

Example 6.4

Determine the level of flying qualities according to Ref. 5 for the longitudinal dynamics of the A-4D aircraft defined by Example 6.1 throughout its design range of c.g. locations.

1) For the phugoid mode, flying qualities are at level 1 because $0.085 \leq \zeta_p \leq 0.087$ for $(x/c)_{cg}$ varying between −0.16 and −0.29 (forward and aft limits) according to Table 6.1 (i.e., $\zeta_p > 0.04$).

2) For the short-period mode, modal damping (cf. Table 6.1) corresponds to level 1 at the aft c.g. limit ($\zeta_{sp} = 0.37$) while meeting level 2 flying qualities at the mid and forward c.g. locations.

The undamped natural frequencies, $\omega_{n_{sp}}$, are shown in Fig. 6.6. as a function of the (n/α) ratio for three c.g. locations. From Fig. 6.6, it can be seen that $\omega_{n_{sp}}$ is consistent with level 1 flying qualities throughout the design range of c.g. values.

The (n/α) ratio was determined from Eq. (3.15) as

$$\frac{dn}{d\alpha} = -\frac{V}{g}\frac{(-M_\alpha Z_\delta + Z_\alpha M_\delta)}{(-M_q Z_\delta + V M_\delta)} \tag{6.7}$$

For the A-4D aircraft at the assumed flight condition, evaluations using Eq. (6.7) are summarized in Table 6.3.

6.3 Modal Approximations

The linearized aircraft longitudinal dynamics considered in Sec. 6.2 identified the two principal modes (i.e., both the eigenvalues and the eigenvectors) including clues as to suitable assumptions to use when considering approximations to simplify the analyses. When a modal approximation produces results in reasonable accord with the complete system analysis, then an insight is gained as to the key elements (i.e., stability derivatives) of the plant's matrix contributing to that mode.

Although the approximations will contain the terms due to longitudinal control, airframe response due to control inputs will not be considered in Sec. 6.3.

6.3.1 Short-Period Approximation

An initial short-period approximation can be made based on the nature of the eigenvector as seen in Sec. 6.2; i.e., the velocity perturbation is negligible by comparison to the α and θ components. These observations are reflected in the following assumptions.

1) The c.g. maintains a constant forward velocity; i.e., $u(t)/V = 0$. Because longitudinal acceleration, du/dt, is governed by the dynamics embodied in the F_x equation, the assumed constraint of no change in forward velocity allows the F_x equation to be dropped.

2) The aircraft is free or unconstrained both to pitch rotation about the c.g. and to vertical motion of the c.g. (z direction). The M_y and F_z dynamic equations will be altered by the deletion of the (u/V) related terms.

3) The perturbation will start from level flight (i.e., $\theta_0 = 0$) with the aircraft reference being the body's stability axes.

The stated constraints, when applied to Eq. (4.42), yield a simplified set of expressions,

$$\begin{aligned}(V - Z_{\dot{\alpha}})\dot{\alpha} &= Z_\alpha \alpha + (V + Z_q)q + Z_\delta \delta \\ -M_{\dot{\alpha}}\dot{\alpha} + \dot{q} &= M_\alpha \alpha + M_q q + M_\delta \delta\end{aligned} \qquad (6.8)$$

Note that the $\dot{\theta} = q$ expression has not been included. This simplification occurred because the assumed initial condition of level flight using stability axes implied that $\sin\theta_0$ was zero with a subsequent removal of the $\theta(t)$ term as a valid state vector component in the F_z equilibrium equation.

The two coupled linear equations (6.8) will be restated into functions solely of α and q by the use of algebraic substitutions, i.e.,

$$\dot{\alpha} = \frac{Z_\alpha}{(V - Z_{\dot{\alpha}})}\alpha + \frac{V + Z_q}{(V - Z_{\dot{\alpha}})}q + \frac{Z_\delta}{(V - Z_{\dot{\alpha}})}\delta$$

$$\dot{q} = \left[M_\alpha + \frac{M_{\dot{\alpha}} Z_\alpha}{(V - Z_{\dot{\alpha}})}\right]\alpha + \left[M_q + \frac{(V + Z_q)M_{\dot{\alpha}}}{(V - Z_{\dot{\alpha}})}\right]q + \left[M_\delta + \frac{M_{\dot{\alpha}} Z_\delta}{(V - Z_{\dot{\alpha}})}\right]\delta$$

A further simplification can be made by recognizing that both $Z_{\dot{\alpha}}$ and Z_q are nearly zero in magnitude and most assuredly are negligible when compared to the freestream velocity V in the preceding equations. The short-period approximation in a commonly used form becomes

$$\begin{Bmatrix}\dot{\alpha} \\ \dot{q}\end{Bmatrix} = \begin{bmatrix} Z_\alpha/V & 1 \\ (M_\alpha + \frac{M_{\dot{\alpha}} Z_\alpha}{V}) & (M_q + M_{\dot{\alpha}})\end{bmatrix}\begin{Bmatrix}\alpha \\ q\end{Bmatrix} + \begin{Bmatrix} Z_\delta/V \\ (M_\delta + \frac{M_{\dot{\alpha}} Z_\delta}{V})\end{Bmatrix}\delta \qquad (6.9)$$

Equation (6.9) is in the standard matrix form (cf. Sec. 5.6),

$$\{\dot{x}\} = [A]\{x\} + \{B\}u \qquad (5.50)$$

Example 6.5

Use the approximate form, Eq. (6.9), to estimate the short-period modal properties for the A-4D aircraft considered in Example 6.1.

```
% The plant matrix [A] in x-dot = [A] x is:
disp(A)
    -0.8185    1.0000
   -12.6811   -1.4240
```

Note that the elements of the 2×2 plant matrix are the same as elements $A(2,2)$, $A(2,3)$, $A(3,2)$, and $A(3.3)$ in the full 4×4 matrix $[A]$ of Example 6.1.

```
% Find the characteristic polynomial and the eigenvalues.
P=poly(A); R=roots(P); disp(P)
    1.0000    2.2425   13.8466
%    (s^2)     (s^1)     (s^0)
disp(R)
    -1.1213  ±3.5482 i
% Find modal damping and undamped natural frequency
[Wn,Z]=damp(R); disp(Z)
    0.3013
disp(Wn); % rad/sec
    3.7211
```

The MATLAB *eig* command could be used to find the eigenvector. However, the components can be found directly by considering the first equation in the eigenvalue statement of Eq. (5.53),

$$\lambda_n \{x^{(n)}\} = [A]\{x^{(n)}\} \tag{5.53}$$

For $n = 1$, one finds that

$$(-1.1213 + 3.5482i)x_1^{(1)} = (-0.8185)x_1^{(1)} + x_2^{(1)}$$

or

$$x_2^{(1)} = (-0.3028 + 3.5482i)x_1^{(1)}$$

where

$$x_1^{(1)} = \alpha \qquad x_2^{(1)} = q = \frac{d\theta}{dt}$$

Although θ was not a state vector component in Eq. (6.9), it can be estimated once its derivative q has been found. The mode shape for the short-period approximation is

$$\{x\}_{sp} = \left\{ \begin{matrix} \alpha \\ q \end{matrix} \right\}_{sp} = \left\{ \begin{matrix} 1.000 \angle 0.0 \text{ deg} \\ 3.5610 \angle 94.88 \text{ deg} \end{matrix} \right\}$$

and the θ estimate by Eq. (5.54) is

$$\theta_{est} = 0.9571 \angle -12.66 \text{ deg}$$

A further approximation to the short-period mode can be made by noting that $\alpha \cong \theta$ and imposing a constraint that the aircraft has only pitch rotation, which corresponds to having no vertical aircraft motion; i.e., the c.g. travels in a straight path. The addition of the no-vertical motion constraint alters Eq. (6.9) by the added assumptions that 1) force equilibrium about the z axis no longer applies and the F_z related equation can be dropped and 2) $\alpha(t) = \theta(t)$.

The resulting dynamics equation is a single second-order differential equation solely based on airframe pitch moment equilibrium.

$$\ddot{\alpha} - (M_{\dot{\alpha}} + M_q)\dot{\alpha} - M_\alpha \alpha = M_\delta \delta \qquad (6.10)$$

which is of the form

$$\ddot{x} + 2\zeta\omega_n \dot{x} + \omega_n^2 x = f(t)$$

The homogeneous equation solution is

$$\omega_n = (-M_\alpha)^{1/2}$$

$$\zeta = -(M_{\dot{\alpha}} + M_q)/(2\omega_n)$$

Example 6.6

Evaluate the short-period mode for the A-4D aircraft of Example 6.1 using the approximation involving the second-order differential equation [Eq. (6.10)]

$$\omega_n = (12.97)^{1/2} = 3.601 \text{ rad/s}$$

$$\zeta = (0.353 + 1.071)/(2 * 3.601) = 0.198$$

$$\omega_d = \sqrt{1 - \zeta^2}\omega_n = 3.530 \text{ rad/s}$$

The approximate eigenvalue is

$$\lambda = -\zeta\omega_n \pm i\omega_d = -0.712 \pm 3.530i \text{ rad/s}$$

A comparison of the two short-period approximations with the full solution for the example aircraft is shown on Table 6.4. The first approximation, where

Table 6.4 A-4D short-period summary

Item	Airframe Example 6.1	Approximation Eq. (6.9)	Approximation Eq. (6.10)
Eigenvalue	$-1.121 \pm 3.547i$	$-1.121 \pm 3.548i$	$-0.712 \pm 3.530i$
Damping, ζ	0.301	0.301	0.198
Eigenvector			
(u/V)	0.015 ∠ 61.33 deg	n/a	n/a
α	1.000 ∠ 0.0 deg	1.000 ∠ 0.0 deg	1.000 ∠ 0.0 deg
q	3.561 ∠ 94.86 deg	3.561 ∠ 94.88 deg	n/a
θ	0.957 ∠ −12.68 deg	0.957 ∠ −12.66 deg	1.000 ∠ 0.0 deg

$(u/V) = 0$, shows very good agreement with the complete airframe solution. This observation implies that the drag force equilibrium relation does not have a strong influence on the short-period mode. The second approximation, where both $(u/V) = 0$ and $\alpha = \theta$, leaves only the pitching moment equation to approximate the short-period mode. As can be seen in Table 6.4, the highly simplified approximation is in error relative to estimating modal damping (34% too low) whereas the characteristic frequency is reasonably close (0.5% too low). From this observation, one may conclude that the slight vertical motion of the c.g. contributes significantly to short-period modal damping while having only a small influence on the modal frequency.

Although the agreement shown on Table 6.4 is airframe specific, the accord with the first approximation [Eq. (6.9)] will normally be quite reasonable for an airframe with ample static stability margin. If the c.g. location were in the neighborhood of the neutral point, which would place the short-period eigenvalues closer to the phugoid mode roots, the agreement from the first approximation could not be expected to be as good.

6.3.2 Long-Period (Phugoid) Approximation

It was apparent in Sec. 6.2 that the velocity perturbation dominated the phugoid mode and that the angle-of-attack variation was orders of magnitude smaller. Consequently, the long-period approximation will assume that the aircraft's phugoid oscillation occurs at a constant angle of attack (i.e., constant C_L). In addition to decreasing the number of time-varying quantities by one, this assumption will remove the influence of c.g. location because the M_α term (a function of static margin) in the moment equation will not make a contribution to the solution. From a constraint standpoint, the approximation will include the following.

1) The pitching moment equation is dropped based on the assumptions that angle of attack $\alpha(t)$ is not a dominant variable and compressibility effects are absent (i.e., $M_u = 0$).

2) Only the F_x and F_z equilibrium relations will be retained.

3) The phugoid perturbation will start from level flight with the body axes assumed as stability axes.

When these constraints are considered, the F_x and F_z equilibrium relations in Eq. (4.42) simplify to

$$(\dot{u}/V) = X_u(u/V) - (g/V)\theta + (X_\delta/V)\delta$$

$$q = -Z_u(u/V) + 0 - (Z_\delta/V)\delta$$

but $q = d\theta/dt$. Therefore, the phugoid mode approximation may be stated in a matrix form as

$$\begin{Bmatrix} (\dot{u}/V) \\ \dot{\theta} \end{Bmatrix} = \begin{bmatrix} X_u & -g/V \\ -Z_u & 0 \end{bmatrix} \begin{Bmatrix} (u/V) \\ \theta \end{Bmatrix} + \begin{Bmatrix} X_\delta/V \\ -Z_\delta/V \end{Bmatrix} \delta \qquad (6.11)$$

which corresponds to the state-space relation

$$\{\dot{x}\} = [A]\{x\} + \{B\}u$$

The characteristic equation comes from $|\lambda I - A| = 0$, i.e.,

$$\begin{vmatrix} (\lambda - X_u) & +g/V \\ Z_u & \lambda \end{vmatrix} = \lambda^2 - X_u\lambda - (gZ_u/V) = 0 \qquad (6.12)$$

which is recognized as having the form

$$\lambda^2 + 2\zeta\omega_n\lambda + \omega_n^2 = 0$$

To interpret the physical meaning of the characteristic equation [Eq. (6.12)], it is appropriate to restate the definitions for the X_u and Z_u dimensional derivatives, cf. Sec. 2.2.1,

$$X_u = -\frac{QS}{mV}\left[2C_D + M\frac{\partial C_D}{\partial M}\right] \text{ (s}^{-1}\text{)} \qquad (2.6)$$

$$Z_u = -\frac{QS}{mV}\left[2C_L + M\frac{\partial C_L}{\partial M}\right] \text{ (s}^{-1}\text{)} \qquad (2.7)$$

First, consider Z_u in low-to-medium subsonic flight where the influence of Mach number can be neglected (i.e., $\partial C_L/\partial M = 0$). Also, recognize that in level flight, $C_L = W/(QS)$. Direct substitution into Eq. (6.12) gives

$$\omega_n^2 = -gZ_u/V = 2(g/V)^2$$

which yields an approximation for the undamped natural frequency of the phugoid mode as

$$\omega_n = \sqrt{2}g/V \qquad (6.13)$$

and the estimated period of a lightly damped phugoid mode becomes

$$T_p = \sqrt{2}\pi(V/g)$$

If the velocity were expressed in miles per hour, an approximation sometimes used for estimating the long period is

$$T_p = 0.2025\, V \cong (1/5)\, V$$

This approximation is easily verified during low-speed general aviation operations using a wristwatch while observing the passage of the aircraft nose through the horizon when the phugoid mode has been excited.

Continuing with the approximation, it may be recognized from Eq. (6.12) that

$$2\zeta\omega_n = -X_u = \frac{\rho VS}{m}C_D$$

Dividing by the aproximation for ω_n [Eq. (6.13)] and using the definition for C_L provides an estimate for the phugoid mode damping as

$$\zeta = (\sqrt{2}C_L/C_D)^{-1} \qquad (6.14)$$

From Eq. (6.14), it may be concluded that the damping of the phugoid mode could be expected to be low for an efficient airframe. Typically, high-performance gliders with glide ratios of 40:1 could be expected to have phugoid mode damping on the order of $\zeta \cong 0.02$, i.e., lightly damped.

Example 6.7

Estimate the properties of the phugoid mode for the A-4D airplane of Example 6.1 by use of the modal approximation from the homogeneous form of Eq. (6.11),

$$\begin{Bmatrix} (\dot{u}/V) \\ \dot{\theta} \end{Bmatrix} = \begin{bmatrix} -0.0129 & -0.0507 \\ 0.104 & 0 \end{bmatrix} \begin{Bmatrix} (u/V) \\ \theta \end{Bmatrix}$$

From MATLAB, one finds that the eigenvalue λ is given by

$$\lambda = -0.0065 \pm 0.0723i \text{ s}^{-1}$$

which yields

$$\omega_n = 0.0726 \text{ rad/s} \quad \text{and} \quad \zeta = 0.0888$$

while the eigenvector is

$$\begin{Bmatrix} (u/V) \\ \theta \end{Bmatrix} = \begin{Bmatrix} 1.000 \angle 0.0 \text{ deg} \\ 1.432 \angle -95.10 \text{ deg} \end{Bmatrix}$$

and the estimate for q by Eq. (5.54) is

$$q_{est} = 0.104 \angle 0.0 \text{ deg}$$

A comparison of the phugoid approximation with the full solution for the example aircraft is shown in Table 6.5. Although the long-period approximation appears to be in good accord with the complete airframe analysis, these results are airframe specific. Similar comparisons made with other representative airframes

Table 6.5 A-4D long-period summary

Item	Airframe Example 6.1	Approx. Eq. (6.11)
Eigenvalue	$-0.0065 \pm 0.0752i$	$-0.0065 \pm 0.0723i$
Damping, ζ	0.0867	0.0888
Eigenvector		
(u/V)	$1.000 \angle 0.0$ deg	$1.000 \angle 0.0$ deg
α	$0.010 \angle -3.91$ deg	n/a
q	$0.112 \angle 0.10$ deg	$0.104 \angle 0.0$ deg
θ	$1.487 \angle -94.87$ deg	$1.432 \angle -95.10$ deg

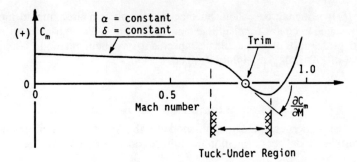

Fig. 6.7 Sketch of C_m variation with Mach number.

(e.g., Appendix B) has resulted in the approximation being considerably in error with respect to the complete aircraft (linearized) phugoid mode.

Compressibility effects at high-subsonic Mach numbers can result in a long-period instability that is described by the term tuck under. The first encounter with tuck under occurred during World War II (WWII) when fighter pilots found their aircrafts' noses continuing to rotate downward with an increasing speed during high-speed dives; hence, the term tuck under. Fatalities did occur due to tuck under when the pilots were unable to recover from the high-speed dive by direct control application due to the high stick forces. One aircraft, the Lockheed P-38, was equipped with a wing lower surface spoiler that caused a corrective trim change when extended in order to avoid the consequences of tuck under.

The tuck-under effect is associated with the onset of local areas of supersonic flow on the aircraft wing during high-subsonic operation with a resultant shift rearward of the wing's centroid of lift loading. Figure 6.7 is a sketch showing a typical pitching moment coefficient variation with Mach number while α and δ (longitudinal control) are constant. These kind of curves are normally obtained from high-subsonic wind-tunnel tests during an aircraft's design evolution. Aircraft with swept-back wings usually have less severe tuck-under behavior than an unswept wing such as found in WWII propeller-driven fighters.

An early U.S. prototype jet transport experienced mild tuck-under behavior during flight tests while in level flight at $M \cong 0.84$, and if left uncorrected, the aircraft would proceed into a gentle dive with stable longitudinal trim re-established at $M \cong 0.92$. Mach-dependent compensation was added to the longitudinal control system of that aircraft to counteract the tuck-under trait. Current high-speed jet aircraft have Mach number-dependent stability augmentation included into the longitudinal control system, which makes the tuck-under effect transparent to the pilot.

A brief study of a jet transport aircraft during cruise (cf. Appendix B.4, DC-8, $M = 0.84$) was made with respect to the long-period mode. Use of the dimensionless coefficients (cf. Table B.4e) provided estimates for the long-period modal properties from Eqs. (6.13) and (6.14), i.e.,

$$\omega_n = \sqrt{2}g/V = 0.055 \text{ rad/s}$$

$$\zeta = (\sqrt{2}C_L/C_D)^{-1} = 0.041$$

Table 6.6 Compressibility effects on the long-period mode DC-8 aircraft, $M = 0.84$, $h = 33,000$ ft

C_{D_M}	C_{m_M}	Phugoid root, s^{-1}	ζ	ω_n, s^{-1}	Remarks
0	0	$-0.002 \pm 0.052i$	0.038	0.052	No Mach effect
0.1005	0	$-0.007 \pm 0.051i$	0.131	0.052	Drag rise only
0.1005	-0.17	$-0.006 \pm 0.023i$	0.244	0.024	Baseline[a]

[a]See Appendix B.

As can be seen on Table 6.6, the estimates were within 10% of the complete aircraft's (4 × 4) plant values when X_u and M_u were modified to have no Mach number dependent derivatives. The effect of the $\partial C_L/\partial M$ term in altering the frequency via the Z_u derivative was not considered because Table B.4e listed C_{L_M} as zero for the stated cruise flight condition. The influence of the drag rise term, $\partial C_D/\partial M$, can be seen on Table 6.6 as increasing the modal damping while having little effect on the natural frequency. This is a trend that is in accord with Eqs. (2.6) and (6.12). As one might expect, the baseline aircraft results shown on Table 6.6, which included all of the Mach number related stability derivatives, were in considerable error relative to the phugoid mode approximation.

The influence of the tuck-under related stability derivative, M_u, on the long-period roots for the DC-8 aircraft at cruise was considered further by varying the $\partial C_m/\partial M (= C_{m_M})$ dimensionless derivative in the M_u expression as defined by Eq. (2.8), i.e.,

$$M_u = +\frac{QSc}{I_y V} M \frac{\partial C_m}{\partial M} \qquad (2.8)$$

A root-locus plot depicting the migration of the complex conjugate root pairs when C_{m_M} assumes values less than zero is shown in Fig. 6.8. The trajectory

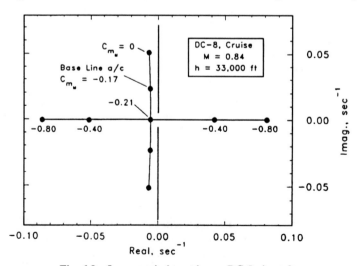

Fig. 6.8 Long-period root locus, DC-8 aircraft.

of the root pairs bifurcate (meet) on the real axis when $C_{m_M} = -0.21$ and then move individually in opposite directions along the real axis with one real root becoming positive to yield a long-period instability. During the root-locus analysis, the corresponding short-period roots remained relatively invariant. Note that this root-locus study is airframe specific.

Stability considerations applied to the long-period roots in an aircraft tuck-under region provide an insight into the cause but may be misleading because the actual tuck-under process is nonlinear. Time-history estimates of an aircraft experiencing tuck under could be made by numerical integration of the equations of motion coupled with a table lookup procedure for describing the variation of the aircraft's dimensionless coefficients. A real-time presentation of the tuck-under process using this method is, in essence, the role of a flight simulator.

6.4 Control Response

Considerations to this point have been oriented toward identifying the airframe's longitudinal dynamic behavior, which has included solving the eigenvalue problem for the modal frequencies, damping, and the respective eigenvectors. However, the actual prediction of the airframe's dynamic motion will be the summation of each dynamic mode's response to control inputs. The longitudinal control input employed when considering static stability and control (cf. Chapter 3) was by either elevator or movable horizontal tail deflections. The same type of control systems will be used when considering either time- or frequency-dependent inputs in the ensuing material. It will be found that the use of modern control theory, combined with matrix oriented computational software (e.g., MATLAB), will make airframe response studies convenient in either the time or frequency domains.

A useful consideration to make before describing the airframe response to longitudinal control inputs relates to the use of the control (e.g., elevator system) either for changing airspeed in unaccelerated flight or to maneuver at constant airspeed. Practical experience tells one that the proper application of a single control allows these change of states to be attained. Because changes in the aircraft state are associated with long- and short-period modes, respectively, it is relevant to understand the criteria requirements for a single control to be able to influence a state change without undue excitation of one or the other of the dynamic modes.

The answer to the criteria question is based on the concept of controllability, a concept introduced by Kalman[6] in 1960. From a heuristic viewpoint, the control should be able to influence each mode of the system. This can be seen by considering the system's state equation [Eq. (5.50)], where the control $u(t)$ is defined as a scalar quantity for illustrative purposes although the equation could be generalized to the multiple input situation,

$$\{\dot{x}\} = [A]\{x\} + \{B\}u \tag{5.50}$$

Although the system's dynamics are related to the eigenvalues, it is timely to note that the eigenvalues remain invariant under a coordinate transformation. With this in mind, let us change the state vector to the modal (canonical) form by the transformation of Eq. (6.2), which uses the eigenvector matrix $[P]$. The state equation becomes uncoupled when it is in the modal form providing that the eigenvalues are distinct. Although the situation of repeated roots (eigenvalues) can occur, most physical systems such as aircraft, ships, and automobiles will have distinct roots. The transformation to modal coordinates is expressed by

$$\{x(t)\} = [P]\{v(t)\} \quad \text{or} \quad [P]^{-1}\{x\} = \{v\} \tag{6.2}$$

Applying the modal transformation to Eq. (5.50) yields

$$\{\dot{v}(t)\} = [P]^{-1}[A][P]\{v\} + [P]^{-1}\{B\}u(t) \tag{6.15}$$

The uncoupling of Eq. (6.15) is due to the plant matrix becoming diagonalized into the eigenvalue matrix by the similarity transformation, Eq. (D.30), which leaves a set of n uncoupled differential equations, i.e.,

$$\begin{aligned}
\dot{v}_1 &= \lambda_1 v_1 + f_1 u(t) \\
\dot{v}_2 &= \lambda_2 v_2 + f_2 u(t) \\
&\vdots \\
\dot{v}_n &= \lambda_n v_n + f_n u(t)
\end{aligned} \tag{6.16}$$

where

$$[P]^{-1}\{B\} = \{f\} \quad \text{with elements } f_1, f_2, \ldots, f_n$$

It is evident from Eq. (6.16) that a requirement for total controllability of a linear, time-invariant system is that the matrix $[P]^{-1}\{B\}$ not have any zero rows.[7] The presence of a zero row in the control matrix would imply that the mode corresponding to that row would not be controllable, and its response would depend on either initial conditions or external disturbance inputs. A more formal definition of controllability is provided by the following statement.

Definition: A system is said to be controllable if it is possible by means of an unconstrained controller to transfer the physical system between two arbitrarily specified finite states in a finite time, i.e., due to $u(t)$ in $t_0 \leq t \leq t_f$

$$\{x(t_0)\} \Rightarrow \{x(t_f)\}$$

A system being controllable does not tell one how to uniquely implement the control(s) in order to change the state vector. Instead, controllability implies that a control scheme can be found to cause a state vector to change in a specified manner and, alternatively, it suggests the possibility of developing a continuous sequence of incremental state vector changes by an appropriate selection of control system inputs. Several problems are provided to illustrate the concept of controllability.

6.4.1 Time-Domain Control Response

The system's control response can be investigated for many forms of control input, such as the typical inputs shown in Fig. 6.9. In a flight test, a control doublet is frequently used to excite the short-period mode whereas the step input will result in an angle-of-attack change that can be viewed as inducing either an initial

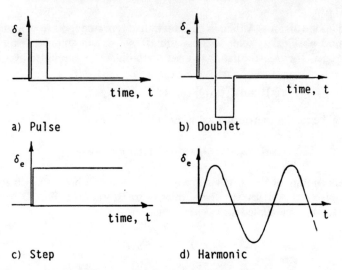

Fig. 6.9 Longitudinal control input choices.

aircraft acceleration change (short-period mode response) or a velocity change (long-period mode response).

The initial consideration will be given to the influence of the longitudinal control on the short-period mode. Consequently, let us consider the short-period approximation [Eq. (6.9)], which includes the assumption that the $Z_{\dot{\alpha}}$ and Z_q dimensional derivatives are negligible in comparison to the freestream velocity V. Equation (6.9) can be restated in a compact form as

$$\begin{Bmatrix} \dot{\alpha} \\ \dot{q} \end{Bmatrix} = \begin{bmatrix} Z_\alpha/V & 1 \\ M'_\alpha & M'_q \end{bmatrix} \begin{Bmatrix} \alpha \\ q \end{Bmatrix} + \begin{Bmatrix} Z_\delta/V \\ M'_\delta \end{Bmatrix} \delta_e(t) \qquad (6.9a)$$

where

$$M'_\alpha = M_\alpha + M_{\dot{\alpha}} Z_\alpha/V$$

$$M'_q = M_q + M_{\dot{\alpha}}$$

$$M'_\delta = M_\delta + M_{\dot{\alpha}} Z_\delta/V$$

which will be noted as having the form of

$$\{\dot{x}\} = [A]\{x\} + \{B\}\delta_e(t)$$

Let us apply the Laplace transformation to Eq. (6.9a) with the understanding of zero initial conditions. Under these assumptions, Eq. (5.55) provides the result that

$$\{X(s)\} = [sI - A]^{-1}\{B\}\delta_e(s) \qquad (6.17)$$

where

$$\{X(s)\} = \mathcal{L}\{x(t)\}$$

and $[sI - A]^{-1}$ is the resolvent matrix.

Equation (6.17) may be alternatively expressed in a transfer function form relating output to input, i.e.,

$$\{X(s)\} = \{G(s)\}\delta_e(s) \tag{6.18}$$

where $\{G(s)\} = [sI - A]^{-1}\{B\}$ is the transfer function matrix.

Because the state vector has only two components when using the short-period approximation, the corresponding two elements of the transfer function matrix can be identified as

$$G_{\alpha\delta}(s) = [(s - M_q')Z_\delta/V + M_\delta']/\Delta(s)$$
$$G_{q\delta}(s) = [(s - Z_\alpha/V)M_\delta' + M_\alpha'Z_\delta/V]/\Delta(s) \tag{6.19}$$

where

$$\Delta(s) = s^2 - (M_q' + Z_\alpha/V)s - M_\alpha' + M_q'Z_\alpha/V$$

It should be recognized that $\Delta(s)$ is the system's characteristic equation, as was illustrated by Example 6.5.

When the control input is a step elevator input of the same form as shown on Fig. 6.9c, i.e.,

$$\delta_e(t) = \delta_0 1(t) = \begin{cases} 0 & \text{for } t < 0 \\ \delta_0 & \text{for } t > 0 \end{cases}$$

then the Laplace transform of the input becomes $\delta_e(s) = \delta_0/s$ and by Eq. (6.18) the output is

$$\{X(s)\} = \delta_0\{G(s)\}(1/s) \tag{6.20}$$

Rather than taking the inverse Laplace transform of the preceding expression to obtain the time response, let us instead consider the results of Sec. 5.6 dealing with state-space fundamentals. From Eq. (5.67), one finds for the elevator step control input that

$$\{x(t)\} = \delta_0[e^{At} - I][A]^{-1}\{B\} \tag{6.21}$$

where

$$\{x(t)\} = [\alpha(t) q(t)]^T$$

and where

$[A]$ = plant matrix
$\{B\}$ = control matrix (vector)
e^{At} = transition matrix
I = unit identity matrix

In Eq. (6.21), the transition matrix e^{At} will vanish when time approaches infinity providing the short-period approximation represents a dynamically stable system (i.e., the characteristic roots have negative real parts). The remaining terms in

190   INTRODUCTION TO AIRCRAFT FLIGHT DYNAMICS

Eq. (6.21) would be the static response, i.e.,

$$\begin{Bmatrix} \alpha/\delta_0 \\ q/\delta_0 \end{Bmatrix}_{\text{stat}} = -[A]^{-1}\{B\} \qquad (6.22)$$

The same result could be obtained by using the final value theorem from Laplace transform theory as described in Sec. 5.2 by Eq. (5.14), i.e.,

$$\lim_{t \to \infty} \{x(t)\} = \lim_{s \to 0} s\{X(s)\} = \delta_0 \lim_{s \to 0} \{G(s)\} \qquad (6.23)$$

Setting $s = 0$ in the transfer function matrix [Eq. (6.19)] gives

$$(\alpha/\delta_0)_{\text{stat}} = (M'_\delta - M'_q Z_\delta/V)/(-M'_\alpha + M'_q Z_\alpha/V)$$
$$(q/\delta_0)_{\text{stat}} = (M'_\alpha Z_\delta/V - M'_\delta Z_\alpha/V)/(-M'_\alpha + M'_q Z_\alpha/V) \qquad (6.24)$$

If the aircraft's short-period response had been modeled as a single second-order differential equation [Eq. (6.10)], which corresponds to having a no vertical motion constraint, then it could be shown that the static response would become

$$(\alpha/\delta_0)_{\text{stat}} = M_\delta/(-M_\alpha) \qquad (6.25)$$

Comparison of Eqs. (6.24) and (6.25) clearly indicates that the vertical motion degree of freedom in the short-period approximation introduces added complexities to the static response relations.

Example 6.8

Consider the A-4D jet aircraft in level flight at $M = 0.6$, $h = 15{,}000$ ft, and c.g. at $0.25\,c$, cf. Appendix B.1. On the assumption that a unit step input was applied to the elevator control at time $t = 0$, estimate using the short-period approximation 1) $\alpha(t)$ and $q(t)$ values at $t = 1.0$ and 2.0 s, respectively, and 2) the static response values when time t tends to infinity.

```
% From Example 6.5, the plant matrix [A] is:
disp(A)
          -0.8185    1.0000
         -12.6811   -1.4240
% The modified control matrix {B} in transposed format is:
disp(B')
          -0.0899  -19.4283
% Find the transition matrix at t = 1.0 second
E1=expm(A); disp(E1)
           0.3108   -0.0364
           0.4617   -0.2889
% Find step response at t = 1.0 second
X1=(E1 - eye(A))*inv(A)*B; disp(X1')
          -1.8900   -0.7232
```

```
%           (rad/rad)  (rad/rad-sec)
% Find the transition matrix at t = 2.0 seconds
E2=E1*E1;  % Using property of transition matrix
disp(E2)
            0.0798    0.0218
           -0.2769    0.0667
% Find step response at t = 2.0 seconds
X2=(E2 - eye(A))*inv(A)*B;  disp(X2')
           -1.2762   -1.3868
%           (rad/rad)  (rad/rad-sec)
% Find static value; i.e., response as t → ∞
Xstat=-inv(A)*B;  disp(Xstat')
           -1.4124   -1.0661
%           (rad/rad)  (rad/rad-sec)
```

MATLAB commands included:

expm(A) = obtain exponential matrix of $[A]$

eye(A) = unit diagonal matrix of same order as matrix $[A]$

Time-history plots of α/δ_0 and q/δ_0 for the aircraft of Example 6.8 are shown on Fig. 6.10 for time varying from $t = 0$ to 10 s. The estimated values at $t = 1$ and 2 s will be seen to agree with the time-history curves. The plot was obtained using matrix computation software (e.g., MATLAB) to calculate the step response for the assumed time span at 0.10-s increments. The 101 sets of the response vector components are shown in Fig. 6.10 as the faired curves.

The exponential decay, as seen in Fig. 6.10, causes a rapid subsidence of the transient to the static values listed subsequently. The negative signs on the static

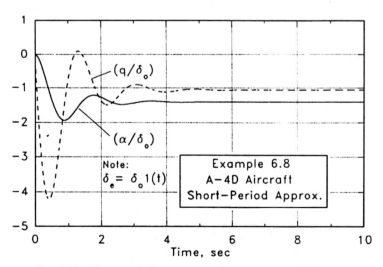

Fig. 6.10 Short-period response to elevator step control input.

values are due to the control deflection sign convention; i.e., positive elevator deflection corresponds to the trailing edge moving down, which results in negative values for both α and q when the aircraft is stable,

$$\{x\}_{\text{stat}} = \begin{Bmatrix} \alpha/\delta_0 \\ q/\delta_0 \end{Bmatrix}_{\text{stat}} = \begin{Bmatrix} -1.412 \text{ rad/rad} \\ -1.066 \text{ rad/rad-s} \end{Bmatrix}_{\text{stat}}$$

It is important to note that Fig. 6.10 is based on the short-period approximation with the result that the long-period mode is suppressed. The full linearized longitudinal airframe dynamic model, which includes the response of both modes, will also show the change in (u/V) and pitch attitude due to an elevator control input. The complete system response for the A-4D aircraft example due to a unit step elevator control input may be seen in Fig. 6.11 for a time span of 25 s. The initial part of the response is shown in order to illustrate the distinctions between the two modes with the understanding that the long-period mode will continue to vary for a time in excess of 100 s.

The α/δ_0 response transient in Fig. 6.11 appears very similar to the corresponding response observed when the short-period approximation was made, cf. Fig. 6.10. Because the short-period response has reached subsidence by approximately 6 s, the remainder of the state vector time-history response represents the reaction of the long-period mode. Other features to note from Fig. 6.11 include the following.

1) The angle of attack is almost constant during the long-period mode's response, a fact that is consistent with the assumptions used in making the long-period approximation.

2) The pitch rate does not attain a nonzero constant value, but instead oscillates as a lightly damped long-period mode response and ultimately attains a value of zero. This is a contrast to the result obtained when using the short-period approximation.

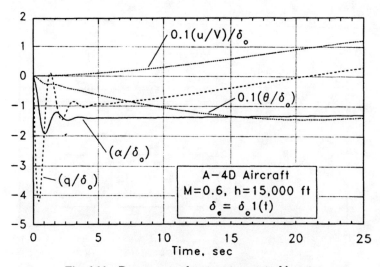

Fig. 6.11 Response to elevator step control input.

3) The pitch attitude response θ/δ_0 shows a lightly damped oscillation, which ultimately will attain a nonzero constant value. A positive elevator step control input will provide a nose-down pitch attitude change with a corresponding increase in velocity. The maintenance of level flight for this situation would require the addition of propulsive thrust.

4) The final system static response after both the short- and long-period modes have subsided was evaluated for the airframe of Example 6.8 using Eq. (6.22) for the complete state vector. It was found that due to $\delta_e = \delta_0 1(t)$,

$$\{x/\delta_0\}_{\text{stat}} = \begin{Bmatrix} (u/V)/\delta_0 \\ \alpha/\delta_0 \\ q/\delta_0 \\ \theta/\delta_0 \end{Bmatrix}_{\text{stat}} = \begin{Bmatrix} 9.501/\text{rad} \\ -1.317 \text{ rad/rad} \\ 0.0 \\ -2.137 \text{ rad/rad} \end{Bmatrix}$$

It should be apparent that application of a control input in the form of a short-duration pulse or a doublet (cf. Fig. 6.9) will primarily result in the short-period mode's response.

6.4.2 Frequency-Domain Control Response

The steady-state harmonic response (gain and phase) due to a harmonic control input is an important way of describing the dynamic behavior of a physical system. Application of the harmonic control input can be considered as starting from an initial static trim flight condition with the result that the response will include a startup transient as well as the long-term steady-state harmonic response (cf. definition of the steady-state harmonic response, Sec. 5.6).

Determination of the steady-state harmonic response at a specified frequency involves the substitution of $s = i\omega$ ($i = \sqrt{-1}$) into the transfer function matrix $\{G(s)\}$. Let us assume that a harmonic elevator control input is applied to a longitudinally stable aircraft, i.e.,

$$\delta_e(t) = \delta_0 e^{i\omega t} = \delta_0(\cos \omega t + i \sin \omega t)$$

The short-period approximation [Eq. (6.9)] led to establishing the transfer function matrix $\{G(s)\}$ as expressed by Eq. (6.19). This matrix can be viewed as made up of complex numerator and denominator functions, i.e.,

$$\begin{aligned} G_{\alpha\delta}(s) &= \text{Num}_{\alpha\delta}(s)/\Delta(s) \\ G_{q\delta}(s) &= \text{Num}_{q\delta}(s)/\Delta(s) \end{aligned} \qquad (6.26)$$

Substitution of $s = i\omega$ into Eq. (6.26) provides

$$\begin{aligned} \text{Num}_{\alpha\delta}(i\omega) &= [M'_\delta - M'_q(Z_\delta/V)] + (i\omega)(Z_\delta/V) \\ \text{Num}_{q\delta}(i\omega) &= [M'_\alpha(Z_\delta/V) - M'_\delta(Z_\alpha/V)] + (i\omega)M'_\delta \\ \Delta(i\omega) &= [M'_q(Z_\alpha/V) - M'_\alpha - \omega^2] - (i\omega)[M'_q + (Z_\alpha/V)] \end{aligned} \qquad (6.27)$$

Each of the complex-natured transfer function terms in Eq. (6.27) can be expressed by a frequency-dependent magnitude and phase quantity using phasor notation, i.e.,

$$(a + ib) = (a^2 + b^2)^{1/2} \angle \Delta\phi$$

where

$$\angle \Delta\phi = \tan^{-1}(b/a)$$

Ultimately, the frequency-response functions may be described by a column vector when a single control is the input, i.e.,

$$\{G(i\omega)\} = [G_{\alpha\delta}(i\omega) \quad G_{q\delta}(i\omega)]^T$$

where the complex quantities are given in phasor notation as

$$G_{\alpha\delta}(i\omega) \Rightarrow G_{\alpha\delta}(\omega) \angle \Delta\phi_{\alpha\delta}(\omega)$$
$$G_{q\delta}(i\omega) \Rightarrow G_{q\delta}(\omega) \angle \Delta\phi_{q\delta}(\omega)$$

Finally, when $\delta_e(t) = \delta_0 \cos \omega t$, the steady-state harmonic-response terms can be viewed as

$$\begin{aligned}\alpha(\omega t) &= \delta_0 G_{\alpha\delta}(\omega) \cos[\omega t + \Delta\phi_{\alpha\delta}(\omega)] \\ q(\omega t) &= \delta_0 G_{q\delta}(\omega) \cos[\omega t + \Delta\phi_{q\delta}(\omega)]\end{aligned} \quad (6.28)$$

It is convenient to summarize Eq. (6.28) by plots of gain and phase angle as a function of frequency. This graphical format is known as a Bode plot when the quantities are expressed in a logarithmic (to base 10) form. For reasons of clarity, the initial presentation will be shown in linear physical units.

Example 6.9

Consider the A-4D jet aircraft in level flight at $M = 0.6$, $h = 15,000$ ft, and c.g. at $0.25c$, cf. Appendix B.1. Estimate the steady-state harmonic response if the elevator control were actuated harmonically at a frequency of $\omega = 1.0$ rad/s. This frequency corresponds to an excitation period of $T = 6.28$ s. Expansion of Eq. (6.27) using the appropriate dimensional derivatives (cf. Example 6.8) gives

$$\text{Num}_{\alpha\delta}(i\omega) = -19.610 - (i\omega)0.090$$

$$\text{Num}_{q\delta}(i\omega) = -14.806 - (i\omega)19.482$$

$$\Delta(i\omega) = (13.847 - \omega^2) + (i\omega)2.243$$

Substituting for $\omega = 1.0$ rad/s and simplifying yields

$$G_{\alpha\delta}(i\omega) = 1.504 \angle +170.4 \text{ deg}$$

$$G_{q\delta}(i\omega) = 1.876 \angle +222.9 \text{ deg}$$

which for an elevator input amplitude described by

$$\delta_e(t) = \delta_0 \cos(\omega t) \quad \text{and} \quad \omega = 1.0 \text{ rad/s}$$

corresponds to

$$\alpha(\omega t) = 1.504 \delta_0 \cos(\omega t + 170.4 \text{ deg})$$

$$q(\omega t) = 1.876 \delta_0 \cos(\omega t + 222.9 \text{ deg})$$

Frequency-response curves due to a harmonic elevator control input (i.e., both gain and phase) are shown in Fig. 6.12 where the aircraft has been modeled by the short-period approximation. The gains are absolute values and reflect the zero-frequency responses combined with the dynamic magnification due to the control application frequency. The plots were obtained by use of the *bode* function in the MATLAB computational software program for a frequency range of 0 to 10.0 rad/s at 0.10 rad/s intervals. The 101 sets of data points are shown as faired curves. The calculated values of gain and phase shift from Example 6.9 agree with the plotted results at $\omega = 1.0$ rad/s. Also note that at $\omega = 0$, the gain and phase values shown in Fig. 6.12 agree with the static values obtained in Example 6.8 for the aircraft's static response due to a step elevator control input. Peak gain values occur near to the short-period resonance frequency of 3.55 rad/s (cf. Examples 6.5 and 6.6). Because the influence of the long-period mode has been suppressed in the short-period approximation, the frequency-response information near to the long-period modal frequency will be in error.

The use of a linear frequency scale in Fig. 6.12 would make it difficult to interpret the responses in the neighborhood of the phugoid mode frequency (0.07 rad/s) while still showing short-period information. For this reason, the presentation of the complete longitudinal plant's frequency response is shown in a Bode plot format (Fig. 6.13). The frequency has been shown using a logarithmic basis from $\omega = 0.01$ to 10.0 rad/s (i.e., a three decade range) in order to illustrate the response behavior of both modes. The gain function for the state vector spans a magnitude range of over four decades in the given frequency range. Consequently, it is convenient to present the gain function in the form of decibels similar to the approach used by electrical engineers, i.e.,

$$\text{gain (decibels)} = 20 \log_{10}(G) \qquad (6.29)$$

It is assumed in Eq. (6.29) that the logarithm function is acting on gain values that have been normalized with respect to the appropriate units, e.g., (u/V), which has units of 1/rad, has been normalized by 1.0 rad^{-1}. An alternate approach of normalizing with respect to the zero-frequency value of the state vector component is not appropriate because the static value of the pitch rate q is zero and would lead to a numerical singularity when calculating decibels.

The Bode plot for the A-4D aircraft (Fig. 6.13) contains information relative to the aircraft longitudinal dynamic behavior that is worthy of note.

1) The (u/V) gain value near the long-period modal frequency of 0.075 rad/s is presented relative to a radian amplitude of elevator control. The asymptotic value as frequency tends to zero is

$$\lim_{\omega \to 0} = 9.501 \text{ rad}^{-1} \Rightarrow 19.56 \text{ dB}$$

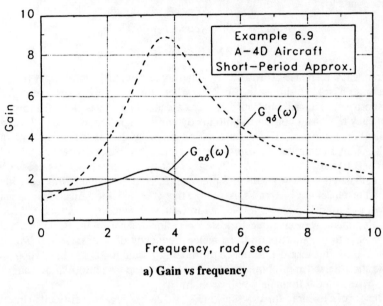

a) Gain vs frequency

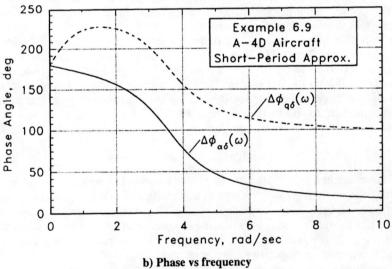

b) Phase vs frequency

Fig. 6.12 Short-period frequency response, elevator input.

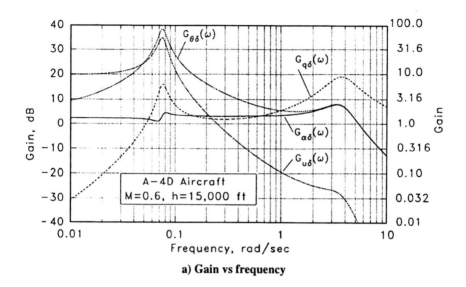

a) Gain vs frequency

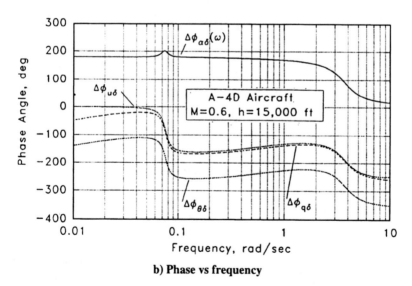

b) Phase vs frequency

Fig. 6.13 Frequency response, harmonic elevator input.

2) The (u/V) gain decreased at frequencies above the long-period frequency at nearly -40 dB per decade. The (u/V) gain was almost 40 dB less in magnitude relative to the α gain near to the short-period modal frequency. This is consistent with the assumption of $(u/V) = 0$ in the short-period approximation.

3) Near to the long-period natural frequency, the pitch attitude gain is of the same magnitude as the velocity perturbation. The angle-of-attack response, in contrast, is in excess of an order of magnitude less (over 20 dB), which is a result consistent with earlier assumptions that $\alpha(t)$ is small in the long-period mode.

4) Near to the short-period natural frequency, the pitch attitude gain and phase approach the corresponding values for α.

5) Pitch rate leads pitch attitude by 90 deg at all excitation frequencies.

6.5 Maneuvering Flight

Maneuvering flight was considered earlier (Sec. 3.2.2) in the context of an airframe experiencing normal acceleration during a steady pull-up. The ensuing material will be broadened in scope to a more general situation such as the normal acceleration transients experienced during a step control input or the frequency response during a harmonic control input. The latter type of control input is occasionally used during flight-test evaluations. Figure 6.14 depicts the airframe's maneuvering perturbations in response to a control input. Assumptions include 1) $u(t)/V = 0$, i.e., response of phugoid mode is absent; 2) the airframe is in level flight at $t = 0$; and 3) body axes are aligned initially with the wind (velocity) vector, i.e., airframe stability axes are used.

The normal acceleration experienced by the airframe is due to the rotation rate of the velocity vector. The velocity vector, tangential to the flight path at any instant, has a perturbation angle γ relative to the initial flight track. Hence the normal acceleration becomes

$$a_n = -\dot{\gamma} V \qquad (6.30)$$

From Fig. 6.14, it is apparent that

$$\gamma = \theta - \alpha$$

which leads to a corresponding angular rate expression

$$\dot{\gamma} = \dot{\theta} - \dot{\alpha} = q - \dot{\alpha}$$

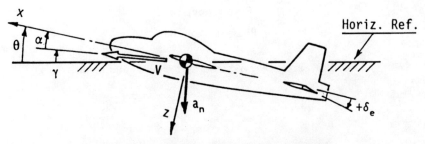

Fig. 6.14 Aircraft during maneuvering flight.

LONGITUDINAL DYNAMICS

which in turn allows the normal acceleration to be defined as

$$a_n = V(\dot{\alpha} - q) \tag{6.31}$$

Note that the earlier considerations of maneuvering flight were applicable to static stability and control topics (cf. Sec. 3.2.2) and the airframe's acceleration response was considered in a steady pull-up. In that situation, the $d\alpha/dt$ term was zero and the rotation rate of the velocity vector simplified to Eq. (3.8), i.e., $a_n = -Vq$.

In accord with assumption 1, the longitudinal dynamics will be described using the short-period approximation [Eq. (6.9)] that includes an added simplification that both $Z_{\dot{\alpha}}$ and Z_q are nearly zero in magnitude and certainly negligible with respect to the velocity V,

$$\begin{Bmatrix} \dot{\alpha} \\ \dot{q} \end{Bmatrix} = [A] \begin{Bmatrix} \alpha \\ q \end{Bmatrix} + \{B\}\delta \tag{6.9b}$$

In spite of the fact that the state vector in Eq. (6.9b) does not contain a component representing the normal acceleration perturbation, it does contain information that can be used to establish a_n. The control theory approach will be used to express the output $y(t)$ as a scalar quantity for the normal acceleration case using a linear combination of the state vector and the control input,

$$y = [C]\{x\} + \{D\}\delta \tag{6.32}$$

For the situation of output $= y(t) = a_n = V(\dot{\alpha} - q)$, the relations of Eqs. (6.9b) and (6.31) can be used to state that

$$a_n = VA_{11}\alpha(t) + V(A_{12} - 1)q + VB_1\delta_e$$

which establishes the terms in Eq. (6.30) as

$$\begin{aligned} [C] = [VA_{11} \quad V(A_{12} - 1)] = [Z_\alpha \quad 0] \\ \{D\} = VB_1 = Z_\delta \end{aligned} \tag{6.33}$$

It will be noted that the maneuvering airframe acceleration response is the first encounter in this text of a situation where the desired output is not a state vector component. This is the reason for using the more general control theory expressions to describe the behavior of a dynamic system as illustrated by Fig. 5.15 and Eqs. (5.50) and (5.63), i.e.,

$$\{\dot{x}\} = [A]\{x\} + [B]\{u\} \tag{5.50}$$

$$\{y\} = [C]\{x\} + [D]\{u\} \tag{5.63a}$$

The transfer functions representing Eqs. (5.50) and (5.63) can be expressed as

$$X(s) = [sI - A]^{-1}BU(s) \tag{5.55a}$$

with an output

$$Y(s) = CX(s) + DU(s) = [C(sI - A)^{-1}B + D]U(s) \tag{6.34}$$

When $Y(s)$ represents the Laplace transformation of the normal acceleration, the $[C]$ and $[D]$ matrices are given by the expressions of Eq. (6.33). In that case, we find that

$$Y(s) = a_n(s) = G_{n\delta}(s)\delta_e(s) \qquad (6.35)$$

where the transfer function $G_{n\delta}(s)$ is of the form

$$G_{n\delta}(s) = Z_\alpha G_{\alpha\delta}(s) + Z_\delta = \text{Num}_{n\delta}(s)/\Delta(s) \qquad (6.36)$$

The $G_{n\delta}(s)$ transfer function in its simplified form includes the $G_{\alpha\delta}(s)$ transfer function of Sec. 6.4.2 due to the influence of the $[C]$ matrix plus an added term due to the $[D]$ matrix. The $\Delta(s)$ quadratic expression in the denominator is the system's characteristic equation as before in Eq. (6.27). The numerator of the transfer function can be determined after a slight bit of algebra as

$$\text{Num}_{n\delta}(s) = Z_\delta\left[s^2 - M'_q s - M'_\alpha + M'_\delta(Z_\alpha/Z_\delta)\right]$$

while the denominator is

$$\Delta(s) = s^2 - (M'_q + Z_\alpha/V)s - M'_\alpha + M'_q Z_\alpha/V$$

The transfer function is second order with respect to s in both the numerator and denominator and can be considered in the form

$$G_{n\delta}(s) = Z_\delta \frac{(s+z_1)(s+z_2)}{(s+p_1)(s+p_2)}$$

where

$-p_1, -p_2 = $ complex conjugate poles representing the short-period eigenvalues
$-z_1, -z_2 = $ zeros of the numerator

The two zeros will usually be real with the added feature that one will be in the right-hand Argand plane. A stable transfer function with a positive zero is described in the literature as a nonminimum phase system.[8,9] A characteristic of a nonminimum phase system is that the initial response to a step control input is in the opposite direction to the steady-state response after all transients have decayed.

The initial acceleration response due to a step control input applied at $t = 0$ for the system described by Eq. (6.35) may be found by using the initial value theorem of Laplace transform theory; i.e., due to a unit elevator control step input, $\delta(t) = \delta_0 1(t)$,

$$\lim_{t \to 0} a_n(t) = \lim_{s \to \infty} s[G_{n\delta}(s)\delta_0(1/s)] = Z_\delta \delta_0 \qquad (6.37)$$

Similarly, the final value theorem may be used to find

$$\lim_{t \to \infty} a_n(t) = \lim_{s \to 0} s\left[G_{n\delta}(s)\delta_0\left(\frac{1}{s}\right)\right] = \frac{Z_\alpha M'_\delta - Z_\delta M'_\alpha}{-M'_\alpha + M'_q(Z_\alpha/V)}\delta_0 \qquad (6.38)$$

LONGITUDINAL DYNAMICS

The recognition that all of the dimensional derivatives are negative in sign and that $(Z_\alpha M'_\delta)$ is greater in magnitude than $(Z_\delta M'_\alpha)$ allows one to conclude from Eqs. (6.37) and (6.38) that 1) the airframe acceleration due to a positive step longitudinal control input will be initially negative in sign and 2) after all transients have decayed, the steady-state response will be positive in sign. This behavior is typical for a nonminimum phase system.

Example 6.10

Consider the A-4D aircraft in level flight at $M = 0.6, h = 15{,}000$ ft ($V = 634$ ft-s^{-1}), and c.g. at $0.25c$. Find the acceleration response due to an elevator step control input of 1 deg.

The following matrices, containing the dimensional stability derivatives, are available from Example 6.8 and Appendix B.1:

$$[A] = \begin{bmatrix} Z_\alpha/V & 1.00 \\ M'_\alpha & M'_q \end{bmatrix} = \begin{bmatrix} -0.8185 & 1.0000 \\ -12.6811 & -1.4240 \end{bmatrix}$$

$$\{B\} = [Z_\delta/V \quad M'_\delta]^T = [-0.0899 \quad -19.4283]^T$$

The output matrices, using Eq. (6.31), are

$$[C] = [-518.9 \quad 0.0] \quad \text{and} \quad [D] = -57.02 \text{ (ft-s}^{-2})$$

1) Find the transfer function $G_{n\delta}(s)$. Use MATLAB to evaluate Eq. (6.36), i.e.,

$$[\text{Num}, \text{Den}] = \text{ss2tf}(A, B, C, D, 1)$$

which provides

$$G_{n\delta}(s) = (-57.02)\frac{s^2 + 1.424s - 164.12}{s^2 + 2.243s + 13.847}$$

Note that the zeros of the numerator are

$$\text{Num}_{n\delta}(s) = (s + z_1)(s + z_2) = (s + 13.54)(s - 12.12)$$

2) Apply Eq. (6.37) to find the initial normal acceleration due to the unit (1-deg) elevator step input,

$$a_n(0) = Z_\delta \delta_0 = (-57.02 \text{ ft-s}^{-2}/\text{rad})(0.01745 \text{ rad/deg})$$

$$= -0.995 \text{ ft-s}^{-2}/\text{deg} \; (= +0.031 \; g/\text{deg})$$

3) Find the steady-state value for normal acceleration due to a 1-deg elevator step input. Instead of evaluating Eq. (6.38), apply the final value theorem to the $G_{n\delta}(s)$ transfer function, i.e.,

$$\lim_{t \to \infty} a_n(t) = (-57.02 \text{ ft-s}^{-2}/\text{rad})\frac{(-164.12)}{(13.847)}(0.01745 \text{ rad/deg})$$

$$= 11.793 \text{ ft-s}^{-2}/\text{deg} \; (= -0.367 \; g/\text{deg})$$

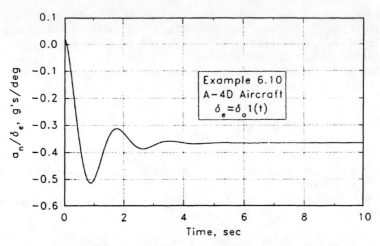

Fig. 6.15 Acceleration response to elevator step input.

The acceleration responses in Example 6.10 are due to a plus (+) 1-deg step elevator input. Because a plus elevator deflection corresponds to moving the trailing edge down, the steady-state acceleration response will be negative in sign. The steady-state acceleration response will agree with the static maneuvering stability results of Chapter 3 inasmuch as that situation corresponds to all of the transient acceleration terms having subsided. A time history of the acceleration response for the aircraft of Example 6.10 is shown in Fig. 6.15. The startup acceleration, as evaluated in Example 6.10, is in the reverse direction to the steady-state response. This is a feature of the step response behavior of a nonminimum phase system. The initial and final responses are in accord with the estimated values.

The steady-state harmonic frequency response for normal acceleration due to harmonic elevator control input is obtained from the $G_{n\delta}(s)$ transfer function by setting $s = i\omega$. The numerator and denominator for the frequency response is given by

$$G_{n\delta}(i\omega) = \text{Num}_{n\delta}(i\omega)/\Delta(i\omega)$$

where

$$\text{Num}_{n\delta}(i\omega) = Z_\delta\{[M'_\delta(Z_\alpha/Z_\delta) - M'_\alpha - \omega^2] - (i\omega)M'_q\}$$

$$\Delta(i\omega) = [M'_q(Z_\alpha/V) - M'_\alpha - \omega^2] - (i\omega)[M'_q + (Z_\alpha/V)]$$

Example 6.11

For the A-4D aircraft considered in Example 6.10, find the gain and phase of the normal acceleration frequency response due to an elevator harmonic amplitude of 1 deg applied at $\omega = \omega_d = 3.548$ rad/s. Substitution of $s = i\omega$ into $G_{n\delta}(s)$ as

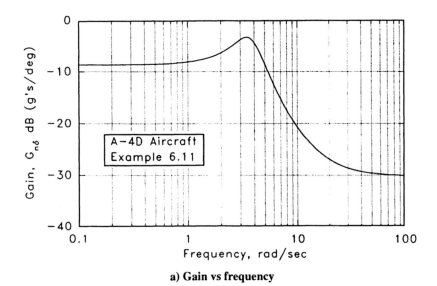

a) Gain vs frequency

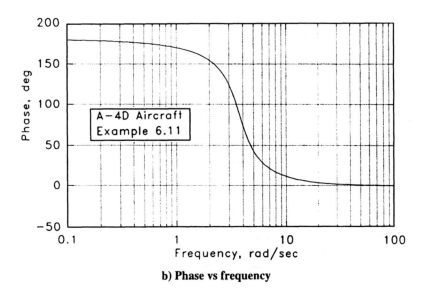

b) Phase vs frequency

Fig. 6.16 Normal acceleration frequency response.

found in Example 6.10 yields for $\omega = 3.548$ rad/s,

$$G_{n\delta}(\omega) = 1251.4 \text{ ft-s}^{-2}/\text{rad}, \qquad \angle = -82.65 \text{ deg}$$

Because elevator amplitude was given as 1 deg and $n_z = -a_n/g$, the gain value at the specified frequency is

$$G_{n\delta}(\omega) = 0.679 \ g/\text{deg} \ (-3.36 \text{ dB}), \qquad \angle = +97.35 \text{ deg}$$

The low-frequency acceleration response will correspond to the result of Example 3.10 part 3 in magnitude and phase, i.e.,

$$G_{n\delta}(0) = 0.367 \ g/\text{deg} \ (-8.71 \text{ dB}), \qquad \angle = +180.0 \text{ deg}$$

A Bode plot (Fig. 6.16) shows the gain and phase for three decades of frequency from 0.1 to 100 rad/s. Items to note include the following.
1) The response values, both gain and phase, are in accord with the estimates of Example 6.11 at the selected frequency of $\omega = \omega_d$.
2) The peak value of acceleration response was $0.683 \ g$ and occurred at $\omega = 3.42$ rad/s, a frequency slightly less than ω_d.
3) The 180-deg phase shift at low frequencies is a result of the sign conventions used for plus elevator deflection (trailing edge down) and plus g ($a_n < 0$).

References

[1] Lanchester, F. W., *Aerodonetics*, Constable and Company, Ltd., London, 1908.
[2] Anon., "NATOPS Flight Manual, Navy Model A-4E/F Aircraft," Naval Air Systems Command, Washington, DC, Feb. 1981.
[3] Anon., "Part 23, Airworthiness Standards: Normal Utility, Acrobatic and Commuter Category Airplanes," *Federal Aviation Regulations*, U.S. Government Printing Office, Washington, DC, Feb. 1991.
[4] Anon., "Part 25, Airworthiness Standards: Transport Category Airplanes," *Federal Aviation Regulations*, U.S. Government Printing Office, Washington, DC, Feb. 1991.
[5] Anon., "Flying Qualities of Piloted Airplanes," MIL SPEC MIL-F-8785C, U.S. Government Printing Office, Washington, DC, Nov. 1980.
[6] Kalman, R. E., "On the General Theory of Control Systems," *Automatic and Remote Control*, London, Butterworths and Co., Ltd., 1961, pp. 481–492.
[7] Gilbert, E. G., "Controllability and Observability in Multivariable Control Systems," *Journal of the Society for Industrial and Applied Mathematics*, (SIAM), Ser. A: Control, Vol. 1, No. 2, 1963, pp. 128–151.
[8] Friedland, B., *Control System Design, An Introduction to State-Space Methods*, McGraw–Hill, New York, 1986, p. 188.
[9] Ogata, K., *Modern Control Engineering*, 2nd ed., Prentice–Hall, Englewood Cliffs, NJ, 1990, pp. 445–446.

Problems

Select an aircraft and flight condition from Appendix B to use in the following analyses, e.g., A-7A attack aircraft at $M = 0.6$, $h = 15,000$ ft, condition 3.

6.1. For the selected aircraft, estimate the location of the stick-fixed neutral and maneuver points.

6.2. Find the short- and long-period characteristics at the specified c.g. value; i.e.:
   a) From the eigenvalues, identify the modal frequencies and damping levels.
   b) Find the mode shapes. Sketch the mode shapes using the phasor representation.

6.3. Determine the effect of 0.05c forward and aft c.g. shifts on the eigenvalues. Do the eigenvectors change significantly due to the c.g. shifts?

6.4. Find the (n/α) ratios and determine the flying quality levels for the natural frequencies. Are the modal damping levels consistent with good flying qualities?

6.5. Estimate the altitude excursion for the phugoid mode due to a (u/V) velocity perturbation of 1%.

6.6. For the short-period approximation, evaluate the state vector at $t = 1$ and 2 s due to a unit step longitudinal control input.
   a) Verify that $e^{2At} = e^{At}e^{At}$.
   b) Apply the final value theorem to obtain the system response as time t tends to ∞.

6.7. Verify the relation for $G_{n\delta}(s)$ [Eq. (6.36)] for both the numerator and denominator.

6.8. Find the normal acceleration response in gravitational acceleration per degree at $t = 1.0$ and 10.0 s for a 1-deg elevator (or stabilizer) longitudinal control input.
   a) Find the zeros of $G_{n\delta}(s)$ to verify that the acceleration response behavior corresponds to that for a nonminimum phase system.
   b) Optional: Plot the time history of the gravitational acceleration per degree for $0 < t < 10.0$.

6.9. Consider an elevator (or stabilizer) harmonic amplitude input of 1-deg applied at the corresponding ω_d frequency.
   a) Find the acceleration response, gravitational acceleration per degree for c.g. located at the nominal, forward, and aft ($\pm 0.05c$) positions. Do the peak acceleration frequency responses change with c.g. position?
   b) Optional: Construct a Bode plot to illustrate the effect of c.g. location on acceleration response and the frequency where maximum gain is attained. It is suggested that the frequency range correspond to three decades from 0.1 to 100.0 rad/s.

6.10. Controllability problem: A well-damped (i.e., real roots) second-order linear system, as depicted by Fig. 5.6, is described by

$$\ddot{y} + 5\dot{y} + 4y = f(t)$$

a) Determine the modal (canonical) form of the governing differential equation.
b) Verify that the system is controllable.
c) Given the condition of the state vector at $t = 0$ and 2.0 s as

$$\{x(0)\} = \begin{Bmatrix} 0 \\ 0 \end{Bmatrix} \quad \text{and} \quad \{x(2)\} = \begin{Bmatrix} 1 \\ 0 \end{Bmatrix}$$

find a control law to satisfy the initial and final values. Hint: The answer will not be unique. One solution is to define $f(t)$ by a pair of 1-s square waves as shown next, followed by an evaluation of their values to satisfy the start and end conditions of the state vector, i.e., assume

$$f_1(t) = f_1 \quad \text{for } 0 < t < 1$$
$$f_2(t) = f_2 \quad \text{for } 1 < t < 2$$

6.11. Perform a controllability check by Gilbert's method to verify that the longitudinal control system for the assumed aircraft can control any of the longitudinal response modes.

7
Lateral-Directional Dynamics

7.1 Background

This chapter considers the aircraft's dynamic behavior about the roll and yaw axis reference frames. The system's dynamics will be developed by viewing the airframe as a multi-degree-of-freedom eigenvalue problem using control theory concepts described in Chapter 5. It will be noted that unlike the longitudinal aircraft responses considered in Chapter 6, the fore and aft c.g. location (i.e., the distance forward from the neutral point) will not play a large role in the dynamic behavior for the airframe's lateral-directional dynamics. In an aircraft, the vertical location of the c.g. is relatively unimportant to roll dynamics because bank angle perturbations from the horizontal do not directly result in restorative roll moments because buoyancy-related forces and moments are negligible. In contrast, surface ship lateral dynamics are most sensitive to the c.g.'s vertical position relative to the center of buoyancy because ship roll motion stability is strongly influenced by the ability of this distance to develop a roll-angle-dependent restorative moment.

From a static stability and control standpoint, aircraft lateral and directional traits influence the ability of an aircraft both to develop sideslip for purposes of maintaining a desired ground track during landing and takeoff flight phases and to maintain a trim condition for straight flight while asymmetries in forces and moments are being experienced such as due to an engine-out condition by a multiengined aircraft.

Dynamic stability considerations for the linearized lateral-directional case involve the ability to develop roll angles and roll rates during maneuvers as well as the ability to control antisymmetric wing span loads induced by turbulence encounters. Atmospheric turbulence can excite a roll (wing drop) along with an ensuing Dutch-roll motion. The latter type of motion, which will be identified in more detail in subsequent sections, relates to a form of aircraft dynamic response that is similar in a colloquial sense to the motion of an ice skater as reportedly observed by an early-age aeronautical engineer when visiting the frozen dikes in Holland.

Whereas a (4 × 4) longitudinal plant frequently yielded two pairs of complex conjugate roots identified as the short-period and phugoid eigenvalues, it will be found that the (4 × 4) lateral-directional plant will most often yield a complex conjugate pair of roots described as the Dutch-roll eigenvalues and two real roots described as the roll response and spiral eigenvalues, respectively.

The spiral root will be found to have a characteristic time to double (or half) amplitude on the order of a minute. Consequently, the application of control feedback to correct an unstable spiral root is not normally a driving concern to the designer. However, as is well known to aircraft pilots, the spiral root can lead to an undesirable condition during instrument flight conditions of an ever tightening spiral, which is described in some circles as the graveyard spiral. After WWII, pilots in high-performance general aviation aircraft occasionally experienced this condition during instrument flight. For this reason, a wing leveler was introduced as a simple feedback control to assist the single pilot during actual instrument flight. This has proven especially helpful when the pilot might be in a busy controlled airspace such as found in the neighborhood of New York City.

The evolution of jet aircraft with swept-back wings for high-speed flight operation introduced aircraft with lightly damped (and even unstable) Dutch-roll modes. The light damping is a result of the apparent dihedral effect due to sweepback, with a rule of thumb being that each 5–7 deg of sweepback is equivalent aerodynamically to 1 deg of dihedral. The increase in dihedral effect leads to light damping in the Dutch-roll mode. Compensation for the low Dutch-roll damping led to the introduction of the yaw damper as a form of stability augmentation where yaw rate was sensed by a rate gyro and the electrical signal generated was used to drive either the rudder or rudder control (or trim) tab. About the same time as jet aircraft were experiencing light Dutch-roll damping (ca. 1950), a few high-performance, single-engine general aviation aircraft were experiencing similar problems in handling qualities. As a result, general aviation aircraft were introduced to the yaw damper concept in the same time frame as high-performance jet aircraft.

The approach to be followed in the subsequent material will be, first, to consider pure roll motion as an approximation for the purpose of introducing the concept of a roll response time constant. Following this, the full (4 × 4) linearized lateral-directional plant will be analyzed from the standpoint of identifying its dynamic properties (i.e., solve the eigenvalue problem). Then a Dutch-roll approximation will be considered based on insights gained from analysis of the complete (4 × 4) plant. Last, an introduction to feedback principles for stability augmentation will be provided by considering an aircraft yaw damper to improve Dutch-roll modal damping.

It should be recognized that lateral-directional dynamics in the linearized form serves the role of introducing the reader to the subject of roll and yaw dynamics. Nonlinear influences also have a large effect on aircraft dynamics, and advanced topics in the area would include inertial cross coupling, wing rock, stall departure, spin entry, and spin recovery. A recent surge of interest in nonlinear aerodynamics has developed because of desires to control maneuvering aircraft at high angles of attack (i.e., $\alpha > 30\text{–}40$ deg).

7.2 Pure Rolling Motion

The idealized example of an aircraft in pure roll about its x stability axis implies that the dynamic equilibrium relations about the z and y axes are assumed as constrained. The x-axis dynamics will involve maintaining a conservation of roll angular momentum with respect to the aerodynamic moments arising from lateral control inputs and rate-dependent roll damping. These considerations will allow one to recognize the following.

1) Lateral control is a roll rate producing device.
2) Roll rate response is similar in a dynamic sense to the low-pass filter as defined by electrical engineers.
3) A time constant appears in the solutions, which is related to the inverse of the aircraft's dimensional roll damping derivative.
4) The complete solution for roll rate response may be interpreted as the scalar form of the more general matrix solution for a state vector application.

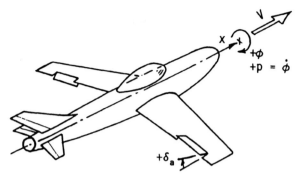

Fig. 7.1 Sketch of aircraft in pure rolling motion.

Figure 7.1 is a sketch representing an aircraft operating at constant airspeed V and constrained to rotate about the x stability axis in response to lateral control inputs. The effects of sideslip and yaw rate are removed from the model by assumption. The conservation of angular momentum about the x roll axis, subject to the stated constraints, may be expressed as

$$I_x \dot{p} = \left[C_{\ell_p}(pb/2V) + C_{\ell_{\delta_a}} \delta_a \right] QSb$$

Dividing both sides of this relation by the mass moment of inertia I_x will yield a first-order, ODE with the constants represented by dimensional stability derivatives as defined in Chapter 4 (cf. Table 4.2), i.e.,

$$\dot{p} = L_p p(t) + L_{\delta_a} \delta_a(t) \tag{7.1}$$

where

L_p = roll damping derivative, $(QSb/I_x)(b/2V)C_{\ell_p}$, s^{-1}
L_{δ_a} = roll control derivative, $(QSb/I_x)C_{\ell_{\delta_a}}$, s^{-2}
p = roll rate, $d\phi/dt$, s^{-1}
δ_a = lateral control input, rad normally

Multiplication of both sides of Eq. (7.1) by the $e^{-L_p t}$ integrating factor allows a solution to be readily obtained in the form of

$$p(t) = e^{L_p t} p(0) + e^{L_p t} \int_0^t e^{-L_p \tau} L_{\delta_a} \delta_a(\tau) d\tau \tag{7.2}$$

It will be noted that Eq. (7.2) is the scalar equivalent of the more general matrix-oriented solution, Eq. (5.61), found in Chapter 5. The exponential term $e^{L_p t}$ serves the scalar role of a transition matrix e^{At}.

Three types of lateral control input will be considered using the simplified roll rate response equation. In the time domain, control inputs will include 1) step input applied at $t = 0$; i.e., $\delta_a = \delta_0 1(t)$ and 2) pulse input applied for the purpose of changing bank angle. In the frequency domain, a harmonic lateral control input will be investigated in order to demonstrate that roll rate dynamics has parallels to the low-pass filter, which is a well-known circuit in the electrical engineering community.

7.2.1 Time-Domain Control Response

It will be assumed that the roll rate is initially zero [i.e., $p(0) = 0$] and that the roll rate response will be due solely to a step aileron control input of amplitude δ_0 applied at $t = 0$. Under this form of control input, Eq. (7.2) simplifies to

$$p(t) = (L_{\delta_a}\delta_0)e^{L_p t} \int_0^t e^{-L_p \lambda} 1(\lambda)\,d\lambda = p_{\text{stat}}(1 - e^{L_p t}) \quad \text{for } t > 0 \quad (7.3)$$

where

$$p_{\text{stat}} = -L_{\delta_a}\delta_0/L_p$$

The steady (static) roll rate, after all transients have subsided, may also be recognized from Eq. (7.1) as the roll rate value remaining when roll acceleration (dp/dt) is zero. In Eq. (7.3), the roll damping dimensional derivative will normally be negative in sign and have dimensional units of (1/time). Therefore, a time constant may be defined by

$$\tau = -1/L_p$$

with the result that Eq. (7.3) can be presented in a more classical form as

$$p(t) = p_{\text{stat}}(1 - e^{-t/\tau})1(t) \quad (7.4)$$

This relation is similar to Eq. (5.31) where now the variable $x(t)$ is replaced by the roll rate $p(t)$. The first-order system response is shown in Fig. 5.4 with emphasis made that the response is 63% of the static value when time t is equal to the system's time constant.

Example 7.1

Estimate the roll rate experienced by the DC-8 during a cruise-flight condition 1 s after a +5-deg aileron deflection was applied as step control input. Also estimate the time constant for the aircraft and the steady-state roll rate for the assumed modest control input.

From Appendix B.4, applicable dimensional derivatives for the DC-8 flying at $M = 0.84$, $h = 33{,}000$ ft are

$$L_p = -1.184 \text{ s}^{-1} \quad \text{and} \quad L_{\delta_a} = 2.120 \text{ s}^{-2}$$

The pure rolling motion governing relation [Eq. (7.1)] provides

$$\dot{p} = -1.184\,p(t) + 2.120\delta_0\,1(t)$$

Because of the +5 deg (0.0873 rad) of aileron control input, the steady-state roll rate is estimated as

$$p_{\text{stat}} = -(2.120 \text{ s}^{-2})(0.0873 \text{ rad})/(-1.184 \text{ s}^{-1})$$

$$= 0.156 \text{ rad/s} \ (= 8.96 \text{ deg/s})$$

LATERAL-DIRECTIONAL DYNAMICS

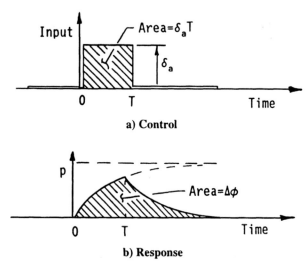

Fig. 7.2 Response to an aileron control pulse.

Next, the time constant for the assumed roll response relation is

$$\tau = -1/L_p = -1/(-1.184 \text{ s}^{-1}) = 0.845 \text{ s}$$

Estimated roll rate at $t = 1.0$ s, from Eq. (7.3), is

$$p(1) = p_{\text{stat}}(1 - e^{-1.0/0.845}) = 0.694 p_{\text{stat}} = 6.22 \text{ deg/s}$$

It should be apparent from Eq. (7.1) that lateral control is a rate producing device. The following strategy will be assumed for changing aircraft bank angle. Initially, a positive control input step will be applied at $t = 0$ followed by an equal negative control input step at a later time T. As shown in Fig. 7.2a, the control strategy represents a pulse input. It may be noted in Fig. 7.2b that the input pulse results in an exponential roll rate growth followed by an exponential subsidence to a zero value. The area under the curve represents the change in bank angle.

The first (initial) control input step yields a roll rate by Eq. (7.4) of

$$p_1(t) = p_{\text{stat}}(1 - e^{-t/\tau}) \qquad \text{for } t > 0$$

whereas the negative second step input applied at $t = T$ gives

$$p_2(t) = -p_{\text{stat}}[1 - e^{-(t-T)/\tau}] \qquad \text{for } t > T$$

Because

$$\phi(t) = \int p(t) \, dt$$

the bank angle change due to the lateral control input pulse can be estimated by linearity principles as

$$\phi_1(t) = p_{\text{stat}} \int_0^t (1 - e^{-\lambda/\tau}) \, d\lambda = p_{\text{stat}}[t + \tau(e^{-t/\tau} - 1)] \qquad (7.5)$$

whereas the second negative step input applied at $t = T$ yields

$$\phi_2(t) = -p_{\text{stat}}[(t-T) + \tau(e^{-(t-T)/\tau} - 1)]$$

The bank angle change due to the lateral control pulse of duration T is

$$\Delta\phi(t) = \phi_1(t) + \phi_2(t) = p_{\text{stat}}[T + \tau e^{-t/\tau}(1 - e^{T/\tau})] \qquad \text{for } t > T \qquad (7.6)$$

In the limit, as time t tends to infinity, the roll angle change due to the application of a lateral control pulse becomes

$$\lim_{t \to \infty} \Delta\phi(t) = p_{\text{stat}} T = (L_{\delta_a} T \tau)\delta_0 \qquad (7.7)$$

The term $(L_{\delta_a}\delta_0)T$ in Eq. (7.7) corresponds to the magnitude of the angular roll moment impulse applied to the system. From a practical standpoint, it can be recognized from Eq. (7.6) that an aircraft with a roll time constant on the order of 1 s will be within 1% of a new steady bank angle approximately 4–5 s following the control pulse application.

Example 7.2

A 5-deg aileron control input is applied as a pulse for a time duration of 2.0 s to the DC-8 aircraft while in a cruise-flight condition (cf. Appendix B.4). Find the change in aircraft roll attitude due to the pulse application.

From Example 7.1, it was determined that 5 deg of aileron control input produced a steady roll rate of

$$p_{\text{stat}} = 0.156 \text{ rad/s } (= 8.96 \text{ deg/s})$$

By application of Eq. (7.7), the final bank angle change due to the specified aileron control pulse input may be estimated as

$$\Delta\phi = (0.156 \text{ rad/s})(2.0 \text{ s}) = 0.312 \text{ rad } (= 17.9 \text{ deg})$$

7.2.2 Response to Control in the Frequency Domain

The following material relates to identifying the steady harmonic response of an aircraft under a pure rolling motion constraint. The validity of the pure roll assumption will be considered in a later section dealing with the full set of linearized lateral-directional equations of motion. As mentioned in Chapter 5, the term steady harmonic response implies that all startup transients have subsided and only the long-term vehicle response to a harmonic input is under consideration. The solution has a direct equivalence to obtaining a transfer function in the Laplace s domain and substituting $s = i\omega$ to obtain a frequency-response function.

The approach considered here will be to assume that the lateral control input was applied in an exponential form vice a sine or cosine input, i.e.,

$$\delta_a(t) = \delta_0 e^{i\omega t} = \delta_0(\cos \omega t + i \sin \omega t)$$

LATERAL-DIRECTIONAL DYNAMICS

Substitution of the control input into the governing roll rate relation [Eq. (7.1)] provides

$$\dot{p} = L_p p + L_{\delta_a} \delta_0 e^{i\omega t} \quad (7.8)$$

Next, assume that the harmonic solution is of the following form:

$$p(t) = P(i\omega) e^{i\omega t}$$

then

$$\dot{p} = i\omega P(i\omega) e^{i\omega t}$$

Substitution of the frequency-dependent response terms into Eq. (7.8) followed by cancellation of the common $e^{i\omega t}$ term gives

$$P(i\omega) = (i\omega - L_p)^{-1} (L_{\delta_a} \delta_0)$$

Using the earlier definitions of p_{stat} and time constant τ allows the complex-natured roll rate to be simplified to

$$P(i\omega) = p_{\text{stat}} (1 + i\omega\tau)^{-1}$$

This result may be viewed in the context of a gain and phase shift, both of which are frequency dependent (cf. Sec. 5.3), i.e.,

$$p(t) = p_{\text{stat}} G(\omega) e^{i(\omega t + \phi)} \quad (7.9)$$

where

$G(\omega)$ = gain function, $[1 + (\omega\tau)^2]^{-1/2}$
$\phi(\omega)$ = phase angle, $\tan^{-1}(-\omega\tau)$

Equation (7.9) includes both the cosine (real part) and sine (imaginary part) responses due to cosine and sine lateral control inputs, respectively. As pointed out in Sec. 5.3, it should be noted that when the applied frequency $\omega = 1/\tau$, the gain amplification factor $G(\omega)$ has a value of $1/\sqrt{2} = 0.707$ while the corresponding phase shift $\phi(\omega)$ is $-\pi/4$ rad (-45 deg). The stated value of frequency is described as the corner frequency, which is a term arising from Bode plot usage.

Example 7.3

A test pilot rocks the wings harmonically of the DC-8 aircraft in a cruise-flight configuration (cf. Appendix B.4). Assuming that the pilot is applying control wheel motion to obtain a 5-deg sinusoidal amplitude of aileron motion at an applied period of $T = 4.0$ s, estimate the roll amplitude and phase angle relative to the control input.

From Example 7.1, $p_{\text{stat}} = 0.156$ rad/s and $\tau = 0.845$ s. Applied frequency $\omega = 2\pi/T = 2\pi/(4.0 \text{ s}) = 1.571$ rad/s. The gain and phase from Eq. (7.9) is

$$G(\omega) = \{1 + [(1.571 \text{ s}^{-1})(0.845 \text{ s})]^2\}^{-1/2} = 0.602$$

$$\phi(\omega) = \tan^{-1}[(1.571 \text{ s}^{-1})(0.845 \text{ s})] = -0.925 \text{ rad} (= -53.0 \text{ deg})$$

Therefore,
$$p(\omega t) = 0.094 \sin(\omega t - 0.925) \text{ rad/s}$$

for a control input of
$$\delta_a(\omega t) = 0.0873 \sin \omega t \text{ rad}$$

However, aileron control produces roll rate and results in roll angle variation by an integration process, i.e.,

$$\Delta\phi(\omega t) = \int p(\omega t)\,dt = (1/\omega) \int p(\omega t)\,d(\omega t)$$
$$= (-1/\omega)\, p_{\text{stat}}\, G(\omega) \cos(\omega t + \phi)$$

By using the following trigonometry identity

$$-\cos(A) = \sin(A - \pi/2)$$

one finds that when

$$\delta_a = \delta_0 \sin \omega t$$

then

$$p(\omega t) = p_{\text{stat}}\, G(\omega) \sin(\omega t + \phi)$$

and

$$\Delta\phi(\omega t) = p_{\text{stat}}\,[G(\omega)/\omega]\sin(\omega t + \phi - \pi/2)$$

For the example case, substitution yields

$$\Delta\phi(\omega t) = (0.060 \text{ rad}) \sin(\omega t - 2.496)$$

at $\omega = 1.571$ rad/s. In degree measure, the estimate would provide a roll angle amplitude of 3.43 deg with roll angle lagging the aileron control input by a 143.0-deg phase angle.

7.3 Lateral-Directional Dynamics

The pure rolling motion approximation was described prior to consideration being given to the full (4×4) linearized lateral-directional system because the roll motion approach provided a realistic representation of the specific response mode, similar by analogy to the short-period approximation for longitudinal dynamics. As a starting point for the (4×4) system, the relation, Eq. (4.48), from Sec. 4.6.2 will be used. The primed lateral-directional moment stability derivatives will be considered as defined in Appendix B in order to correct for the cross product of inertia terms. In this way, the inertial matrix $[I_n]$ becomes a unit diagonal matrix. Therefore, the system will be defined as

$$\{\dot{x}\} = [A]\{x\} + [B]\{u\} \qquad (7.10)$$

where the state vector is given as

$$\{x\} = [\beta \quad p \quad \phi \quad r]^T$$

and the control input, a column matrix of order (2×1), is

$$\{u\} = [\delta_r \quad \delta_a]^T$$

If only rudder or aileron control were under consideration, then the control input would simplify to a single (scalar) value. The plant matrix, which involves primed dimensional moment derivatives, is expressed as

$$[A] = \begin{bmatrix} \frac{Y_\beta}{V} & \frac{Y_p}{V} & \frac{(g\cos\theta_0)}{V} & \frac{(Y_r - V)}{V} \\ L'_\beta & L'_p & 0 & L'_r \\ 0 & 1 & 0 & 0 \\ N'_\beta & N'_p & 0 & N'_r \end{bmatrix} \quad (7.11)$$

whereas the control sensitivity matrix is

$$[B] = \begin{bmatrix} \frac{Y_{\delta_r}}{V} & \frac{Y_{\delta_a}}{V} \\ L'_{\delta_r} & L'_{\delta_a} \\ 0 & 0 \\ N'_{\delta_r} & N'_{\delta_a} \end{bmatrix} \quad (7.12)$$

The characteristic equation formed from Eq. (7.11) will be a quartic of the form

$$|\lambda I - A| = \lambda^4 + a_3\lambda^3 + a_2\lambda^2 + a_1\lambda + a_0$$

which in most cases will factor into a complex conjugate pair of roots and two distinct real roots. The factoring may be expressed as

$$\left(\lambda^2 + 2\zeta\omega_n\lambda + \omega_n^2\right)_{DR}(\lambda + \lambda_{roll})(\lambda + \lambda_{spiral}) = 0$$

where

$$\lambda_{DR} = (-\zeta\omega_n \pm i\omega_d)_{DR}$$

The Dutch-roll mode shape comes from solving the eigenvalue problem, i.e.,

$$\{\dot{x}\}_{DR} = \lambda_{DR}\{x\}_{DR} = [A]\{x\}_{DR}$$

Because the Dutch-roll mode shape (eigenvector) involves solving four, complex-natured, simultaneous equations, computer methods (e.g., MATLAB) are suggested for convenience of evaluation.

It will be found useful, when considering the Dutch-roll mode shape, to include an estimate of the ψ perturbation term even though ψ is not an eigenvector component. This estimation process involves using the relationship between an eigenvector and its derivative as explained in Sec. 5.6 when describing a phasor representation. The same relation holds for a perturbation term and its derivative, i.e.,

$$(\dot{\psi})_{DR} = \lambda_{DR}(\psi) \tag{7.13}$$

The relationship between ψ and $\dot{\psi} = d\psi/dt$ may be recognized by a gain and phase change, i.e.,

$$|\dot{\psi}|_{DR} = (\omega_n)_{DR}|\psi|_{DR}$$

and

$$(\Delta\phi)_{DR} = \tan^{-1}\left[\sqrt{1-\zeta^2}\Big/(-\zeta)\right]$$

The estimated ψ perturbation for the Dutch-roll mode will be almost the same in value as the β perturbation term whereas the phase angle will be almost 180 deg out of phase relative to the β phase angle. This relationship between β and ψ during a Dutch-roll oscillation corresponds to the vehicle's c.g. nearly traveling in a straight flight path (when looking down upon the aircraft's trajectory in the x–y plane). Recognition of this aspect of Dutch-roll modal dynamics will be a key feature later when making modal approximations.

The roll response mode, the fastest of the two real modes, will be found to have small magnitudes of sideslip β and yaw rate r when comparison is made to the roll and roll rate eigenvector components, respectively. The slower spiral mode will be dominated by the roll angle component, which will normally be the largest term in that mode's eigenvector. In case the root of the spiral mode were unstable, a spiral divergence would ensue as a wing bank angle change that slowly builds up in value with time as the aircraft starts into a gradual spiral turn.

Example 7.4

Estimate the lateral-directional dynamic behavior of the DC-8 aircraft in a cruise-flight configuration (cf., Appendix B.4), which corresponds to $M = 0.84$, $h = 33{,}000$ ft, and $V = 825$ ft-s^{-1} (TAS).

Using the primed lateral-directional dimensional derivative information from Table B.4c, the plant matrix $[A]$ according to Eq. (7.11) is

$$[A] = \begin{bmatrix} -0.0869 & 0.0 & 0.0390 & -1.0 \\ -4.424 & -1.184 & 0.0 & 0.335 \\ 0.0 & 1.0 & 0.0 & 0.0 \\ 2.148 & -0.021 & 0.0 & -0.228 \end{bmatrix}$$

A matrix-oriented software program (i.e., MATLAB) was used to identify the following dynamic properties in a manner similar to the longitudinal dynamics listing provided in Example 6.1.

1) The characteristic equation from $|\lambda I - A| = 0$ is

$$\lambda^4 + 1.4989\lambda^3 + 2.5477\lambda^2 + 2.8327\lambda + 0.0113 = 0$$

which may be factored as

$$(\lambda^2 + 0.2368\lambda + 2.2437)(\lambda + 1.2580)(\lambda + 0.0040) = 0$$

2) The four characteristic roots (eigenvalues) are as follows.
Dutch-roll roots, s^{-1}:

$$\lambda_{1,2} = -0.1184 \pm 1.4932i$$

Roll response root, s^{-1}:

$$\lambda_3 = -1.2580$$

Spiral response root, s^{-1}:

$$\lambda_4 = -0.0040$$

3) The information contained in the Dutch-roll eigenvalue may be interpreted as

$$\zeta_{DR} = 0.0791, \qquad (1 - \zeta^2)^{1/2} = 0.9969$$
$$\omega_d = 1.4932 \text{ s}^{-1}, \qquad \omega_n = 1.4979 \text{ s}^{-1}$$

and Dutch-roll period

$$T_d = 2\pi/\omega_d = 4.208 \text{ s}$$

4) The Dutch-roll mode shape may be identified by solving

$$\{\dot{x}\}_{DR} = \lambda_{DR}\{x\}_{DR} = [A]\{x\}_{DR}$$

and using λ_1 with the + value for ω_d to yield the Dutch-roll eigenvector of interest, normalized with respect to β, as

$$\{x\}_{DR} = \begin{Bmatrix} \beta \\ p \\ \phi \\ r \end{Bmatrix} = \begin{Bmatrix} 1.000 \angle 0.0 \text{ deg} \\ 2.412 \angle 131.84 \text{ deg} \\ 1.610 \angle 37.30 \text{ deg} \\ 1.457 \angle -86.79 \text{ deg} \end{Bmatrix}$$

It will be noted in the eigenvector listing that the relationship in gain and phase between p and ϕ is consistent with the concept expressed by Eq. (7.13). From the eigenvector, one finds

$$|p| = \omega_n|\phi| = 1.4979|\phi|$$

and $\Delta\phi = 94.54$ deg, which is the phase lead between rate and displacement.
An estimate of the ψ perturbation term in degrees using the information contained in the eigenvector is

$$\psi = 0.973 \angle 178.67 \text{ deg}$$

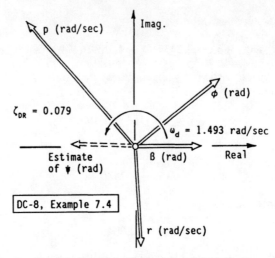

Fig. 7.3 Dutch-roll phasor interpretation.

based on $r = 1.457 \angle -86.79$ deg. Although ψ is not an eigenvector component in the linearized formulation, the airplane will be oscillating about the heading axis when excited in the Dutch-roll mode. The estimate for ψ suggests that the heading changes are almost the same magnitude as the sideslip β variations except for being about 180 deg out of phase. This observation supports an earlier statement that the c.g. travels in nearly a straight flight path for the Dutch-roll mode when observed in the x–y plane. The Dutch-roll mode may be viewed in the context of a phasor diagram (Fig. 7.3) in order to provide an alternate perspective of the mode.

5) The roll response real root corresponds to a roll time constant of

$$\tau_{\text{roll}} = (-1/\lambda_3) = 0.795 \text{ s}$$

which may be compared with $\tau = 0.845$ s in Example 7.1 where a single-degree-of-freedom roll approximation was used. The mode shape, obtained from solving

$$\{\dot{x}\}_{\text{roll}} = \lambda_{\text{roll}}\{x\}_{\text{roll}} = [A]\{x\}_{\text{roll}}$$

gave, when normalized to the roll rate component,

$$\{x\}_{\text{roll}} = \begin{Bmatrix} \beta \\ p \\ \phi \\ r \end{Bmatrix} = \begin{Bmatrix} 0.016 \\ 1.000 \\ -0.795 \\ -0.013 \end{Bmatrix}$$

As stated, the roll mode is dominated by the p component with a minimal participation by the β and r components of the eigenvector.

6) The spiral (real) root corresponds to a time constant of

$$\tau_{\text{spiral}} = (-1/\lambda_4) = 250.0 \text{ s}$$

The mode shape, when normalized with respect to bank angle ϕ, was estimated as

$$\{x\}_{\text{spiral}} = \begin{Bmatrix} \beta \\ p \\ \phi \\ r \end{Bmatrix} = \begin{Bmatrix} 0.004 \\ -0.004 \\ 1.000 \\ 0.039 \end{Bmatrix}$$

The spiral mode in Example 7.4, as illustrated by the eigenvector, showed that the roll angle component was dominant with a slight participation by the r (yaw rate) component. The estimated ψ component, when r is considered in light of Eq. (7.13), becomes approximately -9.75 when compared to the ϕ component. The minus sign is due to the example situation of a stable spiral mode. This observation implies that the aircraft heading angle is a major participant in the spiral mode while sideslip angle perturbations remain negligible. Because the example mode had a quite large (stable) time constant, its impact on the airplane dynamics would normally be considered as imperceptible by the pilot.

If the spiral root were unstable, the pilot would have little difficulty in correcting for the aircraft's tendency to enter into a slow, apparently coordinated (because of $\beta \cong 0$), spiral turn with a gradual buildup of bank angle due to the large time period involved to double an initial amplitude. Consider, for example, that if the spiral root were unstable with $\lambda_{\text{spiral}} = 0.010$ s^{-1}, the time constant would be 100 s and the time to double amplitude would be

$$\{x(t_2)\} = e^{0.01 t_2} \{x(0)\} = 2\{x(0)\}$$

Solving for the time to double amplitude, t_2, gives

$$t_2 = \ell_n(2.0)/0.01 = 69.3 \text{ s} \, (= 1.16 \text{ min})$$

7.3.1 Dutch-Roll Approximation

It was observed during the identification of the Dutch-roll mode in Example 7.4 that, although the heading perturbation was not included in the equations of motion, it was possible to estimate a heading perturbation term from the yaw rate (r) component. The Dutch-roll approximation will be based on the observation that the heading angle is nearly the same as the sideslip perturbation with almost a 180-deg phase difference.

It will be assumed that the aircraft, while experiencing a Dutch-roll type of modal oscillation, is traveling in a straight flight track. The c.g.'s lack of lateral motion imposes a side-force constraint on the dynamic system. This assumption is equivalent to stating 1) $\psi = -\beta$ and $\dot{\psi} = +r = -\dot{\beta}$ and 2) the linearized roll and yaw moment equations are retained while the side-force equation is omitted. Removal of the side-force equation in the complete set of lateral-directional equations, Eq. (7.11), simplifies the relations to

$$\begin{Bmatrix} \dot{p} \\ \dot{r} \end{Bmatrix} = \begin{bmatrix} L'_p & L'_r & L'_\beta \\ N'_p & N'_r & N'_\beta \end{bmatrix} \begin{Bmatrix} p \\ r \\ \beta \end{Bmatrix}$$

which is almost in the form of

$$\{\dot{x}\} = [A]\{x\}$$

When the assumption of $r = -\dot{\beta}$ is included in the formulation, the resulting approximation, including the control terms, becomes

$$\begin{Bmatrix} \dot{p} \\ \ddot{\beta} \\ \dot{\beta} \end{Bmatrix} = \begin{bmatrix} L'_p & -L'_r & L'_\beta \\ -N'_p & +N'_r & -N'_\beta \\ 0 & 1 & 0 \end{bmatrix} \begin{Bmatrix} p \\ \dot{\beta} \\ \beta \end{Bmatrix} + \begin{bmatrix} L'_{\delta_r} & L'_{\delta_a} \\ -N'_{\delta_r} & -N'_{\delta_a} \\ 0 & 0 \end{bmatrix} \begin{Bmatrix} \delta_r \\ \delta_a \end{Bmatrix} \quad (7.14)$$

where

$$\{x\} = [p \; \dot{\beta} \; \beta]^T$$

is the state vector for the approximation.

The characteristic equation can be obtained from Eq. (7.14) in the cubic form (cf. Problem 7.6 for the expansion of a_0, a_1, and a_2) as

$$\lambda^3 + a_2\lambda^2 + a_1\lambda + a_0 = 0$$

which factors into a form to provide estimates for both the Dutch-roll and roll response modes, i.e.,

$$(\lambda^2 + 2\zeta\omega_n\lambda + \omega^2)_{\text{DR}}(\lambda + \lambda_{\text{roll}}) = 0$$

Example 7.5

Use the Dutch-roll approximation to estimate the lateral-directional dynamics of the DC-8 aircraft in the cruise-flight configuration. Applicable dimensional stability derivatives from Example 7.4 (cf. Appendix B.4) will be applied to Eq. (7.14) for the analysis. The plant matrix is

$$[A] = \begin{bmatrix} -1.184 & -0.335 & -4.424 \\ 0.021 & -0.228 & -2.148 \\ 0.0 & 1.0 & 0.0 \end{bmatrix}$$

1) The cubic characteristic equation from $|\lambda I - A| = 0$ is

$$\lambda^3 + 1.4120\lambda^2 + 2.4250\lambda + 2.6361 = 0$$

which may be factored as

$$(\lambda^2 + 0.2026\lambda + 2.1798)(\lambda + 1.2093) = 0$$

2) The three characteristic roots are as follows.
Dutch-roll roots, s^{-1}:

$$\lambda_{1,2} = -0.1013 \pm 1.4730i$$

Roll response root, s^{-1}:

$$\lambda_3 = -1.2093$$

3) The Dutch-roll eigenvalues may be interpreted as

$$\zeta_{DR} = 0.0686, \quad (1 - \zeta^2)^{1/2} = 0.9976$$

$$\omega_d = 1.4730 \text{ s}^{-1}, \quad \omega_n = 1.4764 \text{ s}^{-1}$$

The eigenvector for the Dutch-roll approximation, when normalized with respect to β, is

$$\{x\}_{DR} = \begin{Bmatrix} p \\ \dot{\beta} \\ \beta \end{Bmatrix} = \begin{Bmatrix} 2.417 \angle 132.73 \text{ deg} \\ 1.476 \angle 93.94 \text{ deg} \\ 1.000 \angle 0.0 \text{ deg} \end{Bmatrix}$$

The p component compares well with the result obtained with the full (4 × 4) plant in Example 7.4 (i.e., 2.412 ∠ 131.84 deg). Recognition that $\dot{\beta} = -r$ in the approximation allows comparison with the r component in Example 7.4 (i.e., 1.457 ∠ −86.79 deg).

4) The roll response root corresponds to a time constant of

$$\tau_{roll} = (-1/\lambda_3) = 0.826 \text{ s}$$

The mode shape for the roll mode, when normalized relative to the roll rate component, is

$$\{x\}_{roll} = \begin{Bmatrix} p \\ \dot{\beta} \\ \beta \end{Bmatrix} = \begin{Bmatrix} 1.000 \\ -0.008 \\ 0.006 \end{Bmatrix}$$

The roll angle component can be estimated by using Eq. (7.13), i.e.,

$$\phi = (1/\lambda_3) p = -0.826$$

It is apparent from Table 7.1 that the roll time constant for the (3 × 3) model is about midway between the more exact value of Example 7.4 and the pure roll approximation of Example 7.1. Although the sideslip component is small for this mode, the value for the (3 × 3) model again is approximately midway between the full (4 × 4) plant result and $\beta = 0$, which is an assumption in the pure rolling airframe model.

A very simplified approximation for the Dutch-roll mode may be obtained by assuming that the c.g. travels in a straight line and that bank angle does not vary. These assumptions imply that constraints are applied to both the airframe's side-force and rolling moment equations. The resulting airframe motion, sometimes colloquially referred to as a snaking mode, is described by the single, second-order yawing moment equation, i.e.,

$$\ddot{\beta} - N_r \dot{\beta} + N_\beta \beta = -N_{\delta_r} \delta_r \tag{7.15}$$

Table 7.1 DC-8 roll mode summary

Item	Airframe (4 × 4)	Approx. (3 × 3)	Approx. (pure roll)
τ_{roll}, s	0.795	0.826	0.845
Eigenvector			
β	0.016	0.006	n/a
p	1.000	1.000	1.000
ϕ	−0.795	−0.826	−0.845
r	−0.013	n/a	n/a

where it can be seen that

$$\omega_n^2 = N_\beta \quad \text{and} \quad 2\zeta\omega_n = -N_r$$

Example 7.6

Estimate the response dynamics for the DC-8 aircraft during cruise flight (cf. Appendix B.4) using the snaking mode approximation. From Eq. (7.15), one may estimate

$$\omega_n = \sqrt{N_\beta} = \sqrt{2.175} = 1.4748 \text{ s}^{-1}$$

$$\zeta = 0.5(-N_r)/\sqrt{N_\beta} = 0.078$$

and

$$\omega_d = \sqrt{1-\zeta^2}\,\omega_n = 1.4703 \text{ s}^{-1}$$

The Dutch-roll approximations illustrated the influence of motion constraints on the dynamic behavior. A summary is shown on Table 7.2. Although the imposition of side-force and rolling motion constraints did not have a major effect on the natural frequency and the mode shape components, the side-force constraint from

Table 7.2 DC-8 Dutch-roll mode summary

Item	Airframe (4 × 4)	Approx. (3 × 3)	Approx. (snaking)
ω_n, s^{-1}	1.498	1.476	1.475
ζ	0.079	0.069	0.078
Eigenvector			
β	1.000 ∠ 0.0 deg	1.000 ∠ 0.0 deg	1.000 ∠ 0.0 deg
p	2.412 ∠ 131.8 deg	2.417 ∠ 132.7 deg	n/a
ϕ	1.610 ∠ 37.3 deg	1.637 ∠ 38.8 deg	n/a
r	1.457 ∠ −86.8 deg	1.476 ∠ −86.1 deg	1.475 ∠ −85.5 deg

Example 7.5 did reduce the modal damping by $\Delta\zeta = 0.01$. The snaking mode does not represent the Dutch-roll motion properly due to the addition of roll moment and side-force constraints; however, its use in modeling airframe responses will be invaluable in a later discussion (Chapter 8) dealing with inertial cross coupling. It also should be remembered that the results of Tables 7.1 and 7.2 are airframe specific with results based on the dimensional stability derivative information of Appendix B.4.

7.4 Control Response

The response of an aircraft to both rudder and aileron control inputs will be considered in this section. The preceding sections of this chapter were focused on identifying the modal properties using the homogeneous form of the governing state equations. Initially, the responses will be in the time domain and will reflect either rudder or aileron inputs. The rudder input will be in the form of an impulse. An impulse input will correspond to a rudder kick, a form of input frequently used to excite the Dutch-roll mode. The aileron input will be in the form of a pulse, with a reason being that lateral control is primarily used to change bank angle during maneuvers. Because aileron deflection produces roll rate, a control pulse of a finite time length is required to achieve a desired bank angle change. In a later section, it will be observed that lateral control design requirements for an airframe are oftentimes linked to the time required to change a bank angle in support of a heading change. Both the steady-state roll helix angle and the time to change bank angle are used as design criteria when establishing a lateral control system.

The lateral-directional frequency response due to both harmonic rudder and aileron control will be considered and presented in Bode plot format. The viewpoints gained from these latter studies will prove of value in a later section when the elements of a yaw damper design are discussed.

7.4.1 Rudder Control in the Time Domain

The lateral-directional system behavior will be considered initially due to a rudder kick applied at $t = 0$. This control input may be viewed as a constant rudder deflection applied as a pulse for a time period of 50–100 ms. The area defining the pulse (i.e., $\int \delta_r \, dt$) of time duration T is approximately equal to $\delta_r T$ and may be replaced by an equivalent impulse I_δ. The impulse I_δ may effectively be represented by a Dirac delta function providing the pulse duration were several orders of magnitude less than the Dutch-roll period.

The system is described by Eq. (5.50) in state equation format as

$$\{\dot{x}\} = [A]\{x\} + \{B\}u \qquad (5.50)$$

The response solution to a unit impulse control input due to $u(\tau) = \delta(\tau)$ was also described in Chapter 5 by Eq. (5.65), i.e.,

$$\{x(t)\} = e^{At}\{B\} \qquad (5.65)$$

When considering an impulse input due to a control deflection, care should be taken relative to the units used for both input and output, i.e., $|I_{\delta_r}|$ = rad-s, whereas $|\beta|$ and $|\phi|$ are in radians, and $|p|$ and $|r|$ are in radians per second. Alternatively, the numerically determined values would scale equally well to degree vice radian angular measure.

Example 7.7

Estimate the response at $t = 1.0$ and 2.0 s due to a rudder pulse (kick) of 5.0 deg applied for 200 ms as an input to the DC-8 aircraft at the cruise condition (cf. Appendix B.4). Dimensional stability derivatives needed, in addition to those considered in earlier examples, include Y_{δ_r}/V, L'_{δ_r}, and N'_{δ_r}, which are used to form the control vector as

$$\{B_{\delta_r}\} = [0.0223 \quad 0.547 \quad 0.0 \quad -1.169]^T$$

1) For $t = 1.0$ s, the transition matrix may be found using matrix-oriented software, e.g., MATLAB,

$$[e^A] = \begin{bmatrix} 0.1083 & 0.0161 & 0.0257 & -0.5717 \\ -1.2865 & 0.2842 & -0.0447 & 1.2149 \\ -1.1458 & 0.5794 & 0.9817 & 0.5480 \\ 1.2540 & 0.0025 & 0.0318 & 0.0437 \end{bmatrix}$$

The rudder pulse under consideration is $I_\delta = 1.0$ deg-s. Application of Eq. (5.65) with suitable scaling for the impulse magnitude yields

$$\{x(1)\} = e^A\{B_{\delta_r}\} = \begin{Bmatrix} \beta \\ p \\ \phi \\ r \end{Bmatrix} = \begin{Bmatrix} 0.68 \text{ deg} \\ -1.29 \text{ deg-s}^{-1} \\ -0.35 \text{ deg} \\ -0.02 \text{ deg-s}^{-1} \end{Bmatrix}$$

2) For $t = 2.0$ s,

$$\{x(2)\} = e^{2A}\{B_{\delta_r}\} = e^A\{x(1)\} = \begin{Bmatrix} 0.06 \text{ deg} \\ -1.25 \text{ deg-s}^{-1} \\ -1.88 \text{ deg} \\ 0.84 \text{ deg-s}^{-1} \end{Bmatrix}$$

These solutions were validated using the *impulse* function in MATLAB. A time-history plot of β, p, and ϕ due to the assumed rudder control impulse (Example 7.7) for a time span of $t = 0$–20 s is shown in Fig. 7.4.

The exponential decay of the β response in Fig. 7.4 to an impulselike rudder input shows the properties of a lightly damped Dutch-roll mode. However, the rudder impulse input excited all of the modes including the roll and spiral modes. The pure Dutch-roll mode behavior could have been obtained if the initial condition at $t = 0$ represented the Dutch-roll mode shape, a concept that was used in Chapter 6 to illustrate the short-period modal response behavior. Although Fig. 7.4 illustrated the Dutch-roll response to the assumed rudder kick of Example 7.7, the short-time scale relative to the period of the stable spiral mode

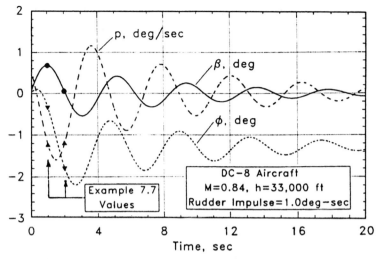

Fig. 7.4 Aircraft response due to a rudder impulse input.

(i.e., 20 s vs $T_{\text{spiral}} = 250$ s) masked the true roll and spiral mode responses as would be evident in the bank angle time history. The impulse response of Fig. 7.4 was replotted for a time scale of $t = 0$–1000 s to illustrate that the bank angle ultimately tended to a zero value, cf. Fig. 7.5. The range of time considered in Fig. 7.5 corresponded to several spiral mode time periods in order to show the ultimate bank angle behavior, i.e., well after the Dutch-roll oscillation had subsided.

An important feature to note in Fig. 7.4 is that the bank angle initially moved in a direction opposite to the apparent dominant bank angle response, i.e., $\phi > 0$

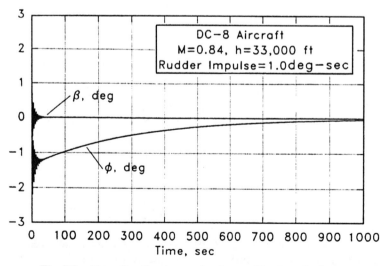

Fig. 7.5 Aircraft roll response due to a rudder impulse input.

for $0 < t < 1.0$ s whereas $\phi < 0$ for $t > 1.0$ s until subsidence occurred. This behavor suggests that the $G_{\phi\delta_r}(s)$ transfer function represents a nonminimum phase system similar in principle to the aircraft's normal acceleration response to elevator control as described in Chapter 6. As will be apparent by Example 7.8, the roots of the numerator (i.e., the zeros) of the $G_{\phi\delta_r}(s)$ transfer function include a positive valued member, which is responsible for the bank angle's initial response behavior.

Example 7.8

Determine the zeros of the $G_{\phi\delta_r}(s)$ transfer function for the DC-8 aircraft in a cruise-flight configuration. The poles of the transfer function are the roots of the denominator, which correspond to the system's eigenvalues as determined in Example 7.4.

The transfer function can be expressed in a matrix form as

$$\{G(s)\}_{\delta_r} = [sI - A]^{-1}\{B_{\delta_r}\}$$

Because the state vector is

$$\{x(t)\} = [\beta \quad p \quad \phi \quad r]^T$$

the third row in the transfer function matrix will correspond to $G_{\phi\delta_r}(s)$. The *ss2tf* (state-space to transfer function) function from MATLAB made possible the conversion of the plant from the state space to the transfer function format. The lateral-directional plant definition for the Example 7.7 aircraft was used to identify the numerator of the transfer function as

$$\text{Num}_{\phi\delta_r} = 0.547s^2 - 0.318s - 4.026$$

which factors to

$$\text{Num}_{\phi\delta_r} = 0.547(s + 2.438)(s - 3.019)$$

The zero at $z_1 = +3.019$ contributes to the nonminimum phase behavior of bank angle response due to a rudder impulse input.

It will be noted in Fig. 7.4 that the bank angle due to the 1.0 deg-s rudder impulse (kick) input resulted in the roll response mode tending toward a value of approximately -1.3 deg before the stable spiral mode reacted to return the bank angle to zero, as is indicated in Fig. 7.5. Application of the final value theorem to Example 7.8 provides an estimate of the bank angle due to a unit rudder impulse input as time becomes large, i.e.,

$$\lim_{t \to \infty} \phi(t) = \lim_{s \to 0} (s) \frac{(0.547s^2 - 0.318s - 4.026)}{(s^3 + 1.495s^2 + 2.542s + 2.823)(s + 0.004)} \quad (7.16)$$

In the limit, as s tends to 0, Eq. (7.16) states that the bank angle would tend to zero in accord with the time history of Fig. 7.5. However, the denominator can be factored (as shown) into a cubic polynomial representing the Dutch-roll and roll response modes along with the first-order polynomial representing the stable spiral mode. If one were to assume that an approximate pole-zero cancellation took place in Eq. (7.16) before the limit operation took place, the pole due to the very slow spiral mode would be canceled by the s in the numerator from the final value theorem leaving an estimate for the limit bank angle after the Dutch-roll oscillation subsided. On this basis, Eq. (7.16) could be interpreted as stating that a bank angle due to the 1.0-deg-s rudder impulse in the limit would be

$$\lim_{t \to \text{large}} \Delta\phi = (-4.026)/(2.823) = -1.43 \text{ deg}$$

The Dutch-roll approximation based upon the c.g. traveling in a straight flight path provides an alternate lateral-directional plant that does not contain the spiral mode response. Application of the final value theorem to the $G_{\phi\delta_r}(s)$ transfer function, as estimated from the (3×3) plant, provides a result close in value to that obtained by the pole-zero cancellation approach applied to the $G_{\phi\delta_r}(s)$ transfer function from Eq. (7.16).

7.4.2 Aileron Control in the Time Domain

As stated in Sec. 7.2, lateral control is a rate producing device. Although aileron control is frequently used to obtain roll rate during maneuvering operations, a very common use of the aileron is to change bank angle. Bank angle change can be obtained by providing a control pulse input to the lateral control system because $\Delta\phi = \int p \, dt$. In Example 7.2, an aileron control pulse was applied when the airframe was modeled by pure rolling motion. The complete airframe model, which will include the presence of the Dutch-roll and spiral modes along with the roll response mode, will be considered to illustrate the impact of the individual modes on the airframe's response to lateral control input.

Example 7.9

Consider the DC-8 aircraft in the cruise-flight condition (cf. Appendix B.4) when lateral control similar to that used in Example 7.2 is applied. Find the airframe sideslip and bank angle responses due to a 5-deg aileron control input applied as a pulse for a time duration of 2.0 s.

Estimate the state vector for time = 3 s after completion of the pulse input, i.e., at $t = 5$ s. Because the roll mode time constant was identified as 0.8 s in Example 7.4, the 3-s time represents approximately four roll mode time constants following completion of the lateral control input.

The lateral-directional plant is the same as that used in Example 7.4. The lateral control dimensional stability derivatives (cf. Appendix B.4) include Y_{δ_a}/V, L'_{δ_a}, and N'_{δ_a}, which are used to form the control vector as

$$\{B_{\delta_a}\} = [0.0 \quad 2.120 \quad 0.0 \quad 0.065]^T$$

The response may be obtained using the MATLAB *lsim* function in combination with a control input representing the 2-s pulse. A file entitled pulse.M is listed in Appendix E, the use of which is convenient for the determination of the input as a vector. Function *lsim* provides a numerical evaluation for the time history that is equivalent to the matrix solution

$$\{x(t)\} = e^{At} \int_0^t e^{-A\tau} \{B_{\delta_a}\} u(\tau) d\tau$$

where $u(\tau)$ may be established using the pulse function.

For 5 deg of aileron control input as a 2-s pulse, the solution obtained at $t = 5$ s was

$$\{x(5)\} = \begin{Bmatrix} \beta \\ p \\ \phi \\ r \end{Bmatrix} = \begin{Bmatrix} -0.048 \text{ deg} \\ 0.060 \text{ deg/s} \\ 17.072 \text{ deg} \\ 0.870 \text{ deg/s} \end{Bmatrix}$$

The time-history response of β and ϕ for $t = 0\text{--}20$ s is shown in Fig. 7.6. The start of the aileron pulse was delayed until $t = 1$ s so that the aileron time history would be clear. The aileron control input does not induce a major response from the Dutch-roll mode, a fact attributable to the small value of N'_{δ_a}. The sideslip time-history response was increased by a factor of 10 in order to illustrate the weak Dutch-roll activity due to lateral control.

The lateral response did not reach a steady value in the 20 s shown in Fig. 7.6, primarily due to the influence of the stable spiral mode, which caused the bank angle to slowly return to a zero value. An estimate for the steady-state bank angle value without the presence of the spiral mode was made by using a pole-zero cancellation procedure similar in concept to the method used when considering the bank angle response to a rudder control input (cf. Example 7.8).

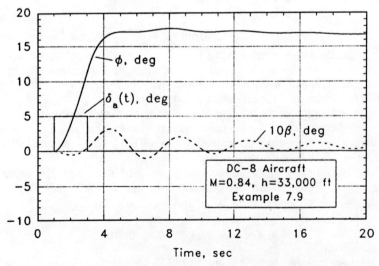

Fig. 7.6 Roll response due to aileron input.

LATERAL-DIRECTIONAL DYNAMICS

Table 7.3 DC-8 roll response summary: input $\delta_a = 5$ deg for $0 < t < 2.0$ s

Item	Airframe (4 × 4)	Approx. (3 × 3)	Approx. (pure roll)
$\Delta\phi$, deg	17.30	18.26	17.9

The $G_{\phi\delta_a}(s)$ transfer function due to lateral control actuation was estimated for the Example 7.9 situation using the MATLAB *ss2tf* function, i.e.,

$$G_{\phi\delta_a}(s) = \frac{(2.120s^2 + 0.689s + 4.885)}{(s^3 + 1.495s^2 + 2.542s + 2.823)(s + 0.004)} \quad (7.17)$$

The control input pulse has a Laplace transform of

$$\mathcal{L}\{1(t) - 1(t - T)\} = (1/s)(1 - e^{-Ts}) = (1/s)[(Ts) - (Ts)^2/2! + \cdots]$$

Applying the final value theorem to obtain the steady-state bank angle estimate yields

$$\lim_{t \to \infty} \phi(t) = \lim_{s \to 0}(s)\frac{(2.120s^2 + 0.689s + 4.885)[(Ts) - (Ts)^2/2! + \cdots]}{(s^3 + 1.494s^2 + 2.542s + 2.823)(s + 0.004)(s)}$$

Application of the pole-zero cancellation approximation to remove the spiral-mode pole in the denominator provides an estimate for the bank angle when time is large. The bank angle change due to the assumed 5 deg of lateral control as a 2-s pulse input becomes

$$\lim_{t \to \text{large}} \Delta\phi = (5.0 \text{ deg})(4.885)(T = 2.0)/(2.823) = 17.30 \text{ deg}$$

The (3 × 3) lateral-directional approximation was also considered from the standpoint of estimating the bank angle change due to the 5-deg aileron pulse input. This approximation suppressed the spiral response mode due to the assumption that the c.g. travels in a straight path. An estimate for the steady bank angle change was $\Delta\phi = 18.26$ deg, an estimate left as a problem exercise.

The preceding comparisons of bank angle response estimates under different modeling assumptions are airframe specific. However, Table 7.3 does illustrate the influence of making simplifying assumptions on the lateral control performance. In actual practice, the influence of the spiral mode to gradually remove the bank angle would be suppressed by the pilot's use of appropriate rudder control inputs.

7.4.3 Lateral Acceleration Response in the Time Domain

The lateral acceleration a_y in body-axis coordinates is normally evident as the lateral displacement of the ball in the needle-ball indicator, which is information available on the aircraft instrument panel for the purpose of assisting the pilot to maintain coordinated flight. The sensing of a_y can be useful for other purposes such as airframe stability augmentation.[1] The governing relation for lateral acceleration will be considered as an output variable from the plant (i.e., aircraft) in a manner similar to the normal acceleration described in Sec. 6.5. Lateral acceleration at the c.g. is defined as

$$a_y(t) = V(\dot{\beta} + r) - g\cos\theta_0 \sin\phi \qquad (7.18)$$

In a steady coordinated turn, when $\dot{\beta} = 0$, the lateral acceleration at the c.g. is zero thereby simplifying Eq. (7.18) to

$$0 = Vr - g\cos\theta_0 \sin\phi$$

a result that is consistent with the earlier static stability discussion in Sec. 3.2.2 where $\cos\theta_0 = 1$ in a constant altitude turn. Linearization of Eq. (7.18) provides a basis for finding the lateral acceleration during dynamically induced bank angle perturbations, i.e.,

$$a_y(t) = V(\dot{\beta} + r) - g\cos\theta_0 \phi \qquad (7.19)$$

Use of state variable methods to find the lateral acceleration requires the use of the (4 × 4) plant inasmuch as the (3 × 3) plant approximation is based on the assumption that $\dot{\beta} = -r$. The $V\dot{\beta}$ term may be established using the first row of Eq. (7.10),

$$V\dot{\beta} = Y_\beta \beta + Y_p p + g\cos\theta_0 \phi + (Y_r - V)r + Y_{\delta_r}\delta_r(t) + Y_{\delta_a}\delta_a(t)$$

Consequently, the $a_y(t)$ output may be expressed as

$$y = Y_\beta \beta + Y_p p + (g\cos\theta_0 - g\cos\theta_0)\phi + (Y_r - V + V)r + Y_{\delta_r}\delta_r + Y_{\delta_a}\delta_a$$

The Y_p and Y_r terms are usually quite small and will be neglected. The aileron deflection influence on lateral acceleration is relatively minor, especially when compared to rudder deflection (i.e., $|Y_{\delta_r}| \gg |Y_{\delta_a}|$) and, subsequently, will also be neglected. Here,

$$y = [C]\{x\} + D\delta_r \qquad (7.20)$$

where the simplifying assumptions yield that

$$[C] = [Y_\beta \quad 0 \quad 0 \quad 0] \qquad [D] = Y_{\delta_r}$$

If x_p and z_p were the coordinates of the pilot's c.g. relative to the aircraft c.g., the lateral acceleration at the pilot, a'_y, would be

$$a'_y = a_y + x_p \dot{r} - z_p \dot{p} \qquad (7.21)$$

and the determination of the lateral acceleration at the pilot's station would require a corresponding change in the [C] and D matrix expressions of Eq. (7.20) to reflect the translations from the aircraft to the pilot's c.g.

Example 7.10

Estimate the lateral acceleration time history for the DC-8 aircraft in a cruise-flight configuration due to a 1-deg rudder pulse applied for a time duration of 2 s. The procedure for obtaining the time history using MATLAB includes the following.

1) Set a time vector for $t = 0$–20 s by 0.1-s increments.
2) Define the rudder control effectiveness matrix by BR in order to distinguish the control due to rudder from that due to the aileron.
3) Establish a 2-s rudder control input pulse $u(t)$ of unit magnitude using the pulse.M file described in Appendix E.
4) Determine the appropriate $[C]$ and D matrices in accord with Eq. (7.20), i.e.,

$$[C] = [-71.73 \text{ ft-s}^{-2} \quad 0 \quad 0 \quad 0]$$

$$D = 18.38 \text{ ft-s}^{-2}$$

5) Use the MATLAB function *lsim* to obtain a numerical evaluation for the time history of the lateral acceleration output y and the state vector x due to the unit rudder input pulse $u(t)$, i.e.,

$$[Y, X] = \text{lsim}(A, \ BR, \ C, \ D, \ u, \ t)$$

Both the sideslip and lateral acceleration are shown in Fig. 7.7 for the 20-s time span. The rudder control pulse was delayed in Fig. 7.7 until $t = 1$ s in order to show clearly the rudder, δ_r, input deflection. Because β response was either in radian/radian or degree/degree of control, the time-history estimate for β was properly scaled as calculated. The lateral acceleration output from the lsim evaluation was scaled by 0.01745 in order to represent acceleration in feet per second^2 due to the 1 deg (vice 1 rad) of assumed rudder pulse amplitude.

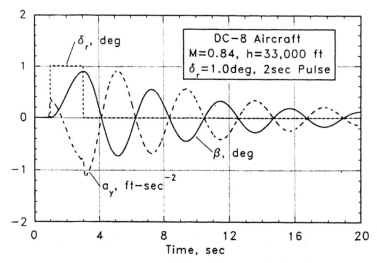

Fig. 7.7 Lateral acceleration due to a rudder impulse input.

The initial lateral acceleration response implies that the transfer function $G_{n\delta}(s)$ represents a nonminimum phase system. The first acceleration response is the acceleration reaction to the lateral force induced by the plus rudder deflection followed by an acceleration buildup in the other direction as sideslip develops followed by the Dutch-roll oscillation. The $G_{n\delta}(s)$ transfer function corresponding to the Example 7.10 case was determined using the MATLAB *ss2tf* program as

$$G_{n\delta}(s) = (18.38)\frac{(s^4 + 1.412s^3 - 2.137s^2 - 2.721s + 0.052)}{s^4 + 1.499s^3 + 2.548s^2 + 2.833s + 0.011}$$

The four roots of the numerator (i.e., zeros of the transfer function) are $z = 1.421$, 0.019, -1.125, and -1.727. The first two roots are positive, which provides agreement as to the nonminimum phase property. It can be seen from the transfer function, when properly scaled for an 0.01745-rad rudder pulse input, that the initial acceleration as found using the initial value theorem agrees with the response shown in Fig. 7.7,

$$a_y(0) = (0.01745)\lim_{s\to\infty}(s)G_{n\delta}(s)(1/s) = 0.321 \text{ ft-s}^{-2}$$

7.4.4 Frequency-Domain Control Response

The lateral-directional transfer function matrices due to either rudder or aileron inputs contain the frequency-response information when the substitution of $s = i\omega$ is made. The aircraft considered by example in this chapter (i.e., the DC-8 at a cruise configuration) can be summarized with respect to these transfer functions as follows.

1) Numerators of $G_{(-)\delta_r}(s)$ due to the harmonic rudder input are

$$\text{Num}_{\beta\delta_r}(s) = (0.022)(s + 52.62)(s + 1.220)(s - 0.007)$$

$$\text{Num}_{\phi\delta_r}(s) = (0.547)(s - 3.019)(s + 2.438)$$

$$\text{Num}_{p\delta_r} = s\,\text{Num}_{\phi\delta_r}$$

$$\text{Num}_{r\delta_r}(s) = (-1.169)(s + 1.279)(s^2 - 0.040s + 0.104)$$

2) Numerators of $G_{(-)\delta_a}(s)$ due to the harmonic aileron input are

$$\text{Num}_{\beta\delta_a}(s) = (-0.065)(s - 1.059)(s + 0.286)$$

$$\text{Num}_{\phi\delta_a}(s) = (2.120)(s^2 + 0.325s + 2.304)$$

$$\text{Num}_{p\delta_a} = s\,\text{Num}_{\phi\delta_a}$$

$$\text{Num}_{r\delta_a}(s) = (0.065)(s + 1.640)(s^2 - 1.054s + 1.771)$$

3) The denominator for all of the lateral-directional transfer functions is

$$\text{Den}(s) = \Delta(s) = s^4 + 1.499s^3 + 2.548s^2 + 2.833s + 0.011$$

A Bode plot showing the harmonic response of the lateral-directional state vector components per unit amplitude of rudder deflection is shown in Fig. 7.8. Items to note include the following.

LATERAL-DIRECTIONAL DYNAMICS

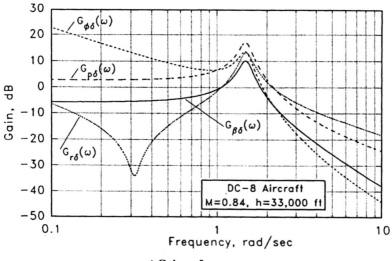

a) Gain vs frequency

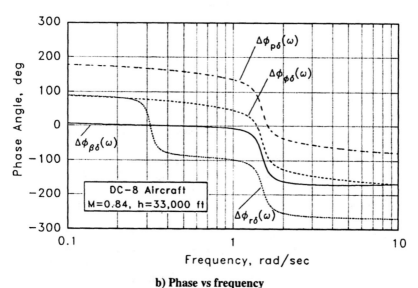

b) Phase vs frequency

Fig. 7.8 Frequency response, harmonic rudder input.

1) The Dutch-roll peak resonance is evident for all of the state vector components in both gain and phase at $\omega_d = 1.493$ rad/s. Sideslip gain at the peak is 16 dB greater than the static response value, which corresponds to a physical increase of 6.3.

An approximation for the Dutch-roll mode was made earlier as a second-order system responding to yawing moments, cf. Eq. (7.15) and the mode summary in Table 7.2. Agreement with this approximation is found by noting that $G_{\beta\delta_r}(\omega)$ has a constant value (0-dB change per frequency decade) until the neighborhood of the modal frequency. Beyond the modal frequency, the gain changes at -40 dB per decade. The phase angle curve also shows a 180-deg change when a comparison is made at frequencies before and after the modal frequency.

2) The yaw rate transfer function indicates a relative gain change of -25 dB as a local negative peak in the frequency region of $\omega = 0.32$ rad/s along with a rapid 180-deg phase change when frequency is varied in that region. This trait in $G_{r\delta_r}(s)$ is a result of the complex conjugate zero pairs in the numerator, which occurs at $\omega_d = 0.32$ rad/s with a damping ratio of 0.06.

3) The phase angle of $G_{p\delta_r}(s)$ leads $G_{\phi\delta_r}$ by 90 deg for all frequencies. This is a direct result of the fact that

$$G_{p\delta_r}(s) = (s) G_{\phi\delta_r}(s)$$

In addition, the gain values are equivalent at $\omega = 1.0$ rad/s while there is a 20-dB difference in gain at $\omega = 0.1$ and 10.0 rad/s, respectively, i.e.,

$$G_{p\delta_r}(0.1) - G_{\phi\delta_r}(0.1) = -20 \text{ dB}$$
$$G_{p\delta_r}(10.0) - G_{\phi\delta_r}(10.0) = +20 \text{ dB}$$

A similar Bode plot that reflects the harmonic response behavior due to aileron deflection is shown in Fig. 7.9.

1) The corner frequency of the roll response mode is at $\omega = 1/\tau_r = 1.26$ rad/s for the full (4 × 4) lateral-directional plant. Its influence on the gain plot may be recognized by noting that $G_{p\delta_a}(\omega)$ is relatively uniform at the static response value below the corner frequency. The $G_{p\delta_a}(\omega)$ gain for frequencies greater than the corner frequency has an approximate slope of -20 dB per frequency decade. This behavior is in close accord with the pure rolling motion approximation described in Sec. 7.2.

2) Since $G_{p\delta_a}(i\omega) = (i\omega) G_{\phi\delta_a}(i\omega)$, the gain of $G_{\phi\delta_a}(\omega)$ shows a 20-dB-per-decade slope increase relative to $G_{p\delta_a}(\omega)$ with their being equal at $\omega = 1.0$ rad/s. Physically, this implies that the bank angle excursion when rocking the wings by harmonic aileron excitation will decrease as aileron input frequency is increased.

3) Both $G_{p\delta_a}(\omega)$ and $G_{\phi\delta_a}(\omega)$ respond slightly to the Dutch-roll mode in the neighborhood of the Dutch-roll frequency. This reflects the fact that the Dutch-roll mode response is more sensitive to yawing rather than rolling moment inputs.

4) In the frequency range shown in Fig. 7.9, $G_{r\delta_a}(\omega)$ shows a gain slope of -20 dB per frequency decade combined with local gain increase in the neighborhood of the Dutch-roll modal frequency. $G_{\beta\delta_a}(\omega)$ shows approximate slopes of -20 and -40 dB per frequency decade below and above the Dutch-roll modal frequency, respectively.

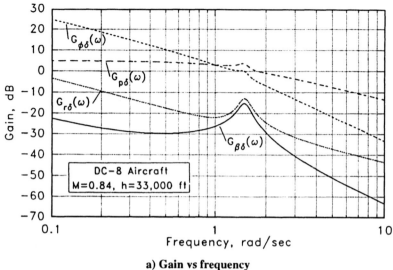

a) Gain vs frequency

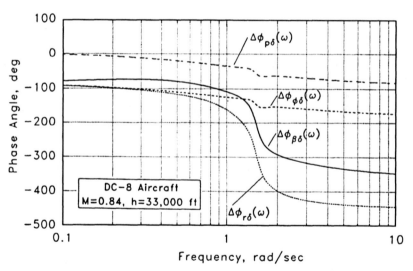

b) Phase vs frequency

Fig. 7.9 Frequency response, harmonic aileron input.

The preceding comments concerning the lateral-directional frequency response behavior due to harmonic control inputs are airframe specific. Other airframes with different modal properties including frequencies, damping, and mode shapes will differ with respect to details from Figs. 7.8 and 7.9; however, their Bode plot behaviors will, in general, display similar traits to those found in the preceding examples.

7.5 Design Guidelines

The preceding sections in this chapter have shown how to identify the lateral-directional modal properties of an aircraft (considered as a plant) along with the aircraft's response to either lateral or directional control inputs. No comments were made relative to the airframe's dynamic properties being suitable for flight operation. Appropriate guidelines for airframe design purposes to yield satisfactory flying qualities are contained in the certification requirements for U.S. civilian aircraft, FAR Parts 23[2] and 25,[3] and the U.S. military specifications, MIL-F-8785C,[4] dealing with aircraft flying qualities.

The civilian requirements[2] address the Dutch-roll mode by stating that any oscillation occurring in the normal flight envelope must be damped to 1/10 amplitude in 7 cycles. This provides a minimum acceptable value of Dutch-roll damping as $\zeta = 0.052$ where the damping level estimate is based on using the log-decrement procedure, Eq. (5.40). No mention is made regarding the Dutch-roll frequency. Transport category requirements[3] specify that any Dutch-roll oscillation must be positively damped and must be controllable with normal use of the primary controls without requiring exceptional pilot skill. Roll capability is described by the minimum time to roll the airplane from a steady 30-deg banked turn through an angle of 60 deg so as to reverse the direction of the turn in either takeoff or landing approach configurations. Typical times given[2] are 4 s as a minimum for the 60-deg bank angle change.

The military specifications[4] are more specific with regard to airframe dynamics and will yield acceptable aircraft flying qualities when used as a guideline during airframe design. Flying qualities are graded by levels as defined by Table 6.2 in Chapter 6 (note: level 1 is good).

Minimum Dutch-roll requirements make a distinction relative to class of aircraft and category of the flight operation. Let us consider T and F as referring to trainer and fighter aircraft classes whereas L and M refer to large and medium weight aircraft in the utility and transport classes. A summary of the requirements[4] for minimum Dutch-roll properties during a precision class of flight operations is shown in Table 7.4.

It is presumed that stability augmentation may be used to improve Dutch-roll damping from a level 2 or 3 to a level 1 flying quality, i.e., use a yaw damper to meet the requirements.

The roll mode time constant during precision category flight operations should not exceed 1.0 and 1.4 s for level 1 and 2 flying qualities, respectively, by the T and F class of aircraft. Not surprisingly, the roll time constant maximums are a little less demanding for the L and M aircraft classes. In their case, the maximum roll mode time constant should not exceed 1.4 and 3.0 s for level 1 and 2 flying qualities, respectively.

Table 7.4 Minimum Dutch-roll properties

Level	Class	Min. ζ	Min. $\zeta\omega_n$, rad/s	Min. ω_n, rad/s
1	T, F	0.19	0.35	1.0
	L, M	0.19	0.35	0.4
2	All	0.02	0.05	0.4
3	All	0.0	——	0.4

Spiral stability requirements apply to unstable roots and specify the minimum time to double amplitude as approximately 12 and 8 s for level 1 and 2 flying qualities, respectively.

Roll performance specifications[4] are similar to the civilian requirements in that the time to make a bank angle change is defined in terms of flying qualities. For the situation of precision category flying with aileron control input only and rudder control free, the T class aircraft should not exceed a time of 1.3, 1.7, and 2.6 s to achieve a 60-deg bank angle change for flying quality levels of 1, 2, and 3, respectively. Similar requirements for the L and M aircraft class to achieve a less demanding 45-deg bank angle change is 1.4, 1.9, and 2.8 s for flying quality levels of 1, 2, and 3, respectively.

The military lateral-directional flying quality requirements stated earlier are just part of a more comprehensive set of specifications that considers both static and dynamic stability and control from the standpoint of flying qualities. References 2 and 3, like Ref. 4, are continually upgraded to reflect increased understanding in the technical community, based on lessons learned, of the essential qualities in an airframe for safe and efficient flight operations.

7.6 Yaw Damper for Stability Augmentation

7.6.1 Background

The principle of a yaw damper for the purpose of improving aircraft Dutch-roll modal damping will be illustrated using state variable feedback concepts. A yaw damper may be viewed as a form of stability augmentation with its goal the improvement of aircraft handling qualities. Rudder deflection proportional to aircraft yaw rate will act to augment the N_r and L_r dimensional stability derivatives in the aircraft's plant matrix. Aileron control feedback can alternatively be employed, but its use requires considerably more complex gain laws to achieve the same results as with rudder control feedback.

The use of a yaw damper as initially described will introduce potential difficulties at low-frequency values inasmuch as pilot application of rudder and/or aileron control to initiate a turn command would be countermanded by the negative yaw rate feedback of the yaw damper. The supplementary use of a washout filter in the feedback loop to correct for this influence will be described. Its use in yaw damper design is well established, i.e., since early developments (ca. 1950) to improve Dutch-roll damping on large swept-wing military aircraft.

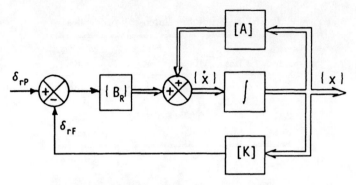

Fig. 7.10   Block diagram of a closed-loop system.

7.6.2 Basic Yaw Damper Concepts

An elementary form of the state vector block diagram, similar to Fig. 5.15, except modified by the addition of negative state vector feedback via a gain matrix is shown in Fig. 7.10. The use of rudder deflection for yaw damping may be implemented by means of a separate control tab on a manual rudder control or by introduction of the feedback signal as information at a summing junction in a power control system. In any event, command information to the rudder is assumed as coming from either pilot input or state vector feedback. The single lines in Fig. 7.10 represent scalar terms whereas the double lines represent vector quantities.

The elements shown in Fig. 7.10 are defined as follows:

$\{x\}$ = $(n \times 1)$ state vector
$[A]$ = unaugmented $(n \times n)$ plant matrix
$\{B_R\}$ = $(n \times 1)$ rudder control effectiveness vector
$[K]$ = $(1 \times n)$ feedback row (matrix) vector
δ_{r_P} = pilot input command to rudder control
δ_{r_F} = feedback input command to rudder control

The (4×4) lateral-directional plant is based on use of the primed stability derivatives (e.g., L'_p) in the manner used in other sections of Chapter 7. The state vector is of order (4×1), therefore the feedback matrix $[K]$ will be a (1×4) row vector in order to yield a scalar output, as shown in Fig. 7.10. The feedback matrix will initially be assumed as consisting only of a yaw rate entry K_r. Therefore, $[K]$ will have the form

$$[K] = [0 \quad 0 \quad 0 \quad K_r]$$

The dual rudder inputs alter the state equation slightly to

$$\{\dot{x}\} = [A]\{x\} + \{B_R\}(\delta_{r_P} + \delta_{r_F}) \qquad (7.22)$$

where the rudder deflection due to yaw damper feedback is defined as

$$\delta_{r_F} = -[K]\{x\} \qquad (7.23)$$

Combining the rudder feedback term into the state equation provides an augmented plant matrix in the form of

$$\{\dot{x}\} = [A - B_R K]\{x\} + \{B_R\}\delta_{r_p} \quad (7.24)$$

where the augmented plant matrix is $[A - B_R K]$.

Example 7.11

Find the value of yaw rate feedback, K_r, in the gain matrix of $[K]$ in order to increase the Dutch-roll modal damping to $\zeta = 0.30$ from the unaugmented value ($\zeta = 0.079$) as determined earlier in Example 7.4 for the DC-8 aircraft in a cruise configuration.

Solution involves using the unaugmented plant $[A]$ and rudder control effectiveness column vector $\{B_R\}$ from Examples 7.4 and 7.7. The augmented matrix can be established using MATLAB once a feedback row vector $[K]$ has been assumed followed by verification that the desired damping ratio has been obtained, i.e.,

```
% Establish the augmented matrix:
Kr = -0.580; K=[0 0 0 Kr]; AAUG=A-BR*K;
[Wn,Z]=damp(AAUG); disp(Z')
0.3000; % Dutch roll damping ratio
Disp(Wn')
1.4821; % Dutch roll natural frequency, rad/sec
-1.2502; % Roll mode real root, 1/sec
-0.0370; % Spiral mode real root, 1/sec
```

In finding the appropriate feedback constant, a manual iteration was done on the first two lines of the listing until the desired damping level was determined.

Note that K_r is negative in value (i.e., $K_r = -0.580$ s), which is primarily a result of the N'_{δ_r} stability derivative also being negative in value.

The Dutch-roll eigenvalue and eigenvector were also evaluated for the Example 7.11 case corresponding to the augmented yaw damper system with the feedback gain that provided the desired level of modal damping (i.e., $\zeta = 0.30$). A comparison summary of the Dutch-roll mode with and without yaw damping is listed in Table 7.5. The eigenvector for the airframe with yaw damper augmentation shows a slight change in mode shape when comparison is made with the original airframe. Because of the yaw damper installation, the ϕ/β ratio changed from 1.61 to 1.80 whereas the phase angle between ϕ and β was altered by approximately 14 deg. The effect of the improved Dutch-roll modal damping would be readily apparent in dynamically oriented handling qualities although the slight changes in mode shape would not be noticeable.

The roll response and spiral roots were altered only slightly by the addition of the yaw damper, and the impact on handling qualities might not be apparent to the pilot.

The movement of the complex-natured Dutch-roll roots with change in feedback gain is shown in Fig. 7.11 for the DC-8 aircraft considered in Example 7.11. In the range of K_r feedback gains shown on the root-locus plot, the roots traveled almost

Table 7.5 Effects of yaw damper on Dutch-roll: DC-8 aircraft, cruise condition, $M = 0.84$, $h = 33,000$ ft

Item	Basic Example 7.4	Augmented Example 7.11
Eigenvalue, s^{-1}	$-0.118 \pm 1.493i$	$-0.445 \pm 1.414i$
Damping, ζ	0.079	0.300
ω_n, rad-s^{-1}	1.497	1.482
Eigenvector		
β	1.00 ∠ 0.0 deg	1.000 ∠ 0.0 deg
p	2.412 ∠ 131.8 deg	2.661 ∠ 130.4 deg
ϕ	1.610 ∠ 37.3 deg	1.795 ∠ 22.6 deg
r	1.457 ∠ −86.8 deg	1.469 ∠ −73.1 deg

on a constant radius centered about the origin, corresponding to $\omega_n = 1.49$ rad/s for the example. The dimensionless damping ratio increased as the magnitude of the feedback gain was increased. This near circular-arc behavior of the root locus plot is very similar to the root migration for a second-order, single-degree-of-freedom harmonic oscillator as damping ζ is increased. The implication of this observation is that the N_r stability derivative in the yaw moment equation has a dominant influence on the root migration. Although the example is airframe specific, the traits exhibited in Fig. 7.11 are generic in nature.

The gain value considered in Example 7.11 to meet the stated goal of improving Dutch-roll damping applied at only that flight condition. An indication of how the gain could be approximately adjusted or varied as the flight environment changed can be obtained by considering the simplifed second-order yaw moment relation

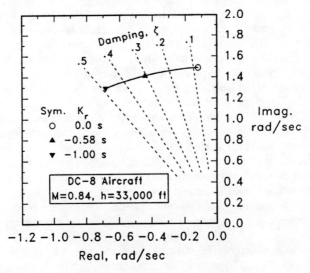

Fig. 7.11 Yaw damper root locus.

of Eq. (7.15) with a modification to reflect rudder command from both pilot and stability augmentation,

$$\ddot{\beta} - N_r \dot{\beta} + N_\beta \beta = -N_{\delta_r}(\delta_{r_P} + \delta_{r_F}) \tag{7.25}$$

Feedback would be modeled as

$$\delta_{r_F} = -K_r r = +K_r \dot{\beta}$$

which yields an approximation for the augmented system as

$$\ddot{\beta} - (N_r - N_{\delta_r} K_r)\dot{\beta} + N_\beta \beta = -N_{\delta_r} \delta_{r_P} \tag{7.26}$$

If the Dutch-roll damping change $\Delta \zeta \omega_n$ were to be kept constant, then

$$K_r N_{\delta_r} \cong \text{const}$$

or

$$K_r \propto 1/Q \tag{7.27}$$

If the improvement in dimensionless damping ratio, $\Delta \zeta$, due to augmentation were to be kept invariant, then

$$\Delta \zeta \cong \frac{+N_{\delta_r} K_r}{2\sqrt{N_\beta}} \cong \text{const}$$

or

$$K_r \propto 1/\sqrt{Q} \tag{7.28}$$

The actual determination of the gain variation with flight condition is aircraft specific and will depend on many effects that influence the variation of the stability derivatives, e.g., angle of attack, Mach number, altitude, and airframe configuration. In addition, the yaw rate sensor in Example 7.11 was assumed as aligned with the airframe's stability axis. This is a condition that can be satisfied at only that flight condition. Changes in angle of attack will introduce roll rate as well as yaw rate for a rate gyro with fixed alignment relative to the airframe. Consequently, the influence of gyro orientation on feedback gain selection must also be considered.

7.6.3 Application of a Washout Filter

There is a potential for a conflict with the simplifed yaw damper model described in Sec. 7.6.1. Pilot inputs will normally occur in a frequency range considerably lower than the Dutch-roll modal frequency; however, the yaw damper considered up to this point will provide rate-dependent feedback to the rudder regardless of the frequency. A pilot command to the rudder or a turn rate established during a steady turn (e.g., while in a holding pattern) will incur an augmentation command to suppress the airframe's yaw rate. A Bode plot (Fig. 7.12) depicts the influence of the yaw damper on the gain and phase of the $G_{r\delta_r}(s)$ transfer function. It is readily

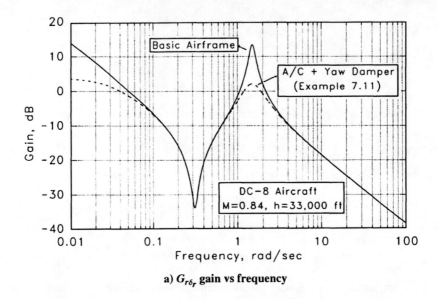

a) $G_{r\delta_r}$ gain vs frequency

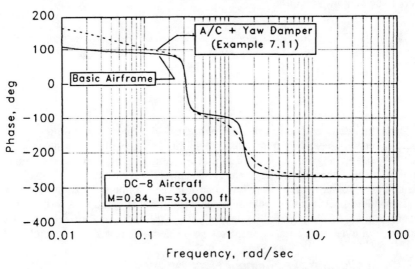

b) $\Delta\phi_{r\delta_r}$ phase vs frequency

Fig. 7.12   Bode plot showing yaw damper influence.

LATERAL-DIRECTIONAL DYNAMICS

apparent that the modal damping has been improved in the neighborhood of the Dutch-roll frequency ω_d in accord with the design goal of Example 7.11. However, the ability of the rudder to develop yaw rate at frequencies below approximately $0.25\,\omega_d$ has been lessened.

A suitable approach to avoid the potential difficulty of using a single control surface for multiple purposes (i.e., both control and stability augmentation) is to include an appropriate filter in the feedback circuit. For the case of a yaw damper, the influence of state vector feedback can be effectively removed at lower frequencies by introducing a washout (or high-pass) filter in the feedback circuit.

The washout filter resembles a low-pass filter in the denominator while containing an s in the numerator to decrease the gain below the corner frequency. The transfer function is defined as

$$G(s) = \frac{s}{1+Ts} \qquad (7.29)$$

with the corresponding frequency response being

$$G(i\omega) = \frac{i\omega}{(1+i\omega T)} = \frac{\omega}{[1+(\omega T)^2]^{1/2}} e^{i(\phi+\pi/2)} \qquad (7.30)$$

where

$$\phi = \tan^{-1}(-\omega T)$$

The s in the numerator causes a $+\pi/2$ difference in phase angle when compared to the low-pass filter. The corresponding gain plot for the washout filter with an assumed time constant of $T = 0.5$ s is shown in Fig. 7.13 as a buildup that combines the numerator and denominator behaviors into a net gain value in accord with well-known Bode plot construction rules (cf. Ogata[5]).

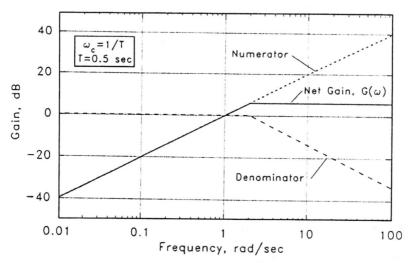

Fig. 7.13 Gain plot for a washout filter.

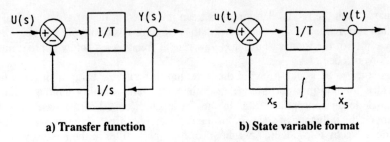

a) Transfer function b) State variable format

Fig. 7.14 Washout filter block diagrams.

At the corner frequency, when $\omega = 1/T$, the true gain value will be 3 dB less than the straight line fairing approximation. The net gain has a slope of +20 dB per frequency decade when the frequency is less than the corner frequency. It is this trait which provides the high-pass character of the washout filter. The gain remains approximately constant at frequencies above the corner frequency, thereby allowing the effective passage of the input information in that frequency range without any frequency-dependent gain change. From Eq. (7.30), it is evident that the approximate gain value is $(1/T)$ when $\omega \gg 1/T$.

Block diagrams showing the principle of a washout filter from two different viewpoints are shown in Fig. 7.14. Figure 7.14a shows in Laplace transform notation the relation between input $Y(s)$ and output $U(s)$ that corresponds to Eq. (7.29), i.e.,

$$Y(s) = -(1/Ts)Y(s) + (1/T)U(s)$$

which factors to

$$Y(s) = \frac{s}{(1+Ts)} U(s) = G(s)U(s) \tag{7.31}$$

In the time domain, Eq. (7.31) readily translates to

$$(\dot{y}) + (1/T)y = (1/T)(\dot{u}) \tag{7.32}$$

The problem with Eq. (7.32) is that it is not in a convenient state vector arrangement due to the derivative action on the input. A way to circumvent this difficulty is to define y in terms of u and a new state variable x_5, i.e.,

$$y = (1/T)u - (1/T)x_5 \tag{7.33}$$

along with

$$\dot{x}_5 = -(1/T)x_5 + (1/T)u \tag{7.34}$$

It is straightforward to show that Eqs. (7.33) and (7.34) satisfy Eq. (7.32). The added state variable x_5, which in general will not have a physical meaning, is a technique used to preserve a state variable solution procedure in the presence of inputs with time derivatives. Figure 7.14b is the state variable format of the washout filter and is consistent with Eqs. (7.33) and (7.34).

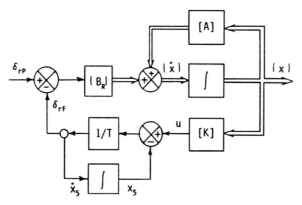

Fig. 7.15 Block diagram, yaw damper plus washout filter.

The use of a washout filter in the feedback circuit of the yaw damper is shown in Fig. 7.15. The state variable format for the washout filter has been sketched into the feedback loop between the output of the gain matrix $[K]$ and the negative feedback summation of δ_{r_F}. An intermediate change in terminology will be used to illustrate the procedure for including a washout filter into the state equation formulation.

First, let

$$\delta_{r_F} = -y \quad \text{and} \quad u = [K]\{x\}$$

But

$$y = (1/T)[K]\{x\} - (1/T)x_5$$

Substitution of these relations into Eq. (7.22) as

$$\{\dot{x}\} = [A]\{x\} + \{B_R\}\delta_{r_F} + \{B_R\}\delta_{r_P} \tag{7.22a}$$

alters the state equation relations to

$$\{\dot{x}\} = [A - (1/T)B_R K]\{x\} + (1/T)\{B_R\}x_5 + \{B_R\}\delta_{r_P} \tag{7.35}$$

plus the added term to account for x_5

$$\dot{x}_5 = (1/T)[K]\{x\} - (1/T)x_5$$

The presence of the washout filter has increased the size of the state vector from (4×1) to (5×1) due to the addition of the x_5 term. The state equation can be viewed in its partitioned form as equivalent to Eq. (7.35). This approach preserves the matrix algebra tools used up to now to solve the eigenvector problem. Here,

$$\left\{\begin{array}{c}\{\dot{x}\} \\ \hline \dot{x}_5\end{array}\right\} = \left[\begin{array}{c|c} A - (1/T)B_R K & (1/T)B_R \\ \hline (1/T)K & -(1/T) \end{array}\right] \left\{\begin{array}{c}\{x\} \\ \hline x_5\end{array}\right\} + \left\{\begin{array}{c}B_R \\ \hline 0\end{array}\right\}\delta_{r_P} \tag{7.36}$$

Equation (7.36) makes possible a lateral-directional analysis using software tools such as MATLAB for the purpose of improving Dutch-roll damping by identifying a suitable feedback matrix [K] plus preserving the airframe's low-frequency behavior by the use of a washout filter with a time constant T.

Example 7.12

Add a washout filter to the yaw damper feedback circuit considered for the DC-8 aircraft in Example 7.11 and find the change in Dutch-roll damping due to its presence.

1) Select the corner frequency of the washout filter at 0.33 of the Dutch-roll natural frequency, i.e.,

$$\omega_c = (0.33)(1.482 \text{ s}^{-1}) = 0.489 \text{ rad/s}$$

2) The yaw damper feedback gain in Example 7.11 was $K_r = -0.580$ s to yield a value for Dutch-roll damping of $\zeta = 0.30$.

3) Apply Eq. (7.36) to find the effect of the washout filter on modal damping. Using MATLAB, one finds

```
K = [0  0  0  -0.580]; % Yaw damper feedback
(1/T) = 0.489; % Washout filter corner frequency
% Establish plant with feedback and washout, Eq. (7.36)
AWO = [A-(1/T)*BR*K, (1/T)*BR; (1/T)*K, -(1/T)]
[AWn,AZ] = damp(AWO); disp(AZ')
0.1793 ; % Dutch-roll damping ratio
disp(AWn')
1.438 ; % Dutch-roll undamped modal frequency, rad/sec
-1.250 ; % Roll mode real root, 1/sec
-0.003 ; % Spiral mode real root, 1/sec
-0.549 ; % Fifth real root due to the washout filter
```

In Example 7.12, a washout filter was added to the yaw damper feedback circuit described in Example 7.11. The corner frequency was selected as approximately 0.33 of the Dutch-roll natural frequency as a compromise between the low-frequency domain where the feedback to the rudder control should be negligible and the Dutch-roll frequency domain where the yaw damper is expected to provide stability augmentation.

The Bode plot (Fig. 7.16) shows the influence of the washout filter in achieving the stated goals. The washout filter altered the high-pass gain by approximately $(1/T)(= \omega_c = 0.49$ rad/s), which in turn decreased the effectiveness of the yaw damper gain. It can be seen in Fig. 7.16a, that the reduction in Dutch-roll damping from $\zeta = 0.30$ to $\zeta = 0.18$ due to the presence of the washout filter is evident as an increase in decibels of the gain $G_{r\delta_r}$ in the neighborhood of the Dutch-roll frequency. However, in the low-frequency range, corresponding to aircraft operations in steady turns and/or slow pilot control inputs, both the $G_{r\delta_r}$ gain and the phase angle functions for the yaw damper plus washout filter are nearly identical to those for the basic airframe.

The Dutch-roll modal damping could be increased to the original stated goal of Example 7.11 by altering the feedback constant K_r to account for the high-pass

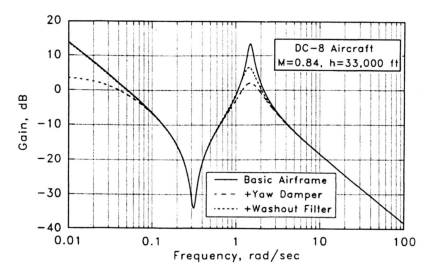

a) $G_{r\delta_r}$ gain vs frequency

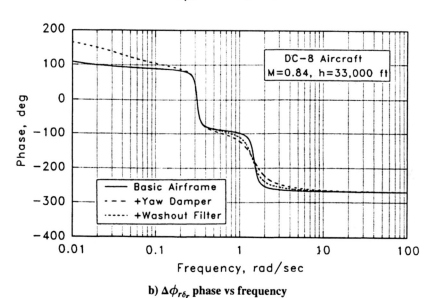

b) $\Delta\phi_{r\delta_r}$ phase vs frequency

Fig. 7.16 Bode plot showing washout filter influence.

gain $(1/T)$ from the washout filter, which is a series member in the feedback loop. In this case, Dutch-roll damping of approximately $\zeta = 0.30$ could be re-established by increasing K_r from -0.58 to -1.18 s, a factor corresponding to the reciprocal of the filter's corner frequency. These comments are airframe specific but are mentioned in order to explain elementary principles.

In summary, the presence of the washout filter has made possible the dual use of rudder control, i.e., both for near-steady control inputs by the pilot and for stability augmentation to improve the Dutch-roll modal damping.

7.7 Controllability and Pole Placement

7.7.1 Background

When the concept of a yaw damper was presented in Sec. 7.6 to improve Dutch-roll damping, only the rudder was considered as a controller. A feedback law based on using control input that was proportional solely on yaw rate was considered. The gain constant was found to meet the desired damping goals using an elementary root locus study technique, as illustrated by Example 7.11. A question arises as to whether the aileron could have been used as an alternate input controller for the yaw damper feedback law. However, inasmuch as improvements in yaw damping are primarily dependent on increasing the magnitude of the N_r stability derivative (when designing a stability augmentation system), one would expect that the aileron might not be the control of choice because usually $|N_{\delta_a}| \ll |N_{\delta_r}|$.

Although the preceding remarks are intuitively correct, it is appropriate to recognize the following.

1) Both aileron and rudder controls can be shown as satisfying the lateral-directional dynamics with regard to controllability criteria, a principle introduced in Sec. 6.4.

2) Inasmuch as the aileron will be found as satisfying controllability criteria, then it should be possible by pole placement methods to find a gain law to achieve the same yaw damper characteristics as was done for the rudder when using $K = [0 \quad 0 \quad 0 \quad K_r]$ in Sec. 7.6.2.

3) Finally, if a gain law could be found using the aileron as a controller, it might turn out that the gain values required were not physically practical or realizable.

The following material will expand on the concept of controllability in a more formal manner than the approach by Gilbert,[6] which is based on showing modal participation from each controller (cf. description in Sec. 6.4). Following this, the concept of pole placement for a single-input/single-output (SISO) system will be considered using Ackerman's formula.[5] Application of Ackerman's formula will finally be made to find a feedback gain law for an aileron controlled yaw damper to achieve the same improvement in Dutch-roll damping as was attained with the rudder in Sec. 7.6. Remarks concerning the required gain values will confirm the intuitive feeling that the use of a rudder as a yaw damper controller is preferred over the aileron in spite of both controls satisfying controllability criteria.

7.7.2 Controllability

A formal definition of system controllability, stated earlier in Sec. 6.4, is as follows.

Definition: A system is said to be controllable if it is possible by means of an unconstrained controller to transfer the physical system between two arbitrarily specified finite states in a finite time, i.e.,

$$\{x(t_0)\} \Rightarrow \{x(t_f)\} \text{ due to } u(t) \text{ in } t_0 \leq t \leq t_f$$

Initially, let us assume that a single control $u(t)$ is being applied to a physical system to change its condition of state. It has been shown previously [cf. Eq. (5.61)] that for a linear, time invariant (LTI) system as described by

$$\dot{x} = Ax + Bu$$

the response is given by

$$\{x(t)\} = e^{At}\{x(0)\} + e^{At}\int_0^t e^{-A\tau}Bu(\tau)d\tau \qquad (7.37)$$

Inasmuch as the states are arbitrary, the final state can be selected as $\{x(t)\} = \{0\}$, which then simplifies Eq. (7.37) to

$$\{x(0)\} = -\int_0^t e^{-A\tau}Bu(\tau)d\tau \qquad (7.38)$$

The question still remains whether a control law exists to satisfy Eq. (7.38), and if it does, what are the requirements for its existence.

The transition matrix under the integrand, which is an infinite series, can be re-expressed into a finite series by the Cayley–Hamilton theorem (cf. Appendix D). Under the assumption that the plant matrix $[A]$ is of order $(n \times n)$, then the transition matrix becomes a series in increasing powers of $[A]$ from $[A]^0$ up to $[A]^{n-1}$, i.e.,

$$e^{-A\tau} = \sum_{k=0}^{n-1} f_k(\tau) A^k \qquad (7.39)$$

The time-dependent variables, $f_k(\tau)$, are understood as existing but, as will be seen, need not be determined in order to establish the controllability concept. Substitution of Eq. (7.39) into Eq. (7.38) provides an expression for $\{x(0)\}$ based on a finite series,

$$\{x(0)\} = -\sum_{k=0}^{n-1} A^k B \int_0^t f_k(\tau) u(\tau) d\tau \qquad (7.40)$$

Next consider the integral expressions as having solutions given by

$$\int_0^t f_k(\tau) u(\tau) \, d\tau = g_k \qquad \text{for } k = 0, 1, \ldots, n-1$$

Then Eq. (7.40) can be viewed in a symbolic form as

$$\{x(0)\} = -\sum_{k=0}^{n-1} A^k B g_k = -[B \ AB \ \cdots \ A^{n-1}B] \begin{Bmatrix} g_0 \\ g_1 \\ \vdots \\ g_{n-1} \end{Bmatrix} \quad (7.41)$$

For a solution of the $\{g\}$ column vector to exist, the $(n \times n)$ controllability matrix in Eq. (7.41) must have a rank of n. This criteria (cf. Appendix D) assures linear independence of the n columns in the controllability matrix, a requirement for a solution to exist. If there were m controls under consideration, then the $(n \times 1)\{g\}$ column vector in Eq. (7.41) would increase in size to an $(nm \times 1)$ column vector. The corresponding controllability matrix would become rectangular in form with order of $(n \times nm)$. The rank of the controllability matrix must still be n for the controllability criteria to be satisfied by the m controllers.[5]

In summary, when a plant described by the $(n \times n)$ matrix A is acted on by m controllers through its interaction with a B matrix of order $(n \times m)$, controllability is assured if the rank of the controllability matrix is n, where

$$[B \ AB \ A^2B \ \cdots \ A^{n-1}B] = \text{controllability matrix} \quad (7.42)$$

7.7.3 Pole Placement

The use of state variable feedback to alter the location of a system's eigenvalues is a common approach for improving a system's dynamic performance. If the system were of low order (i.e., $n \leq 3$), it is possible by direct substitution of the gain matrix into the characteristic polynomial to find suitable gain constants. The gain row matrix for a (3×3) system is

$$K = [k_1 \ k_2 \ k_3] \quad (7.43)$$

which leads to the augmented (compensated) plant matrix as

$$A_c = A - BK \quad (7.44)$$

The characteristic equation for A_c takes the form of

$$|sI - A + BK| = (s - \mu_1)(s - \mu_2)(s - \mu_3) = 0 \quad (7.45)$$

which corresponds to polynomials in powers of s from 0 to 3. By equating coefficients with like powers, the gain matrix may be found. This method becomes labor intensive when the order of the plant matrix exceeds $n = 3$.

The feedback law to place the poles of a SISO system from an initial pole location of $\lambda_1, \lambda_2, \ldots, \lambda_n$ to a second location $\mu_1, \mu_2, \ldots, \mu_n$ may be found by use of Ackerman's formula.[5] It will be seen that plant controllability, in the sense as described in Sec. 7.7.2, is a requirement for pole placement by Ackerman's formula. The characteristic equation [Eq. (7.45)] when expanded in powers of s

has the form of

$$|sI - A + BK| = |sI - A_c| = s^n + a_{n-1}s^{n-1} + \cdots + a_1 s + a_0 = 0$$

Cayley–Hamilton's theorem (cf. Appendix D), which states that the compensated plant matrix A_c satisfies its characteristic equation in a matrix sense, can be applied to the characteristic equation to yield

$$A_c^n + a_{n-1}A_c^{n-1} + \cdots + a_1 A_c + a_0 I = [0] \quad (7.46)$$

The polynomial matrix expression [Eq. (7.46)] also can be expressed in a matrix functional notation as

$$\alpha_c(A_c) = [0]$$

Continuing, it is clear that because $A \neq A_c$,

$$\alpha_c(A) = A^n + a_{n-1}A^{n-1} + \cdots + a_1 A + a_0 I \neq [0]$$

Before considering Eq. (7.46) to derive Ackerman's formula, let us list several powers of the compensated plant in terms of the feedback law, Eq. (7.45), i.e.,

$$A_c = A - BK$$
$$A_c^2 = (A - BK)^2 = A^2 - ABK - BKA_c$$
$$A_c^3 = (A - BK)^3 = A^3 - A^2BK - ABKA_c - BKA_c^2$$

The characteristic equation of the compensated plant, under the assumption of $n = 3$, alternatively can be expressed as

$$[0] = \left(A^3 - A^2BK - ABKA_c - BKA_c^2\right) + a_2(A^2 - ABK - BKA_c)$$
$$+ a_1(A - BK) + a_0 I$$

which can be arranged to an alternate form of

$$\alpha_c(A) = a_1 BK + a_2(ABK + BKA_c) + A^2 BK + ABKA_c + BKA_c^2$$

to then become

$$\alpha_c(A) = [B \ \ AB \ \ A^2 B] \begin{bmatrix} a_1 K + a_2 K A_c + K A_c^2 \\ a_2 K + K A_c \\ K \end{bmatrix}$$

The controllability matrix will be recognized in the preceding matrix factoring. As the single-input system has been assumed to be controllable, an inverse of the controllability matrix exists. Therefore, one finds that

$$[B \ \ AB \ \ A^2 B]^{-1} \alpha_c(A) = \begin{bmatrix} a_1 K + a_2 K A_c + K A_c^2 \\ a_2 K + K A_c \\ K \end{bmatrix}$$

After premultiplying both sides of this solution by [0 0 1], the K gain matrix is obtained as

$$K = [0 \ 0 \ 1][B \ AB \ A^2B]^{-1}\alpha_c(A) \tag{7.47}$$

Ackerman's formula for finding the gain of a controllable SISO system is Eq. (7.47) for the case of $n = 3$. The extension of Eq. (7.47) to higher orders of n is straightforward.

7.7.4 Aileron as a Yaw Damper

Although the ability of the aileron to act as a controller is not in doubt, it is useful to illustrate by example that the aileron satisfies controllability criteria.

Example 7.13

Determine the rank of the controllability matrix for the aileron of the DC-8 aircraft in a cruise configuration.

1) The aileron control column vector, using information from Example 7.9 (and Appendix B.4), is

$$\{B\} = [0.0 \ 2.120 \ 0.0 \ 0.065]^T$$

The lateral-directional plant matrix from Example 7.4 is

$$[A] = \begin{bmatrix} -0.0869 & 0.0 & 0.0390 & -1.0 \\ -4.424 & -1.184 & 0.0 & 0.335 \\ 0.0 & 1.0 & 0.0 & 0.0 \\ 2.148 & -0.021 & 0.0 & -0.228 \end{bmatrix}$$

for a state vector corresponding to

$$\{x\} = [\beta \ p \ \phi \ r]^T$$

2) Form the controllability matrix in accord with Eq. (7.42)

```
C = [B  AB  A²B  A³B]
% Find the rank of matrix C
k = rank(C); disp(k)
4.0 ; % Verifies that rank is 4
```

This result confirms that the lateral-directional system is controllable by the use of the aileron. The rank of the controllability matrix was found using the *rank* function in MATLAB.

Inasmuch as controllability of the lateral-directional system by the aileron has been confirmed by Example 7.13, Ackerman's formula can be used to identify a feedback law using the aileron as a controller, which will provide matching pole placement to that when using the rudder for a yaw damper (cf. Example 7.11).

Example 7.14

Determine a feedback law using the aileron as a controller to match the yaw damper design of Example 7.11. Airframe is the DC-8 in a cruise configuration.

1) The characteristic polynomial corresponding to the roots of Example 7.11 that provided Dutch-roll modal damping of $\zeta = 0.300$ is

$$P = s^4 + 2.1769s^3 + 3.3883s^2 + 2.8690s + 0.1017$$

This polynomial can be converted to a matrix relation using the Cayley–Hamilton theorem.

2) Form the $\alpha_c(A)$ matrix (F) by

```
F = A^4 + P(2)*A^3 + P(3)*A^2 + P(4)*A + P(5)*eye(A);
% C = Controllability matrix from Example 7.13
% Gain matrix when using the aileron as a controller is:
K = [0 0 0 1]*inv(C)*F; disp(K)
-12.670  0.154  -0.139  5.401
```

3) The steps involved in the use of Ackerman's formula to find the gain matrix of a SISO system could also have been obtained using the *acker* function in MATLAB, i.e.,

```
K = acker(A,B,R);  % R = desired pole locations
disp(K)
-12.670  0.154  -0.139  5.401 ; % Agreement is shown.
```

Example 7.14 confirmed that the aileron, when used as a yaw damper controller, could theoretically produce the same degree of modal damping as was shown in Example 7.11 when using the rudder.

Comparison of the gain constants when using the rudder (KR) and the aileron (KA) will illustrate why the rudder is the preferred controller for stability augmentation as a yaw damper.

From Example 7.11

$$KR = [0 \quad 0 \quad 0 \quad -0.580 \text{ s}]$$

and from Example 7.14

$$KA = [-12.670 \quad 0.154 \text{ s} \quad -0.139 \quad 5.401 \text{ s}]$$

where the state vector is

$$\{x\} = [\beta \quad p \quad \phi \quad r]^T$$

and the feedback law is either

$$\delta_{r_F} = -KR\{x\}$$

or

$$\delta_{a_F} = -KA\{x\}$$

Either feedback law will result in the same degree of stability augmentation as a yaw damper. However, the large values in the KA gain matrix are not physically realizable as can be seen by noting that a 2-deg sideslip excursion during a Dutch-roll oscillation would require 26 deg of aileron deflection combined with an added 16 deg of aileron deflection lagging by approximately 70 deg due to an estimated 3-deg/s yaw rate. These estimates are based on the KA gain matrix interacting with the augmented eigenvector information as shown in Table 7.5.

In summary, the aileron theoretically could be used as a controller for a yaw damper. However, the large gain values required to achieve the same degree of stability augmentation as provided by a rudder controller makes its consideration for sole use as a yaw damper not practical. As has been stated previously, the results obtained are airframe specific; however, the conclusions are generic in nature.

References

[1]McRuer, D., Ashkenas, I., and Graham, D., *Aircraft Dynamics and Automatic Control*, Princeton Univ. Press, Princeton, NJ, 1973, p. 483.

[2]Anon., "Part 23, Airworthiness Standards: Normal Utility, Acrobatic and Commuter Category Airplanes," *Federal Aviation Regulations*, U.S. Government Printing Office, Washington, DC, Feb. 1991.

[3]Anon., "Part 25, Airworthiness Standards: Transport Category Airplanes," *Federal Aviation Regulations*, U.S. Government Printing Office, Washington, DC, Feb. 1991.

[4]Anon., "Flying Qualities of Piloted Airplanes," MIL SPEC MIL-F-8785C, U.S. Government Printing Office, Washington, DC, Nov. 1980.

[5]Ogata, K., *Modern Control Engineering*, 2nd ed., Prentice–Hall, Englewood Cliffs, NJ, 1990, pp. 782–784.

[6]Gilbert, E. G., "Controllability and Observability in Multivariable Control Systems," *Journal of the Society for Industrial and Applied Mathematics* (SIAM), Ser. A: Control, Vol. 1, No. 2, 1663, pp. 128–151.

Problems

7.1. Consider an airplane in a constant altitude, steady turn to the left with a 30-deg bank angle ($\phi = -30$ deg). A 5-deg lateral control deflection is applied as a step input at $t = 0$ and maintained until a bank angle change of $\Delta\phi = +60$ deg takes place. At that time, the pilot neutralized the lateral control system to stop the rolling action. If $L_p = -1.516$ s^{-1} and $L_{\delta_a} = 21.2$ rad-s^{-2}, what would be the time required in seconds (± 0.01 s) for the bank angle change? Assume pure rolling motion.

7.2. The aircraft of Problem 7.1 is initially in trimmed wings-level flight. The pilot initiates a wing rocking motion by a harmonic aileron deflection of 2-deg amplitude with a period of 2 s. The pilot continues the control oscillation to attain a steady harmonic response. Under the assumption of pure rolling motion, what is the bank angle response relative to the input?

7.3. Using the phasor of Fig. 7.3 as an initial condition to excite only the Dutch-roll mode, develop a time-history plot for $t = 0$–20 s to show the homogeneous response of sideslip (β) and bank angle (ϕ).

7.4. Apply a rudder kick of 1.0 deg-s to the (3 × 3) modal approximation for the DC-8 at a cruise configuration. Find the resultant steady-state bank angle. Note: Although bank angle ϕ is not a state vector component in the (3 × 3) approximation, the $G_{\phi\delta_r}(i\omega)$ transfer function is related to $G_{p\delta_r}(i\omega)$ by the scaling factor of $(i\omega)^{-1}$.

7.5. Consider a 5-deg aileron deflection applied as a 2-s pulse to the DC-8 aircraft in a cruise-flight condition. Using the (3 × 3) modal approximation to suppress the spiral mode's influence, estimate the resulting steady-state bank angle.

7.6. Confirm that the coefficients in the characteristic polynomial equation corresponding to the (3 × 3) lateral-directional approximation, Eq. (7.14), are given as

$$\lambda^3 + a_2\lambda^2 + a_1\lambda + a_0 = 0$$

where

$$a_2 = -(L'_p + N'_r)$$
$$a_1 = -(L'_p N'_r - L'_r N'_p + N'_\beta)$$
$$a_0 = -(L'_\beta N'_p - L'_p N'_\beta)$$

7.7. For the washout filter depicted by Fig. 7.14, confirm that Eqs. (7.33) and (7.34) satisfy the state equation [Eq. (7.32)] where the input function is given as a time derivative du/dt.

7.8. Consider the A-4D aircraft in trimmed level flight, $M = 0.6$, and $h = 15{,}000$ ft as defined by condition 3 in Appendix B.
  a) Design a yaw damper using yaw rate feedback to the rudder to increase Dutch-roll modal damping to $\zeta = 0.30$.
  b) Compare the improvement in the $G_{r\delta_r}(i\omega)$ transfer function on a Bode plot due to the presence of the yaw damper.
  c) Modify the yaw damper by including a washout filter in the feedback loop to improve the low-frequency behavior while still retaining a modal damping level of $\zeta = 0.30$. Confirm the design solution on an appropriate Bode plot.

7.9. Consider a second-order harmonic oscillator system with real roots of $\lambda_1 = -1$ and $\lambda_2 = -2$. The control vector is given by

$$\{B\} = [0 \quad 1]^T$$

Find a state vector feedback law to move the characteristic roots to $\mu_1 = -3$ and $\mu_2 = -4$.

a) Find the gain matrix by comparing information contained in the characteristic polynomials.

b) Use Ackerman's formula to verify the result of Problem 7.9a.

Remark: Analysis of a system with real roots, where the matrix order is (2 × 2), by hand calculations will reinforce the remarks of Sec. 7.7.3 where mention is made of software algorithms such as *acker* in MATLAB becoming appropriate when the order of a plant matrix being considered for the pole placement of a SISO system is greater than (3 × 3).

8
Nonlinear Dynamics

8.1 Background

Before their first flight, much concern was given by the Wright brothers to the problem of controllability (cf. Wolko[1]). As a result, their first powered vehicle had inherently poor static stability characteristics on the assumption that a man-in-the-loop could maintain control of the aircraft. They were correct for their era, and the rest is history. However, forethought in design evolution does not always follow such a path. The advent of slender, thin-winged transonic and supersonic aircraft introduced new flight mechanic concerns such as tuck under and pitch up being routinely encountered and subsequently corrected by either aerodynamic tailoring or stability augmentation. A like term entitled inertial cross coupling became evident as fighter aircraft progressed into the Century series during the 1950s and, in particular, with the F-100 aircraft where several renowned test pilots were killed while performing high-performance rolling pull-up maneuvers. An early treatment by Phillips[2] demonstrated the existence of critical rolling velocities where the coupled pitching and yawing oscillations became unstable. The explanation of the subject is now a classical topic in flight mechanic references (e.g., Blakelock,[3] Etkin,[4] and Babister[5]). At the present time, aircraft are experiencing a dynamic response at high angles of attack, called wing rock, which although generally not fatal, is disconcerting and deserving of simple explanations. Possibly 20 years from now, wing rock will be well understood in a manner similar to inertial cross coupling today.

Material in Chapters 4, 6, and 7 related to aircraft flight mechanics where the equations of motion were linearized and small perturbation analyses ensued. The initial considerations in the following sections will be related to the equations of motion being broadened to include nonlinear coupling while retaining linear aerodynamics where feasible.

In Sec. 8.2, flight constraints will be applied to the airframe during a pilot-controlled roll maneuver, the inertial coupling terms in Euler's equations will be included, and a stability analysis based on the migration of the eigenvalues (a form of a root locus study) as roll rate is varied will ensue. In this manner, it will be possible to develop a roll-rate-dependent stability boundary for the onset of an unstable, inertial cross coupling between the longitudinal and directional modes of the airframe.

Section 8.3 will describe the wing-rock phenomena experienced by many high-performance jet aircraft. There are several explanations for the motion, but the treatment considered here will relate to a limit-cycle wing-rock motion due to coupling between the stable longitudinal short-period mode and a lightly unstable Dutch-roll mode. The coupling will occur due to the inertial terms in the equations of motion. These inertial influences were considered as higher order terms and assumed as being negligible when a small perturbation analysis was applied followed by a linearization process to describe airframe flight dynamics in Chapter 4.

Solution of the nonlinear sets of coupled equations will be obtained by numerical analysis using a Runge–Kutta integration method in order to identify the onset of a limit cycle that is stable motion in the sense of Lyapunov (cf. Sec. 5.7).

The principles of aircraft stall dynamics will be introduced in Sec. 8.4. Unlike the nonlinear analyses of Secs. 8.2 and 8.3, which involve linear aerodynamics, this treatment will use a nonlinear aerodynamic model. Some aircraft encounter an oscillation during a stall demonstration that is explained as the aircraft's wing stalling and unstalling while the longitudinal control is held fixed. The modeling will yield a limit cycle (known as stall porpoising), which like wing rock, is also considered as stable motion in the sense of Lyapunov.

These three examples of nonlinear dynamics are meant to be an introduction to flight dynamics beyond the realm of the idealized form represented by the coverage in Chapters 6 and 7. Although the majority of aircraft flight can be adequately modeled by linearity considerations, many portions of the flight envelope involve more complex nonlinear principles. It is hoped that one's understanding of the mechanisms leading to the various dynamic behaviors is broadened by making simple studies as presented in this chapter.

8.2 Inertial Cross Coupling

As the design of high-performance aircraft evolved, wings became thinner and more of the mass became concentrated in the fuselage. Consequently, the roll mass moment of inertia I_x tended to decrease in relative value whereas the pitch and yaw mass moments of inertia, I_y and I_z, respectively, tended to increase. As will be seen, this design trend allowed the coupling between the lateral-directional and longitudinal equations of motion to become important at large values of roll rate. To develop this thought further, let us examine Euler's moment equations, from Chapter 4,

$$\begin{Bmatrix} M_x \\ M_y \\ M_z \end{Bmatrix} = \begin{Bmatrix} I_x \dot{p} - I_{xz} \dot{r} \\ I_y \dot{q} \\ I_z \dot{r} - I_{xz} \dot{p} \end{Bmatrix} + \begin{Bmatrix} qr(I_z - I_y) - pq I_{xz} \\ pr(\underline{I_x - I_z}) + (p^2 - r^2) I_{xz} \\ pq(\underline{I_y - I_x}) + qr I_{xz} \end{Bmatrix} \qquad (4.30)$$

When linearization occurred with respect to the inertial related expressions (cf. Sec. 4.5), the second column vector on the right-hand side of Eq. (4.30) was dropped. Stepping back now, reconsider the second column vector and note that the underlined terms relating to $(I_x - I_z)$ and $(I_y - I_x)$ inertia differences in the equation will increase in magnitude as I_x becomes proportionately smaller. The underlined nonlinear quantities are of small import at low roll rate values. It will also be seen in Eq. (4.30) that a weak coupling exists through the I_{xz} terms.

As the state vector term of roll rate p increases, such as during a full authority rolling pull-up maneuver, the underlined quantities would increase in magnitude and develop strong cross coupling between the pitch and yaw moments. When roll rate is zero, the equations essentially become uncoupled, thereby supporting a common assumption that aircraft longitudinal dynamics (short-period and phugoid modes) is isolated or distinct from the lateral-directional dynamics (Dutch-roll, roll, and spiral modes).

The equations relating aerodynamic forces to body-axis-oriented linear momentum conservation principles were also described in Chapter 4 by

$$\begin{Bmatrix} F_x \\ F_y \\ F_z \end{Bmatrix} = mV \begin{Bmatrix} \frac{d(u/V)}{dt} \\ \frac{d\beta}{dt} \\ \frac{d\alpha}{dt} \end{Bmatrix} + mV \begin{Bmatrix} q\alpha - r\beta \\ r - p\alpha \\ p\beta - q \end{Bmatrix} \quad (4.29a)$$

The velocity perturbations are defined by $u(t)/V$ in the axial x direction, $\beta(t) = v(t)/V$ in the lateral y direction, and $\alpha(t) = w(t)/V$ in the vertical z direction. The body-axis angular rate perturbations are defined as p, q, and r for rates about the x, y, and z body-axis directions, respectively.

In the treatment to follow, several simplifying assumptions will be made with respect to the force and moment equation relations [Eqs. (4.29) and (4.30)] to keep the problem tractable while retaining the essentials. In this manner, the underlying physical principles that control the stability traits of inertial cross coupling will be preserved. Assumptions will include the following.

1) The airplane body axes are principal axes, i.e., $I_{xz} = 0$.
2) Aircraft freestream velocity V is constant while rolling.
3) Roll rate p will be constant at p_0 during the consideration of a stability analysis.

Assumption 1 will simplify the cross coupling considerations between the pitch and yaw moment equations without seriously compromising the nature of the final conclusions. The F_x chord force relation will not be considered because it is tacitly assumed by assumption 2 that the aircraft forward velocity is constant during the stability analysis at a specified value of roll rate p_0. Assumption 3 is based on the premise that the applied rolling moment is pilot controlled such that the roll rate is maintained at a constant value in spite of the influence of the inertial $(I_z - I_y)$ term in the roll moment equation. This latter assumption will remove roll rate as a time-dependent variable in the subsequent stability analysis. It will be seen that roll rate values p_0 will influence the characteristic roots (i.e., eigenvalues) of a coupled set of equations in a manner akin to varying feedback gain in the analysis of a closed-loop system. Finally, the constraints imposed by assumptions 2 and 3 permit the dropping of both the roll moment M_x and chord force F_x equilibrium conditions in the subsequent analyses.

Next let us restate the equilibrium equations, Eqs. (4.29) and (4.30), subject to assumptions 1–3, using the dimensional derivative notation as defined in Sec. 4.6. In the discussion to ensue, we will assume that a stability analysis will be considered in an open-loop form; i.e., no control derivatives will be included. Therefore, the four relations are as follows.

1) For normal force, define $Z_w = Z_\alpha/V$ and neglect the $Z_{\dot\alpha}$ and Z_q derivatives. Hence, we obtain

$$\dot\alpha = Z_w \alpha + q - p_0 \beta$$

2) For pitching moment, define $\mu_1 = (I_z - I_x)/I_y$. Therefore,

$$-M_{\dot\alpha}\dot\alpha + \dot q = M_\alpha \alpha + M_q q + p_0 \mu_1 r$$

3) For side force, define $Y_v = Y_\beta/V$ and neglect the Y_p and Y_r derivatives. The side-force equation then becomes

$$\dot{\beta} = p_0 \alpha + Y_v \beta - r$$

4) For yawing moment, define $\mu_2 = (I_x - I_y)/I_z$. Consequently,

$$\dot{r} = p_0 \mu_2 q + N_\beta \beta + N_r r$$

The preceding four equations may be expressed in matrix form as

$$[I_n]\{\dot{x}\} = [A_n]\{x\}$$

which may be simplified to

$$\{\dot{x}\} = [A]\{x\} \tag{8.1}$$

where

$$[A] = [I_n]^{-1}[A_n]$$

whereas the state vector is defined as

$$\{x\} = [\alpha \quad q \quad \beta \quad r]^T$$

The inertia matrix takes the form of

$$[I_n] = \begin{bmatrix} 1 & 0 & 0 & 0 \\ -M_{\dot{\alpha}} & 1 & 0 & 0 \\ 0 & 0 & 1 & 0 \\ 0 & 0 & 0 & 1 \end{bmatrix}$$

whereas the original plant matrix is

$$[A_n] = \begin{bmatrix} Z_w & 1 & -p_0 & 0 \\ M_\alpha & M_q & 0 & p_0\mu_1 \\ p_0 & 0 & Y_v & -1 \\ 0 & p_0\mu_2 & N_\beta & N_r \end{bmatrix}$$

Consequently, the matrix $[A]$ in Eq. (8.1) becomes

$$[A] = \left[\begin{array}{cc|cc} Z_w & 1 & -p_0 & 0 \\ M'_\alpha & M'_q & -p_0 M_{\dot{\alpha}} & p_0\mu_1 \\ \hline p_0 & 0 & Y_v & -1 \\ 0 & p_0\mu_2 & N_\beta & N_r \end{array}\right] = \left[\begin{array}{c|c} A(1,1) & A(1,2) \\ \hline A(2,1) & A(2,2) \end{array}\right] \tag{8.2}$$

where

$$M'_\alpha = M_\alpha + Z_w M_{\dot{\alpha}} \qquad M'_q = M_q + M_{\dot{\alpha}}$$

The partitioned form of Eq. (8.2) makes it evident that when $p_0 = 0$, $A(1, 2) = A(2, 1) = [0]$, and the matrix equation uncouples into the approximations for the longitudinal short-period and lateral-directional Dutch-roll modes, respectively. Also, if the roll rate p_0 were a time-dependent quantity, the equations would be nonlinear due the presence of terms such as $p(t)r(t)$ and $p(t)q(t)$, which would pose a much more difficult problem to solve and understand.

The characteristic equation for the inertially coupled system can be obtained by considering

$$|\lambda I - A| = \lambda^4 + a_3\lambda^3 + a_2\lambda^2 + a_1\lambda + a_0 = 0 \tag{8.3}$$

A typical root locus plot showing the movement of the characteristic roots as roll rate is increased from $p_0 = 0$ is shown in Fig. 8.1. The actual determination of the root migration for an actual aircraft model may be conveniently found using available matrix-oriented software, such as MATLAB. The roots of the fourth-order polynomial are initially complex paired, and as p_0 is increased, a branch will descend to the real axis, bifurcate, and then migrate along the real axis. Typically also, the root migration on the real axis will lead to an instability if the root passes through the origin from the negative side to the positive value. Quite often, if an instability were to exist, the root locus passage through the origin would return back through the origin (as p_0 is increased further), which indicates two critical values of roll rate for the onset of an inertial cross coupling instability. The minimum value of roll rate for root locus passage through the origin would be the critical value from a physical standpoint.

The branch of the root locus plot migrating on the real axis may have been initiated from either a lateral-directional or a longitudinal mode depending on the actual dimensional derivative values used for the particular flight condition under investigation. It is interesting to identify the origin of the modes by looking at the eigenvector using software such as MATLAB. The change in the mode shapes

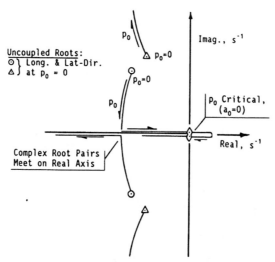

Fig. 8.1 Root locus sketch for inertial cross coupling.

as the roll rate p_0 is varied is illustrative in showing that the effect of the inertial coupling is to introduce a mixed participation of the α, q, β, and r components.

The passage of the root locus through the origin corresponds to the vanishing of the a_0 coefficient in Eq. (8.3), which is a requirement for $\lambda = 0$ to be a root. From a practical, or illustrative standpoint, neglect of the Z_w and Y_v terms in the plant matrix simplifies the determination of the a_0 term in the characteristic equation without losing the essence of the study. Under these assumptions, expansion of the characteristic equation yields

$$a_0 = -p_0^4 \mu_1 \mu_2 + p_0^2(M_q N_r - \mu_1 N_\beta - \mu_2 M_\alpha) - N_\beta M_\alpha \qquad (8.4)$$

As may be noted in Eq. (8.4), the a_0 coefficient is quadratic in the square of roll rate and can be solved to determine a value of p_0^2 for its vanishing. In general, it will be found that $\mu_1 > 0$ for $I_x < I_z$ and $\mu_2 < 0$ for $I_x < I_y$, $N_\beta > 0$ for a directionally stable aircraft and $M_\alpha < 0$ for a longitudinally stable aircraft, and M_q and N_r both <0 for normal pitch and yaw damping. Consequently, it may be seen in Eq. (8.4) that the coefficients on $(p_0^2)^2$ and $(p_0^2)^0$ will normally be positive whereas the $(p_0^2)^1$ coefficient will usually be negative. Under these situations, it is possible to obtain root values of (p_0^2), which are either positive or complex. Complex values indicate that a cross coupling instability does not exist. If one finds a positive root value for (p_0^2), then the value of p_0 will have both a plus and a minus sign, which is consistent with the fact that an inertial cross-coupling instability can occur for either positive or negative valued roll rates. The occurrence of the instability should not depend on the direction of the roll maneuver.

From a stability standpoint, one would be motivated to preclude the occurrence of an inertial cross-coupling instability by increasing either the pitch damping M_q or yaw damping N_r terms by artificial means consistent with the operating roll rate range of the aircraft.

Example 8.1

Consider the A-4D aircraft operating at $M = 0.9$, $h = 35,000$ ft, Appendix B.1, condition 6. Using the inertial cross-coupling approximation [Eq. (8.2)] estimate the mode shapes when $p_0 = 0$ and determine which modal approximation is subject to an instability with respect to inertial cross coupling. Then, using the $a_0 = 0$ approximation of Eq. (8.4), make an estimate of the roll rate where the instability might occur. Finally, identify the root migration as p_0 is varied in order to improve upon identifying the roll rate for the instability.

1) When $p_0 = 0$, the plant matrix for the assumed aircraft flight condition is made up from the partitioned elements as

$$A(1,1) = \begin{bmatrix} -0.670 & 1.00 \\ -14.73 & -1.265 \end{bmatrix} \quad \text{and} \quad A(2,2) = \begin{bmatrix} -0.160 & -1.00 \\ 19.65 & -0.428 \end{bmatrix}$$

and is assembled using MATLAB by

```
A=[ A(1,1), zeros(2); zeros(2), A(2,1) ]
% Eigenvalues are found from the characteristic equation.
```

```
P=poly(A); R=roots(P); disp(R)
λ₁,₂ = -0.967 ±3.826 i  .. Short-period approximation
λ₃,₄ = -0.294 ±4.431 i  .. Dutch-roll approximation
```

The MATLAB *eig*(A) function is used to find the mode shapes; i.e., the short-period mode is

$$\left\{\begin{matrix}\alpha\\q\\\beta\\r\end{matrix}\right\}_{sp} = \left\{\begin{matrix}1.000\angle 0.0\text{ deg}\\3.838\angle 94.5\text{ deg}\\0.0\\0.0\end{matrix}\right\}$$

whereas the Dutch-roll mode shape is

$$\left\{\begin{matrix}\alpha\\q\\\beta\\r\end{matrix}\right\}_{DR} = \left\{\begin{matrix}0.0\\0.0\\1.000\angle 0.0\text{ deg}\\4.432\angle -88.3\text{ deg}\end{matrix}\right\}$$

2) The inertial coupling coefficients that were used to alter the plant matrix in accord with Eq. (8.2) were

$$\mu_1 = 0.821 \ (= MU1) \quad \text{and} \quad \mu_2 = -0.611 \ (= MU2)$$

When $p_0 = 1.0$ rad/s, the eigenvalues were identified as

$$\lambda_{1,2} = -0.731 \pm 3.273i$$

$$\lambda_{3,4} = -0.530 \pm 4.985i$$

The coupled short-period mode shape corresponding to λ_1 is

$$\left\{\begin{matrix}\alpha\\q\\\beta\\r\end{matrix}\right\}_{sp} = \left\{\begin{matrix}1.000\angle 0.0\text{ deg}\\3.827\angle 94.4\text{ deg}\\0.588\angle 112.5\text{ deg}\\2.849\angle 21.6\text{ deg}\end{matrix}\right\}$$

whereas the coupled Dutch-roll mode shape corresponding to λ_3 is

$$\left\{\begin{matrix}\alpha\\q\\\beta\\r\end{matrix}\right\}_{DR} = \left\{\begin{matrix}0.776\angle 111.8\text{ deg}\\2.951\angle -153.2\text{ deg}\\1.000\angle 0.0\text{ deg}\\4.266\angle -88.9\text{ deg}\end{matrix}\right\}$$

Comparison of the mode shapes with and without coupling discloses that for $p_0 = 1$ rad/s, the individual modes did not change significantly; however, the β and r components showed up strongly in the coupled short-period mode whereas the α and q components were clearly evident in the coupled Dutch-roll mode.

Because the coupled short-period mode had a decrease in ω_n whereas the coupled Dutch-roll mode had an increase in ω_n, increasing the roll rate further will drive the coupled short-period mode into a potential region of real roots for this particular example aircraft flight condition.

3) The a_0 polynomial of Eq. (8.4) will be investigated to see if an instability might exist. Substitution of the appropriate dimensional stability derivatives into Eq. (8.4) for the assumed aircraft yielded

$$a_0 = 0.5018 p_0^4 - 24.92 p_0^2 + 294.6 = 0$$

with a solution of

$$p_0^2 = 19.39 \quad \text{and} \quad 30.27 \text{ rad}^2/\text{s}^2$$

or

$$p_0 = \begin{cases} \pm 4.40 \text{ rad/s} & (\pm 252.3 \text{ deg/s}) \\ \pm 5.50 \text{ rad/s} & (\pm 315.3 \text{ deg/s}) \end{cases}$$

4) The migration of the characteristic roots was investigated as roll rate was varied up to $p_0 = 7.0$ rad/s. After a roll rate value, $P0$, was established, the cross coupling matrices became

$$A(1,2) = \begin{bmatrix} -P0 & 0.0 \\ 0.389*P0 & P0*MU1 \end{bmatrix}$$

$$A(2,1) = \begin{bmatrix} P0 & 0.0 \\ 0.0 & P0*MU2 \end{bmatrix}$$

```
% and the [A] matrix is formed by:
A = [A(1,1), A(1,2); A(2,1), A(2,2)]
P=poly(A); R=roots(P); disp(R)
```

Using these MATLAB commands, the critical roll rates that define the unstable inertial cross coupling region were

$$p_0 = \begin{cases} 4.84 \text{ rad/s} & (277.4 \text{ deg/s}) \\ 5.11 \text{ rad/s} & (292.3 \text{ deg/s}) \end{cases}$$

These values differed from the values found in part (3) where Z_w and Y_v were assumed as 0 to make a first estimate tractable.

The migration of the critical roots (from Example 8.1) as roll rate was increased is shown in the three-dimenssional plot (Fig. 8.2) where the third dimension corresponds to roll rate. In this example, the critical root represented the short-period eigenvalue when $p_0 = 0$. The complex-conjugate eigenvalue is omitted for sake of clarity, but can be visualized as a mirror image of the root shown. The two complex modes reached the real axis when $p_0 = 4.25$ rad/s and bifurcated as a pair of real roots with further increases of roll rate. The real roots met (became equal) again when $p_0 = 5.68$ rad/s and then separated into a pair of complex-conjugate roots with further increases of the roll rate variable.

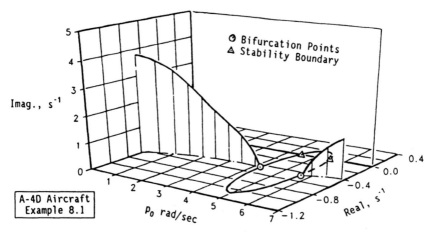

Fig. 8.2 Critical root migration, inertial cross coupling.

The region of interest relates to the roll rate value(s) where the real root trajectory crossed the origin of the real axis and became positive valued, i.e., when $277 < p_0 < 292$ deg/s. Although the first onset of the cross coupling instability would be considered as critical, a parallel concern would relate to the severity of the instability. If the aircraft of Example 8.1 had a roll rate of 285 deg/s, which is midway between the onset to and return from the instability region, the unstable real root ($\lambda = +0.011$ s^{-1}) would result in the instability propagating by about 1% providing the aircraft were constrained to not exceed a 360-deg roll maneuver. It should be noted that many high-performance military jet aircraft are restricted to ± one revolution roll maneuvers.

8.3 Wing Rock

8.3.1 Background

The term wing rock is a colloquial expression that describes the self-induced rocking motion of an aircraft about its longitudinal (roll) axis. Wing rock has been observed in a number of aircraft including the A-4, F-4, T-38, F-5, F-14, F-15, and the F-18, usually when operating at high angles of attack near to a stall or stall departure condition. In some cases, the motion may be suppressed by introducing stability augmentation to the aircraft with careful attention being given in the design such that the cure for wing rock does not become compromised by more serious consequences such as an increase in the tendency for a stall departure or spin entry.

Wing-rock motion, although similar to an aircraft's response of a lightly damped Dutch-roll mode during flight in atmospheric turbulence, should be viewed as a limit cycle considering that the amplitude of the motion tends to a constant value independent of the perturbation from the state vector's static equilibrium. The rationale for considering wing rock as a limit cycle is based on stability principles of nonlinear systems. As an illustration, consider a harmonic oscillator with nonlinear damping, i.e.,

$$\ddot{y} - c\dot{y} + \dot{y}^3 + y = 0 \qquad (8.5)$$

The linear form of Eq. (8.5) will have complex eigenvalues when the damping constant c satisfies $|c| < 2$. When the damping constant c is less than zero, the variable y will return to static equilibrium at $y = 0$. However, when $c > 0$, the eigenvalues of the linear system system cross the imaginary axis and static equilibrium at $y = 0$ will not exist. A small initial condition will result in a divergent oscillation until the system nonlinearity becomes significant with its influence stopping the divergence. Consequently, the oscillation reaches a constant amplitude and is described as a limit cycle. Even though Eq. (8.5) may be statically unstable when $c > 0$, the existence of a limit cycle implies that the nonlinear oscillator system is stable in the sense of Lyapunov (cf. Sec. 5.7).

Equation (8.5) may be put into a state vector format consisting of

$$\begin{Bmatrix} \dot{x}_1 \\ \dot{x}_2 \end{Bmatrix} = \begin{bmatrix} 0 & 1 \\ -1 & c \end{bmatrix} \begin{Bmatrix} x_1 \\ x_2 \end{Bmatrix} + \begin{Bmatrix} 0 \\ 1 \end{Bmatrix} f(x_1, x_2) \qquad (8.6)$$

where the nonlinearity is $f(x_1, x_2) = -x_2^3$.

A time history of the x_1 (i.e., y) term in Eq. (8.6) was obtained by a second-order Runge–Kutta integration scheme using the *ode23* function in MATLAB. The resulting time history, when $c = +0.5$ (shown in Fig. 8.3), reaches a constant amplitude of 0.83. The initial response from the $t = 0$ condition illustrates the divergent motion taking place prior to the influence of the cubic damping term.

A phase-plane plot of x_1 vs x_2 is shown in Fig. 8.4, again for Eq. (8.6) when $c = +0.5$. In addition to the phase-plane trajectory for the initial condition used in Fig. 8.3, a second intitial condition of $x_1(0) = 0$, $x_2(0) = 1.75$ is shown by the dashed trajectory curve. These two trajectories are representative and are shown to illustrate that the final limit-cycle trajectory is independent of the initial condition in either x_1 or x_2, thereby confirming stability in the sense of Lyapunov.

The equilibrium conditions for the dynamic response of the nonlinear oscillator [Eqs. (8.5) and (8.6)] is shown in Fig. 8.5 when the constant c is varied. It will be noted that static equilibrium is stable only when the damping constant satisfies $c < 0$. The crossing of the imaginary axis by the linearized system's eigenvalues

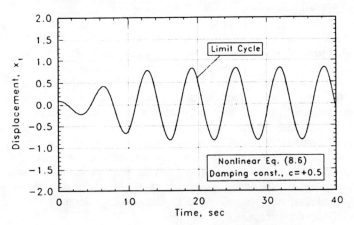

Fig. 8.3 Time history for the nonlinear oscillator.

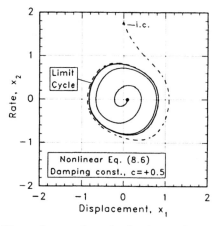

Fig. 8.4 Phase-plane trajectories for the nonlinear oscillator.

when $c = 0$ leads into the development of limit-cycle amplitudes. The unstable static equilibrium is shown in Fig. 8.5 by the dashed line. This type of system dynamic response is known as a Hopf bifurcation, with the bifurcation process taking place in the example nonlinear oscillator when $c = 0$. The appearance of the curves shown in Fig. 8.5 leads to the characterization of the nonlinear system as a pitchfork type of Hopf bifurcation. There are many references describing nonlinear mechanics, e.g., Thompson and Stewart.[6]

The dynamic response of the nonlinear oscillator was presented as an introduction to the more complex situation of aircraft wing rock. Wing rock is frequently described as a pitchfork type of Hopf bifurcation. Of course, the existence and character of a limit cycle when an unstable static equilibrium exists will depend on the nature of the system nonlinearities. Numerous physical traits have been offered as explanations for the existence of aircraft wing rock including the following.

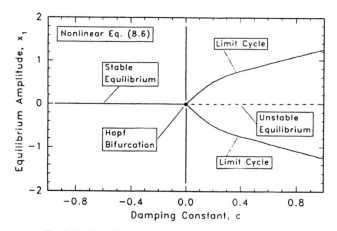

Fig. 8.5 Stability traits of the nonlinear oscillator.

1) Nonlinear yawing moment and side-force variations with sideslip angle resulted in the statically unstable Dutch-roll mode progessing into a limit cycle as the amplitude of the roll ϕ and sideslip β angles introduced the stabilizing influences of the aerodynamic nonlinearities. The analyses by Ross[7] relate to a prototype slender-delta research aircraft that experienced wing rock.

2) Nonlinear roll damping, initially unstable at small sideslip angles followed by becoming stable at larger sideslip angles, resulted in dynamic equilibrium (a limit cycle) being attained when an energy balance was reached between the stable and unstable roll damping effects. Nguyen et al.[8] identified the roll damping influences based on wind-tunnel tests of a slender-delta wing model at a fixed angle of attack and constrained to motion about the roll axis. Sideslip angles were kinematically induced by the roll angle.

3) Other researchers (e.g., Arena and Nelson[9]) have investigated the coupling between roll motion and vortex shedding by wind-tunnel tests on slender-delta wings in order to provide a more detailed explanation for the nonlinear aerodynamic behaviors that lead to the self-induced wing-rock motion.

The preceding examples (Refs. 7–9) are illustrative of wing-rock experiences by airframe configurations incorporating slender delta wings with highly swept leading edges (i.e., 60 deg or greater) where the nonlinear aerodynamics are influenced by the vortex formed as a sheet starting from the wing leading edge. Many other high-performance airframes with more modest wing planforms (e.g., F-4, T-38, and F-5) also are known to exhibit a wing-rock behavior at or near to stall angles of attack.

A candidate wing-rock mechanism[10] will be described in Sec. 8.3.2 that uses linear aerodynamics in order to illustrate that an unstable lateral-directional (Dutch-roll) mode, when inertially coupled with a stable longitudinal (short-period) mode, can attain a stable response amplitude (limit cycle). The modal coupling due to the nonlinear terms in the equations of motion serves in the role of a damping energy transferral that acts to stabilize the otherwise unstable Dutch-roll oscillation.

8.3.2 Nonlinear Equations of Motion for the Wing-Rock Model

The equations of motion are in stability axes and include linear aerodynamics combined with the inertial coupling terms in a form similar to Eq. (4.33). The lateral-directional aerodynamics will be described by the homogeneous form of Eq. (4.48), i.e.,

$$\{\dot{x}\}_{LD} = [A]_{LD}\{x\}_{LD} \qquad (8.7)$$

where

$$[A]_{LD} = \begin{bmatrix} Y_\beta/V & Y_p/V & (g/V)\cos\theta_0 & (Y_r - V)/V \\ L'_\beta & L'_p & 0 & L'_r \\ 0 & 1 & 0 & 0 \\ N'_\beta & N'_p & 0 & N'_r \end{bmatrix}$$

and

$$\{x\}_{LD} = [\beta \quad p \quad \phi \quad r]^T$$

The longitudinal system will only include the short-period modal approximation based on the assumption that the airspeed remains constant during the wing-rock oscillation. The homogeneous form of the linearized short-period equation, similar to Eq. (6.9), is

$$\{\dot{x}\}_{sp} = [A]_{sp}\{x\}_{sp} \tag{8.8}$$

where

$$[A]_{sp} = \begin{bmatrix} Z_\alpha/V & 1 \\ M'_\alpha & M'_q \end{bmatrix}$$

and

$$\{x\}_{sp} = [\alpha \; q]^T$$
$$M'_\alpha = M_\alpha + M_{\dot{\alpha}} Z_\alpha / V$$
$$M'_q = M_q + M_{\dot{\alpha}}$$

The small angle assumption for motion about an initial level-flight condition was removed when analyzing the full set of equations. In addition, Euler angle relations were introduced; i.e., roll angle ϕ became the Euler angle Φ whereas the pitch angle θ became the Euler angle Θ. The Euler angle relation for $\dot{\Theta}$ in terms of the pitch and yaw rates was introduced [cf. Eq. (4.8)] by

$$\dot{\Theta} = qC_\Phi - rS_\Phi \tag{8.9}$$

where trigonometric terms are typically shown as $C_\Phi = \cos\Phi$, $S_\Phi = \sin\Phi$, $T_\Theta = \tan\Theta$, etc.

The linearized equations of motion were modified, in accord with the concepts expressed in Sec. 4.4, to a nonlinear form by adding the corresponding terms of Eq. (8.10) to Eqs. (8.7) and (8.8), which provided six of the seven expressions needed for the analysis. The nonlinear additions are denoted by (NL) with the subscript denoting the element of the linear expressions to which they apply. Thus,

$$\begin{aligned}
(\text{NL})_\beta &= p\alpha + (C_\Theta S_\Phi - C_{\Theta_0}\Phi)(g/V) \\
(\text{NL})_p &= -qr(I_y - I_z)/I_x \\
(\text{NL})_\Phi &= (qS_\Phi + rC_\Phi)T_\Theta \\
(\text{NL})_r &= -pq(I_x - I_y)/I_z \\
(\text{NL})_\alpha &= (C_\Theta C_\Phi - C_{\Theta_0})(g/V) - p\beta \\
(\text{NL})_q &= -pr(I_z - I_x)/I_y + M_{\dot{\alpha}}(\text{NL})_\alpha
\end{aligned} \tag{8.10}$$

Modal cross coupling is provided in Eq. (8.10) by typical terms such as $p\alpha$ in $(\text{NL})_\beta$ and $-p\beta$ in $(\text{NL})_\alpha$. The I_{xz} cross product of inertia terms have not been included in Eq. (8.10) on the basis of their having a small influence.

Finally, the six nonlinear set of motion equations suitable for the wing-rock analysis consist of Eq. (8.10) added onto Eqs. (8.7) and (8.8) plus the seventh Euler angle statement, Eq. (8.9). The state vector for the nonlinear system becomes

$$\{x\}_{NL} = [\beta \quad p \quad \Phi \quad r \quad \alpha \quad q \quad \Theta]^T$$

8.3.3 Analysis of a Wing-Rock Model

The baseline aircraft under consideration is the Navy A-4D attack aircraft operating at sea level with a Mach number of 0.20. Reference values for the aerodynamic stability derivatives, although available from many sources, were selected from Ref. 11. The data corresponded to a dynamically stable aircraft with a c.g. at 0.25 $\bar{c}$. Inasmuch as the purpose of the analysis related to an investigation of nonlinear inertia coupling effects when the Dutch-roll mode was unstable, it was first necessary to alter the dimensional stability derivatives of the baseline aircraft. Although there are many ways to numerically alter the Dutch-roll dynamic stability, the method considered was to apply pole placement to the linear lateral-directional equations using Ackerman's formula (cf. Sec. 7.7.3) and preserve the time constants of the roll and spiral modes while altering only the Dutch-roll mode.

Example 8.2

Apply Ackerman's pole placement formula to the lateral-directional equations of the baseline A-4D aircraft ($M = 0.2$, sea level) to obtain an unstable Dutch-roll eigenvalue corresponding to $\zeta_{DR} = -0.03$ and $\omega_d = 1.40$ rad/s.

The reference plant, in accord with Eq. (8.7), is

$$[A] = \begin{bmatrix} -0.103 & 0.0 & 0.144 & -1.0 \\ -4.470 & -0.396 & 0.0 & 0.146 \\ 0.0 & 1.0 & 0.0 & 0.0 \\ 3.540 & -0.452 & 0.0 & -0.320 \end{bmatrix}$$

with eigenvalues at

$$\lambda_{DR} = -0.095 \pm 1.892i \text{ s}^{-1} \text{ (i.e., } \zeta_{DR} = 0.050)$$

$$\lambda_{roll} = -0.563 \qquad \lambda_{spiral} = -0.065$$

Perform pole placement using the rudder control effectiveness column vector, i.e.,

$$\{B\} = [0.017 \quad 0.617 \quad 0.0 \quad -1.374]^T$$

The desired roots, including the unstable Dutch-roll mode, are

$$\{R\} = \begin{Bmatrix} 0.014 + 1.400i \\ 0.014 - 1.400i \\ -0.563 \\ -0.065 \end{Bmatrix}$$

Apply MATLAB using the *acker* function.

```
K = acker(A,B,R)
disp(K)
-1.192  0.059  -0.015  0.212
% Establish modified lat.-dir. plant
A03=A-B*K
disp(A03)
```

$$\begin{matrix} -0.083 & -0.001 & 0.145 & -1.004 \\ -3.735 & -0.433 & 0.009 & 0.015 \\ 0.0 & 1.0 & 0.0 & 0.0 \\ 1.902 & 0.036 & -0.021 & -0.029 \end{matrix}$$

The [A03] matrix will form the basis of the wing-rock analysis, i.e., a lightly unstable Dutch-roll mode while preserving the original roll and spiral mode time constants.

Initial coupling studies were made with the M_α derivative in $[A]_{sp}$ [Eq. (8.8)] modified such that the short-period to Dutch-roll frequency ratio was 2, i.e., $\omega_{sp}/\omega_{DR} = 2.0$. The technique for integrating the nonlinear equations of motion used a second-order Runge–Kutta numerical solution method, similar to the method described in Sec. 8.3.1, available as a function in MATLAB. Example 8.3 illustrates the technique.

Example 8.3

Find the time history for the example A-4D aircraft with an unstable Dutch-roll mode ($\zeta_{DR} = -0.03$) and the coupling frequency ratio of $\omega_{sp}/\omega_{DR} = 2.0$. A MATLAB M file entitled wrock.m was created (cf. Appendix E) to be used in conjunction with the *ode23* function to perform the Runge–Kutta numerical integration for the system of seven, coupled ODEs.

The ALD matrix statement in wrock.m corresponds to the lateral-directional plant matrix [A03] as determined in Example 8.2.

The ALG matrix in wrock.m is of order (2×3) and contains the following short-period mode information:

$$[ALG] = \begin{bmatrix} Z_\alpha/V & 1.0 & g/V \\ M_\alpha & M_q & M_{\dot\alpha} \end{bmatrix}$$

The MU column vector in wrock.m corresponds to the following mass moment of inertia ratios as described in Eq. (8.10):

$$\{MU\} = \begin{Bmatrix} (I_z - I_y)/I_x \\ (I_y - I_x)/I_z \\ (I_x - I_z)/I_y \end{Bmatrix} = \begin{Bmatrix} 0.360 \\ 0.365 \\ -0.641 \end{Bmatrix}$$

The initial condition for the integration using the *ode23* function was based on

$$\{X0\} = [1.000 \quad 0.082 \quad 1.919 \quad -0.374 \quad 0.0 \quad 0.0 \quad 0.0]^T$$

where the first four components represented the unstable lateral-directional Dutch-roll eigenvector at time $t = 0$.

The MATLAB statements to obtain the time history are

```
% Wing Rock, Zeta = -0.03, Freq. Ratio = 2.0
% i.c. is Beta = 2.0 deg (=0.03749 rad) × mode shape
X0 =[1.00 0.0816 1.919 -0.374 0.0 0.0 0.0]^T
t0=0; tf=160; xin=0.03749*X0
[t,x]=ode23('wrock',t0,tf,xin)
```

The buildup of roll angle to a limit cycle due to the initial condition (i.c.) of Example 8.3 is shown as a time history in Fig. 8.6. The roll angle response of the unstable linear lateral-directional system subject to the same i.c. is also shown in Fig. 8.6 by the dashed-line curve to illustrate the stabilizing influence provided by the inertial coupling of the system with the stable short-period mode. Although not shown, other i.c.s including some greater than the limit-cycle amplitude were evaluated as a confirmation that the limit cycle was stable in the sense of Lyapunov.

A comparative time history of roll angle and angle-of-attack perturbations during the limit cycle of Example 8.3 is shown in Fig. 8.7. It should be noted that the angle-of-attack perturbations show a doubling relative to the roll limit-cycle frequency. This feature, described as kinematic coupling, can be attributed to the product of two sinusoids [i.e., $p(t)\beta(t)$ and $p(t)r(t)$] in the $(NL)_\alpha$ and $(NL)_q$ terms, respectively, of Eq. (8.10).

An alternate perspective of the wing-rock limit cycle of Example 8.3 may be obtained from a cross plot of sideslip β with roll ϕ angle, cf. Fig. 8.8. The phase-plane trajectory appears as an elliptical trajectory corresponding to an amplitude ratio of $|\phi|:|\beta| = 1.92$ and a 15.9-deg phase lag of sideslip relative to roll angle.

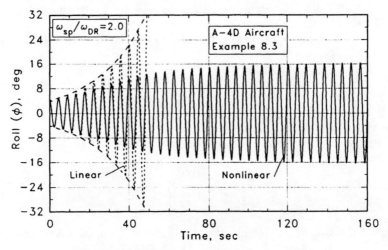

Fig. 8.6 Roll limit cycle buildup, $\zeta_{DR} = -0.03$.

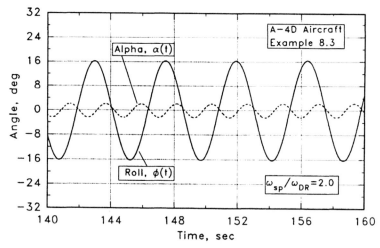

Fig. 8.7 Limit-cycle time histories of ϕ and α, $\zeta_{DR} = -0.03$.

The unstable linear system had a similar $|\phi|:|\beta|$ ratio but with a 20.7-deg phase lag of $\beta(t)$ relative to $\phi(t)$. The elliptical trajectory behavior of β vs ϕ in Fig. 8.8 may be attributed to the low damping value of the Dutch-roll instability combined with the use of linear aerodynamic relations. If the limit-cycle amplitudes involved nonlinear aerodynamic quantities, one would expect distortions from an elliptical trajectory to appear in the phase-plane plots.

The effect of Dutch-roll damping ratio ζ_{DR} on the roll ϕ, sideslip β, and angle-of-attack α limit-cycle amplitudes is shown in Fig. 8.9 subject to the constraint that the frequency ratio (ω_{sp}/ω_{DR}) was 2. The trend is one of increasing amplitude with increasing levels of Dutch-roll modal damping instability (i.e., in a linear context). It should be apparent from Fig. 8.9 that the limit cycles, when $\zeta_{DR} < 0$, correspond to a pitchfork type of Hopf bifurcation as described in Sec. 8.3.1 because static equilibrium does not exist.

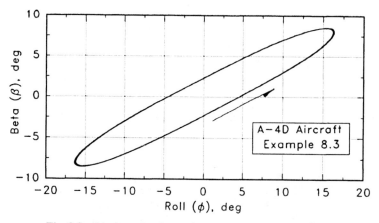

Fig. 8.8 Limit-cycle phase-plane trajectory, $\phi(t)$ vs $\beta(t)$.

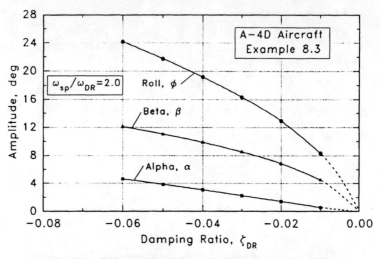

Fig. 8.9 Effects of damping on limit-cycle amplitude.

The doubling of the angle-of-attack perturbation frequency as compared to roll (cf. Fig. 8.7) is recognized as a product of inertial coupling. The ability of the longitudinal motion to couple and transfer energy for the creation of a stable limit cycle could be expected to be near an optimum when the ω_{sp}/ω_{DR} frequency ratio was near to 2.0. Furthermore, when in a situation of near optimum coupling, the resultant limit-cycle amplitudes should be near to a minimum. The variation of limit-cycle amplitudes as a function of frequency ratio, as shown in Fig. 8.10, bears out these traits.

A candidate mechanism for a wing-rock limit-cycle oscillation has been indicated based on modifying available dimensional stability derivatives. However,

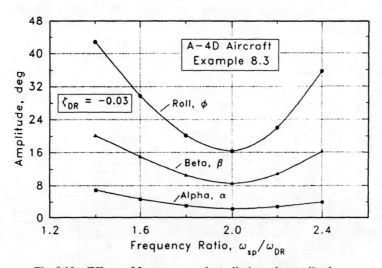

Fig. 8.10 Effects of frequency ratio on limit-cycle amplitudes.

it should be recognized that the existence of a wing-rock limit cycle by an actual aircraft may be due to a multitude of factors including inertial coupling in the equations of motion, nonlinearities in the aerodynamics, and coupling of the aerodynamics with the motion. In the final analysis, each factor results in specific attributes, and the actual cause of the wing-rock motion will depend on an analysis of flight-test and wind-tunnel data for the specific airframe configuration and flight condition. Theoretical approaches serve the role of providing clues as to candidate sources for identifying the nonlinear mechanism leading to the Hopf bifurcation.

8.4 Stall Dynamics

8.4.1 Background

Aircraft stall, which occurs when flight is near to the maximum C_L, sets the criteria for a minimum flight speed. The stall speed, when determined in accord with the procedures described in either the *Federal Aviation Regulations* (Parts 23[12] or 25[13]) or the U.S. military specifications of MIL-F-8785C,[14] establishes many performance factors. Takeoff and landing speeds are usually determined by setting them at a safe margin (i.e., $1.2–1.3V_s$) above the stall speed, where V_s has been determined by flight tests for a particular aircraft configuration (i.e., flap, thrust, and landing gear settings) and weight.

Flying qualities at or near to a stall condition are important with respect to safe aircraft behavior, and desirable traits are described in Refs. 12–14. Typical considerations for wing-level stalls include the following.

1) Following establishment of trimmed flight, use of longitudinal control to reach aircraft stall should require a smooth application of control without reversal in direction. The presence of a pitch-up tendency as stall is approached is undesirable.

2) Stall warning should be evident from a smooth buildup of buffet intensity and/or a warning to the pilot from either an audible sound or a control stick shaker.

3) The aircraft should pitch nose down following stall as part of the stall recovery. Also any tendency for a sudden wing drop at stall should not be excessive (i.e., greater than 20–30 deg bank angle change).

4) Aircraft lateral stability (i.e., dihedral effect) should be adequate at or near to stall such that a bank angle correction to a wing drop may be made using the rudder. In some aircraft, use of lateral control to correct for a wing drop may aggravate the stall by increasing the span load asymmetry.

5) Stall recovery can be done without excessive demands on piloting skills.

The maneuvers for establishing performance related stall speeds are based on establishing trimmed level flight (for a given configuration) at approximately $1.2–1.4V_s$ followed by a smooth deceleration of 1 KEAS/s until stall is reached and then followed by recovery from the stall. Flight-test procedures usually require a series of stall maneuvers at the appropriate configuration and c.g. condition for varying approach rates. Postflight analyses, which include data corrections such as consideration of weight changes due to fuel usage, are made to define the minimum flight speed V_s for the ideal stall approach rate.

Improvement of aircraft stall traits can be made by incorporating design features such as the following.

276   INTRODUCTION TO AIRCRAFT FLIGHT DYNAMICS

1) Provide built-in twist spanwise on the wing to delay tip stall by reducing the span load in the wing tip region. Washout implies that the chord plane at the tip is negative with respect to the root.

2) Leading-edge stall strips may be installed on the wing near to the fuselage in order to have wing stall take place initially in the local wing root area. This technique is frequently used on general aviation aircraft.

3) Leading-edge devices such as movable slats, fixed slots, leading-edge flaps, etc., may be included in the wing design to delay stall onset either locally or for the total wing.

4) Wing fences on the upper surface of swept-wing aircraft may act, when properly configured, to delay the outward spanwise propagation of separated flow on the wing upper surface as stall is approached.

Stall enhancing features are determined by both wind-tunnel testing during aircraft development and fixes as a result of flight-test experiences. The concepts may be used to either improve maximum C_L and/or to improve flying qualities during a stall approach. In keeping with these design principles, a well-known U.S. test pilot once stated during a preflight conference: "One test is worth a thousand expert opinions," A. M. "Tex" Johnston.[15]

In addition to stall maneuvers from an initial level-flight trim condition, accelerated stalls are demonstrated from turning flight while in a banked attitude. However, Sec. 8.4 will address the nonlinear behavior for a wing-level stall in order to illustrate the effects of deceleration rate on minimum flying speed. In addition, a limit-cycle stall porpoising motion will be described with the action attributable to relay action caused by a ΔC_L jump at stall. Earlier (cf. Sec. 8.3), a wing-rock limit cycle was described due to the inertial coupling provided by a stable short-period mode on a lightly unstable Dutch-roll mode. These two nonlinear flight dynamic examples were chosen to illustrate the existence of limit-cycle motions in the lateral-directional and longitudinal frames of reference, respectively.

The aircraft considered in the following examples will be a generic jet trainer with physical properties similar to a Lockheed T-33 trainer. The flaps-up configuration is described by wing area $S = 250$ ft^2, wing span $b = 38.73$ ft, wing aspect ratio AR $= 6.0$, wing taper ratio $\lambda = 0.40$, wing MAC $\bar{c} = 9.22$ ft, aircraft weight $W = 12,500$ lb, and pitch mass moment of inertia $I_y = 20,000$ slug-ft^2.

Summaries of the estimated C_L, C_D, and $(C_m)_{cg}$ variations with angle of attack are shown in Fig. 8.11. The aerodynamic model is based on representative experimental data (i.e., Bihrle et al.[16]), altered slightly to provide an estimated full-scale value of maximum $C_L = 1.40$ when $\delta_e = 0.0$. The dashed C_L vs α curve in Fig. 8.11 shows a ΔC_L option at the stall α, which will be used during the stall-porpoising limit-cycle investigations. The $(C_m)_{cg}$ curve shows an increase in stability when stall α is exceeded in accord with experimental data.[16] Inasmuch as the stall analysis will involve values of the three aerodynamic coefficients vs α in the nonlinear range, MATLAB function files of lift.m, drag.m, and moment.m will be used (cf. Appendix E for software listings). The lift.m and moment.m function files also will include the effects of elevator deflection on C_L and C_m.

In the linear range, the principle coefficient derivatives (in rad^{-1}) may be summarized as

$$C_{L_\alpha} = 4.60; \quad C_{m_\alpha} = -0.48$$

$$C_{L_{\delta_e}} = 0.42; \quad C_{m_{\delta_e}} = -0.90$$

$$C_{m_q} = -8.0; \quad C_{m_{\dot\alpha}} = -4.0$$

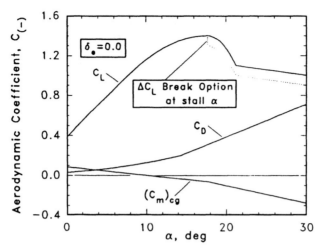

Fig. 8.11 Aerodynamic model, stall dynamic analysis.

These derivatives represent a c.g. location that is $0.104\bar{c}$ forward of the stick-fixed neutral point.

8.4.2 Wings-Level Stall Dynamics

A sketch showing force equilibrium on the aircraft stall model (Fig. 8.12) includes a thrust term constrained to fixed body axes while lift and drag due to C_L and C_D are in stability axes, i.e., oriented with respect to the velocity vector. The u and w components of the velocity vector may be recognized as

$$u = V C_\alpha \quad \text{and} \quad w = V S_\alpha \tag{8.11}$$

which in unyawed flight provides

$$V = [u^2 + w^2]^{1/2}$$

The time derivatives of u and w in Eq. (8.11) may be combined to yield

$$\dot{\alpha} = -\frac{\dot{u}}{V} S_\alpha + \frac{\dot{w}}{V} C_\alpha \tag{8.12}$$

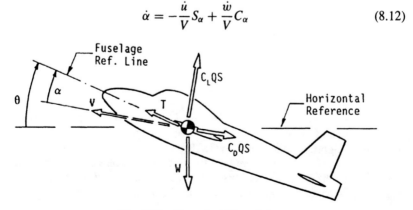

Fig. 8.12 Aircraft stall model.

Force equilibrium using the acceleration components in the airframe body axes, which is considered in a rotating frame of reference, can be expressed [cf. Eq. (4.29)] as

$$\dot{u} = \frac{T}{m} + \frac{QS}{m}[C_L S_\alpha - C_D C_\alpha] - gS_\theta - w\dot{\theta} \quad (8.13)$$

$$\dot{w} = -\frac{QS}{m}[C_L C_\alpha + C_D S_\alpha] + gC_\theta + u\dot{\theta} \quad (8.14)$$

Combining Eqs. (8.13) and (8.14) into Eq. (8.12) simplifies the $d\alpha/dt$ relation to

$$\dot{\alpha} = -\frac{T}{mV}S_\alpha - \frac{QS}{mV}C_L + \frac{g}{V}C_{(\theta-\alpha)} + \dot{\theta} \quad (8.15)$$

The pitch angular acceleration is

$$\ddot{\theta} = \frac{QSc}{I_y}\left[(C_m)_{cg} + \frac{c}{2V}(C_{m_q}\dot{\theta} + C_{m_\alpha}\dot{\alpha})\right] \quad (8.16)$$

Note that the $(C_m)_{cg}$ term in Eq. (8.16) includes the contribution due to elevator control $C_{m_\delta}\delta_e$.

The relations of Eqs. (8.13–8.16) are nonlinear in nature due to expressions such as C_L, C_D, $\cos\theta$, and product terms such as $(-w\dot{\theta})$. For the purpose of the subsequent analyses, consider the state vector as being

$$\{x\} = [\alpha \quad q \quad \theta \quad u \quad w]^T$$

whereas the nonlinear state equations are

$$\dot{x}_1 = -\frac{T}{mV}\sin(x_1) - \frac{QS}{mV}C_L + \frac{g}{V}\cos(x_3 - x_1) + x_2$$

$$\dot{x}_2 = \frac{QSc}{I_y}\left[C_m + \frac{c}{2V}(C_{m_q}x_2 + C_{m_\alpha}\dot{x}_1)\right]$$

$$\dot{x}_3 = x_2 \quad (8.17)$$

$$\dot{x}_4 = \frac{T}{m} + \frac{QS}{m}[C_L \sin(x_1) - C_D \cos(x_1)] - g\sin(x_3) - x_2 x_5$$

$$\dot{x}_5 = -\frac{QS}{m}[C_L \cos(x_1) + C_D \sin(x_1)] + g\cos(x_3) + x_2 x_4$$

The five relations of Eq. (8.17) are described by the MATLAB function *stall.m* (cf. Appendix E) and will subsequently be solved using the second-order Runge–Kutta integration algorithm *ode23* in MATLAB.

Before modeling a stall demonstration, the condition of initial trim must be established. The initial conditions will involve identifying a thrust setting such that level-flight trim occurs (i.e., $\theta = \alpha$ in Fig. 8.12) at a velocity of $V \cong 1.3 V_s$.

NONLINEAR DYNAMICS

The determination of the initial conditions requires an iteration scheme as will be illustrated by the following example.

Example 8.4

For the generic jet trainer, find the i.c.s when $V \cong 1.3V_s$ for level flight at an altitude of 10,000 ft. Considering it is assumed that $\theta = \alpha$, a first estimate for the C_L is

$$(C_L)_1 = \frac{W}{QS} = \frac{(C_L)_{max}}{(1.3)^2} = \frac{1.40}{1.69} = 0.8284$$

From $(C_L)_1$, estimate α_1 when $\delta_1 = 0.0$. The expression for C_L in the linear range is

$$C_L = C_{L_0} + C_{L_\alpha}\alpha + C_{L_\delta}\delta$$

Using the values from *lift.m* in Appendix E, one finds

$$(0.8284) = 0.38 + 4.60(\alpha_1) + 0.42(0.0)$$

which yields a value for α well in the C_L linear range, i.e.,

$$\alpha_1 = 0.09748 \text{ rad } (= 5.59 \text{ deg})$$

Next, using values from *drag.m* in Appendix E, one estimates

$$(C_D)_1 = 0.0188 + 0.0744(0.38 + 4.60\alpha_1)^2 = 0.0699$$

The trim condition in level flight can be used to determine the required thrust value for longitudinal equilibrium; i.e., for $\dot{V} = 0$,

$$(T/QS)_1 \cos\alpha_1 - (C_D)_1 = 0$$

which yields a first trial value for the thrust coefficient as

$$(T/QS)_1 = 0.0702$$

Lift equilibrium for level flight implies that $\dot{w} = 0$, and assume that $S_\alpha \cong \alpha$; hence,

$$(T/QS)_1 \alpha_2 + (0.38 + 4.60\alpha_2 + 0.42\delta_2) = W/QS = 0.8284$$

or

$$4.6702\alpha_2 + 0.42\delta_2 = 0.4484$$

Moment equilibrium (cf. *moment.m* in Appendix E) provides a second relation

$$C_{m_0} + C_{m_\alpha}\alpha + C_{m_\delta}\delta = 0$$

or

$$0.48\alpha_2 + 0.90\delta_2 = 0.084$$

The subsequent 2 × 2 matrix problem is

$$\begin{bmatrix} 4.6702 & 0.42 \\ 0.48 & 0.90 \end{bmatrix} \begin{Bmatrix} \alpha \\ \delta \end{Bmatrix}_2 = \begin{Bmatrix} 0.4484 \\ 0.084 \end{Bmatrix}$$

which yields

$$\begin{Bmatrix} \alpha \\ \delta \end{Bmatrix}_2 = \begin{Bmatrix} 0.09203 \text{ rad} \\ 0.04425 \text{ rad} \end{Bmatrix}$$

Using the second iteration results, go back to restart the iteration process at the C_D relation using values from *drag.m* in Appendix E, as described earlier, and repeat the calculations. After two more iterations, convergence to five decimal accuracy established that

$$(T/QS)_4 = 0.06714$$

$$\alpha_4 = 0.09210 \text{ rad} \quad \text{and} \quad \delta_4 = 0.04422 \text{ rad}$$

The dynamic pressure at level-flight trim is found by considering the $\dot{w} = 0$ relation, Eq. (8.14), i.e.,

$$\dot{w} = 0 = -(QS/m)[C_L C_\alpha + C_D S_\alpha] + gC_\theta$$

or

$$Q = \frac{W}{S} \frac{C_\theta}{[C_L C_\alpha + C_D S_\alpha]}$$

But, $W/S = 50$ psf for the assumed generic trainer while the trim values for lift and drag are $C_L = 0.8222$ and $C_D = 0.0669$. Consequently, the dynamic pressure and velocity at $h = 10{,}000$ ft may be determined; i.e., $Q = 60.36$ psf and trim $V = 262.23$ fps. Based on the results of the iterative solution, the initial conditions for level-flight trim at $V \cong 1.3 V_s$ may be stated as

$$\begin{Bmatrix} \alpha(0) \\ q(0) \\ \theta(0) \\ u(0) \\ w(0) \end{Bmatrix} = \begin{Bmatrix} 0.09210 \text{ rad} \\ 0.0 \\ 0.09210 \text{ rad} \\ 261.12 \text{ fps} \\ 24.12 \text{ fps} \end{Bmatrix}$$

whereas the thrust value to be used in *stall.m* is $T = 1013.1$ lb or $T/m = 2.6077$ ft-s^{-2}.

Elevator control is used to change the velocity from the trim condition as found in Example 8.4 to a minimum speed. A control input based on applying a constant rate change to the trim setting (at $1.3V_s$) can provide a numerical example of a minimum speed. However, the existence of a lightly damped phugoid mode may result in θ oscillations that could mask the true stall process. In actual practice, the pilot applies an appropriate time-dependent elevator control input to achieve a near constant deceleration rate until a minimum flight speed is obtained.

Furthermore, the pilot would automatically suppress the onset of any unwanted phugoid mode oscillations during the stall process by suitable adjustment of the elevator deflection rates. Suppression of unwanted phugoid mode oscillations in the numerical analysis will be achieved by including suitable state vector feedback terms in the elevator control algorithm *elev.m*; cf. Appendix E.

Example 8.5

Find a feedback law at $1.3V_s$ for the generic jet trainer aircraft to be used with *elev.m* that provides suitable suppression of the phugoid mode response. The increase in modal damping will be viewed as an approximate equivalent to elevator control action as provided by a pilot during a stall demonstration.

1) The linear longitudinal state equations at $1.3V_s(V = 262.23$ fps), based on the geometry and aerodynamics of Example 8.4, are for the plant matrix

$$[A] = \begin{bmatrix} -0.0198 & 0.0403 & 0.0 & -0.1227 \\ -0.2436 & -0.6915 & 1.0 & 0.0 \\ 0.0657 & -2.2938 & -0.8098 & 0.0 \\ 0.0 & 0.0 & 1.0 & 0.0 \end{bmatrix}$$

for the control vector,

$$\{B\} = [0.0 \quad 0.0622 \quad -4.6675 \quad 0.0]^T$$

and for the state vector

$$\{x\} = [u/V \quad \alpha \quad q \quad \theta]^T$$

The eigenvalues (i.e., using MATLAB) are

$$\lambda_{sp} = -0.7552 \pm 1.5102i \quad \text{and} \quad \lambda_p = -0.0054 \pm 0.1612i$$

which represents

$$\zeta_{sp} = 0.447 \quad \text{and} \quad \zeta_p = 0.033$$

$$(\omega_n)_{sp} = 1.730 \text{ s}^{-1} \quad \text{and} \quad (\omega_n)_p = 0.157 \text{ s}^{-1}$$

2) Desired phugoid mode damping is $\zeta_p = 0.707$ without changing the natural frequency. Use a pole placement technique, as available in MATLAB.

```
% Desired poles are:
P = { -0.7552 ± 1.5102 i  }
    { -0.1141 ± 0.1141 i  }
% [A] and {B} are available from part (a.)
K = acker(A,B,P); disp(K)
0.0000  0.1458  -0.0446  -0.1814
```

3) Given that δ_0 is the trim value at $1.3V_s$, rad, and $\dot{\delta}$ is the elevator control application rate, rad-s^{-1}, then the elevator control law will be of the form

$$\delta(t) = \delta_0 + \dot{\delta} - K_2\alpha - K_3 q - K_4\theta \tag{8.18}$$

where, from part 2, the gain constants are

$$K_2 = -0.1458, \quad K_3 = +0.0446 \text{ s}, \quad \text{and} \quad K_4 = +0.1814$$

The feedback law [Eq. (8.18)] in Example 8.5 is valid for reducing and/or removing phugoid mode due to ramp type elevator control inputs at a velocity in the neighborhood of $1.3V_s$. The numerical application of the information in Examples 8.4 and 8.5 for a stall demonstration is shown in the next example. It will be assumed that the thrust setting for the jet trainer does not change significantly during the analysis. Also note that the assumed initial condition of trim at level flight can be altered to climbing or descending flight by reconsidering the value of the thrust coefficient.

Example 8.6

Determine the minimum speed attained during a stall demonstration when elevator control is applied from an initial level-flight trim condition at $V \cong 1.3V_s$ for the generic jet trainer.

```
% Stall from 1.3 Vs; DEdot=-0.170 deg/sec
% Elev. smoothing, 0.707 Phugoid damping, Eq. (8.18)
% Input initial condition vector to start analysis
xin=[0.09210; 0.0; 0.09210; 261.12; 24.12];
t0=0.0; tf=80.0;
[t,x]=ode23('stall',t0,tf,xin,1.e-5);
% Use ''whos'' to find size of the time vector, t
[V,nz,De]=velocity(188,t,x);
% Save results for analysis and plotting
Z(:,1)=t; % Time, sec
Z(:,2)=x(:,1)/0.01745; % Alpha, deg
Z(:,3)=x(:,3)/0.01745; % Theta, deg
Z(:,4)=V'; % Velocity, fps
Z(:,5)=nz'; % Load factor, nz, g-s
Z(:,6)=De'/0.01745; % Elev. angle, deg
save STALL170.DAT Z /ascii
% This file is reference data for Figs. 8.13 and 8.14
```

The aircraft attitudes during the numerical stall demonstration of Example 8.6 are shown in Fig. 8.13. The stall maneuver was not initiated until $t = 10$ s in order to confirm that a level-flight trim condition had been achieved using the iterative method described in Example 8.4. Although a $d\delta_e/dt$ value ($= -0.170$ deg/s) was specified, the effects of the control law, which is included in the *elev.m*

NONLINEAR DYNAMICS

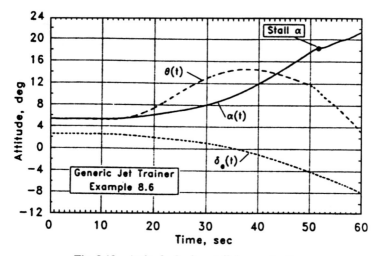

Fig. 8.13 Attitude during stall demonstration.

function (cf. Appendix E), results in a smooth change of elevator angle as might be expected if the stall were conducted by pilot control actuation. The α and θ curves in Fig. 8.13 are smooth progressions without any indication of a phugoid mode oscillation.

During the time period of $t = 15$–42 s, the aircraft shows a climbing tendency with the maximum climb angle (of the velocity vector) being approximately 5 deg and occurring at $t = 32$ s, as shown by the load factor n_z plot in Fig. 8.14. As shown in Fig. 8.13, the aircraft at $t = 43$ s would again be in level flight (i.e., because $\theta = \alpha$). However, the load factor at this time is less than 1.0, which indicates that the aircraft is in curvilinear flight with the instantaneous center of rotation being below the aircraft.

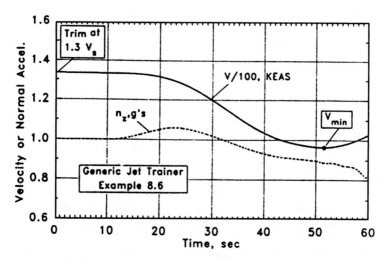

Fig. 8.14 Velocity and n_z during stall demonstration.

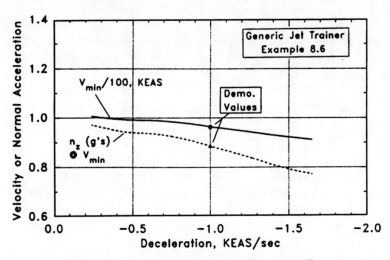

Fig. 8.15 Airframe deceleration effects on stall.

The velocity (cf. Fig. 8.14) smoothly decreased until a minimum took place at $t = 52$ s. Following the stall, continuing application of negative elevator angle (cf. Fig. 8.13) increased the α angle beyond stall while the velocity increased during a pronounced nose-drop tendency. In actual practice, the pilot would initiate stall recovery after reaching the minimum speed. The stall approach, as shown by the velocity curve in Fig. 8.14, corresponds to an approximate deceleration of -1.0 KEAS based on the velocity and time differences between V_s and $1.2V_s$.

Deceleration rate may be expected to alter the minimum speed and load factor at stall; hence, a standard is normally set for the rate in order to establish a safe and consistent definition of minimum aircraft flight speed for use in setting performance related criteria. The investigations of Example 8.6 were repeated for deceleration rates other than -1 KEAS/s by varying $d\delta_e/dt$ in $elev.m$. The cross plot of $V_{\min}$ as a function of $\dot{V}$, as shown in Fig. 8.15, illustrates that increasing the deceleration rate decreases the $V_{\min}$ value attained. Additionally, it will be noted in Fig. 8.15 that the load factor encountered at $V_{\min}$ also decreased as $-\dot{V}$ was increased, which corresponds to rate-entry-dependent changes in the radius of curvature of the aircraft flight path during a stall demonstration.

The demonstration values (i.e., for $\dot{V} = -1$ KEAS/s) shown in Fig. 8.15 correspond to an apparent $(C_L)_{\max} = 1.591$ when $n_z = 0.886g$. This represents an equivalent $n_z = 1.0$ value of $(C_L)_{\max} = 1.409$. When a second correction is made for the thrust coefficient component [i.e., $(T/QS)\sin\alpha$], agreement is found with the static value of trimmed $(C_L)_{\max} = 1.370$. This accord is due to the assumption that $C_{L_{\dot{\alpha}}}$ and C_{L_q} were negligible in the lift model, a fact that may be unrealistic when predicting flight-test results. Also note that stall speeds determined for certification purposes (cf. Refs. 12 and 13) are based on setting thrust at idle rather than that required for level flight at $1.3V_s$ as done in the examples.

The solid C_L vs α curve shown in Fig. 8.11 depicts a gentle stall progression and was used in the analyses of this section as an illustration of the principles experienced by an aircraft during a wings-level stall demonstration. In contrast, the dashed curve shown in Fig. 8.11 would result in a more pronounced stall break and would be representative of an aircraft with wing airfoil sections that experience vestiges of abrupt leading-edge separation at stall. Flight tests in support of the principles described here relative to the effects of stall approach rate have been reported by Kier.[17]

8.4.3 Stall Porpoising Limit Cycle

The term stall porpoising refers to a longitudinal oscillation that may occur when the aircraft is constrained to flight near to or at a wings-level stalled condition. When student pilots encounter these oscillations, the flight instructor explains the motion as being due to the wing alternately stalling and unstalling. Studies have been made of this aspect of stall dynamics and reported by Frederiksen[18] and Schoenstadt.[19]

Assumptions to be made in the following analysis on the generic jet trainer include the three listed next.

1) Aircraft velocity remains constant during the oscillation. Nonlinear analyses[18] without this constraint have shown that the velocity variation is less than 1% during the oscillation.

2) The aircraft is trimmed to the stall attitude at a C_L prior to the action of a ΔC_L jump (cf. Fig. 8.11).

3) A one sinusoid elevator control input of 2-s duration will be used to initiate the aircraft response.

The time-history analysis will be based on a solution technique similar to that used in Sec. 8.4.2. The *stall.m* function will be modified to be compatible with the stated assumptions by removing the phugoid mode related terms, setting the velocity V at a specified value, and introducing elevator angle via a function *elev1.m*, which is shown in Appendix E.

Example 8.7

Find the trim settings for the generic jet trainer for steady flight at the nominal stall α. Also consider the thrust as defined by Example 8.4, i.e., $T = 1013.1$ lb. Given that stall $\alpha = 0.3089$ rad ($= 17.70$ deg), first find δ_e from $C_m = 0$ at $\alpha = 0.3089$ rad [answer: $\delta_e = -0.7141$ rad ($= -4.09$ deg)]. Next find trim $(C_L)_{max}$ at $\alpha = 0.3089$ rad (answer: $C_L = 1.370$). Then find C_D at $\alpha = 0.3089$ rad (answer: $C_D = 0.3107$). From Fig. 8.12, force equilibrium relative to the fuselage reference line provides that

$$Q[C_L C_\alpha + C_D S_\alpha] - (W/S)C_\theta = 0$$

$$Q[C_L S_\alpha - C_D C_\alpha] - (W/S)S_\theta = -T/S$$

Substitution of the known values into the two equations provides a nonlinear relation, i.e.,

$$1.3996Q - 50.0C_\theta = 0$$

$$0.1205Q - 50.0S_\theta = -4.052$$

Convergence was reached after several iterations (answer: $Q = 35.23$ psf and $\theta = 0.1667$ rad), and from Q, find $V = 200.34$ fps at $h = 10,000$ ft.

Because pitch attitude θ is less than angle of attack α, the trim condition corresponds to descending flight. However, the influence of density variations due to altitude changes will be considered as small. The following example will provide a time history of the aircraft's short-period response because the ΔC_L jump will be introduced later as a variable.

Example 8.8

Find the time-history response of the generic jet trainer due to a control disturbance that is provided to the aircraft while trimmed at stall.

```
% In ''stall.m'', set V = 200.34, DE = elev1.m
% and disable xdot(4) and xdot(5) eqns.
xin=[0.3089; 0.0; 0.1667]; t0=0.0; tf=30.0;
[t,x]=ode23('stall',t0,tf,xin,1.e-5);
% Save results for analysis and plotting
Z(:,1)=t; % Time, sec
Z(:,2)=x(:,1)/0.01745; % Alpha, deg
Z(:,3)=x(:,3)/0.01745; % Theta, deg
save STALMAX1.DAT Z /ascii
% This file is reference data for Fig. 8.16
```

A $\Delta C_L = 0.10$ jump at stall was provided by making a $DL = 0.10$ entry into *lift.m*. The procedure shown in Example 8.8 was repeated to develop the stall porpoising limit cycle. Results of the analysis are shown in Fig. 8.16. The short-period response (i.e., when $\Delta C_L = 0$) is well damped even though the local

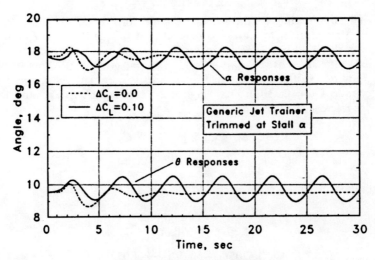

Fig. 8.16 Limit cycle at stall.

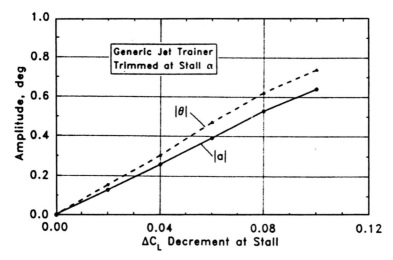

Fig. 8.17   Effects of ΔC_L on limit-cycle amplitudes.

$dC_L/d\alpha$ value was zero. The limit-cycle response, which is occurring near to the short-period frequency, is a result of the ΔC_L jump. It was necessary to increase the velocity to 203.6 fps to accommodate the reduced average value of C_L for the same mean value of descent angle during the limit cycle. For the value of $\Delta C_L = 0.10$, the jump could be considered as relay action described by

$$C_L = 1.365 + 0.05 \, \text{sgn}(\alpha_{\text{stall}} - \alpha) \tag{8.19}$$

where sgn(x) is a sign function, which is $+$ for $x > 0$ and $-$ for $x < 0$.

Rather than perform an integration to describe the stall porpoising limit cycle, Schoenstadt[19] used the relay action concept described by Eq. (8.19) combined with some simplifying assumptions to prove the existence of the limit cycle.

Inasmuch as the ΔC_L jump at stall could be treated as a variable, an appropriate study was made to vary the magnitude of ΔC_L. Results are shown in Fig. 8.17. Although the limit-cycle amplitudes appear to be linear with respect to ΔC_L, the fact that the response is stable when $\Delta C_L = 0$ implies that the limit cycle is a form of a Hopf bifurcation in a manner similar to observations made in Sec. 8.3 with respect to wing-rock motions.

References

[1]Wolko, H. S., *The Wright Flyer, An Engineering Perspective*, Smithsonian Inst. Press, Washington, DC, 1987, pp. 24–30.

[2]Phillips, W. H., "Effect of Steady Rolling on Longitudinal and Directional Stability," NACA TN 1627, June 1948.

[3]Blakelock, J. H., *Automatic Control of Aircraft and Missiles*, 2nd ed., Wiley, New York, 1991, Chap. 5.

[4]Etkin, B., *Dynamics of Atmospheric Flight*, Wiley, New York, 1972, Chap. 10.

[5]Babister, A. W., *Aircraft Stability and Control*, Pergamon, Oxford, England, UK, 1961, Chap. 17.

[6]Thompson, J. M., and Stewart, H. B., *Nonlinear Dynamics and Chaos*, Wiley, New York, 1986, Chap. 7.

[7]Ross, A. J., "Investigation of Nonlinear Motion on a Slender-Wing Research Aircraft," *Journal of Aircraft*, Vol. 9, No. 9, 1972, pp. 625–631.

[8]Nguyen, L. T., Yip, L., and Chambers, J. R., "Self-Induced Wing Rock of Slender Delta Wings," AIAA Paper 81-1883, Aug. 1981.

[9]Arena, A. S., and Nelson, R. C., "Unsteady Surface Pressure Measurements on a Slender Delta Wing Undergoing Limit Cycle Wing Rock," AIAA Paper 91-0434, Jan. 1991.

[10]Schmidt, L. V., and Wright, S. R., "Aircraft Wing Rock by Inertial Coupling," AIAA Paper 91-2885, Aug. 1991.

[11]McRuer, D., Ashkenas, I., and Graham, D., *Aircraft Dynamics and Automatic Control*, Princeton Univ. Press, Princeton, NJ, 1973, pp. 700–704.

[12]Anon., "Part 23, Airworthiness Standards: Normal Utility, Acrobatic and Commuter Category Airplanes," *Federal Aviation Regulations*, U.S. Government Printing Office, Washington, DC, Feb. 1991.

[13]Anon., "Part 25, Airworthiness Standards: Transport Category Airplanes," *Federal Aviation Regulations*, U.S. Government Printing Office, Washington, DC, Feb. 1991.

[14]Anon., "Flying Qualities of Piloted Airplanes," MIL SPEC MIL-F-8785C, U.S. Government Printing Office, Washington, DC, Nov. 1980.

[15]Johnston, A. M., *Tex Johnston, Jet Age Test Pilot*, Smithsonian Inst. Press, Washington, DC, 1991, p. 185.

[16]Bihrle, W., Barnhart, B., and Pantason, P., "Static Aerodynamic Characteristics of a Typical Single-Engine Low-Wing General Aviation Design for an Angle-of-Attack Range of $-8°$ to $90°$," NASA CR-2971, July 1978.

[17]Kier, D. A., "Flight Comparison of Several Techniques for Determining the Minimum Flying Speed for a Large, Subsonic Jet Transport," NASA TN D-5806, June 1970.

[18]Frederiksen, J. T., "An Evaluation of the Longitudinal Dynamic Stability of an Aircraft at Stall," Masters Thesis, U.S. Naval Postgraduate School, Aeronautics Dept., Monterey, CA, June 1972.

[19]Schoenstadt, A. L., "Nonlinear Relay Model for Post-Stall Oscillations," *Journal of Aircraft*, Vol. 12, No. 7, 1975, pp. 572–577.

Problems

8.1. Consider a representative, high-performance jet fighter aircraft such as the F-4C operating at $M = 0.9$, $h = 15,000$ ft, cf. Appendix B.3, condition 2. Investigate the possibility for an inertial cross-coupling instability including the identification of the mode that develops into real roots as the coupling due to roll rate progresses.

8.2. Verfiy Eq. (8.4).

8.3. Consider the F-4C aircraft operating at $M = 0.6$, $h = 35,000$ ft, cf. Appendix B.3, condition 3.

a) Modify the linearized lateral-directional plant to a slightly unstable Dutch-roll mode while retaining the Dutch-roll frequency and both the spiral and roll time constants, i.e.,

$$\zeta_{DR} = -0.002, \qquad \omega_{DR} = 1.823 \text{ s}^{-1}$$

$$1/\tau_r = -0.65 \text{ s}^{-1}, \qquad 1/\tau_s = -0.017 \text{ s}^{-1}$$

b) Investigate the buildup of the wing-rock limit cycle for the c.g. as specified in Appendix B.3. This will correspond to $\omega_{sp}/\omega_{DR} \cong 0.75$.

c) Modify the ω_{sp}/ω_{DR} ratio to approximately 1.50, and find the change in wing-rock limit-cycle amplitudes.

8.4. a) Verify Eq. (8.12).

b) Verify Eq. (8.15). Visualize the physical meaning of this equation by recognizing that $\dot{\gamma} = \dot{\alpha} - \dot{\theta}$ is related to the aircraft normal acceleration (cf. Sec. 6.5).

8.5. Determine the minimum speed and the corresponding normal load factor when performing a stall maneuver from wings-level, horizontal flight at an approximate trim speed of $1.3V_s$ for the aircraft of Example 8.6. The deceleration rate should be -1.0 KEAS/s, and the lift curve considered should have a $\Delta C_L = 0.10$ jump when the stall α is encountered (cf. Fig. 8.11). Note: The nose-drop tendency as indicated by the pitch attitude change when the stall break occurs should be more marked than the situation shown in Fig. 8.13.

8.6. Estimate the altitude changes during the stall maneuver of Problem 8.5.

9
Atmospheric Turbulence

9.1 Background

The aircraft dynamics in the preceding chapters have been focused on the vehicle response to deterministic type of pilot control inputs. However, it is also realistic to consider the effects of external influences (i.e., atmospheric gusts and/or turbulence) on both the aircraft response dynamics and the aircraft structural design. Etkin[1] pointed out that, prior to first flight, the Wright brothers had concerns relating to their aircraft's ability to maintain a wings-level attitude in the presence of lateral gusts. It is also interesting to note that the first technical report by NACA[2] (i.e., provided as a historic reminder by both Etkin[1] and Houbolt[3]) described the contemporary (1915) approach to predict aircraft dynamic behavior in the presence of gusts.

A sharp-edged gust can occur when an aircraft passes over an up- or downdraft in the vicinity of a mountain ridge. Another source of vertical gusts may be found when flying at a low altitude over agricultural fields with varying surface textures. Lateral gusts are commonly encountered during landings when a crosswind may generate unsteady patches due to the wake of the wind blowing over variable structural shapes (i.e., buildings, hangars, etc.). Turbulence inputs, which can be described by spectral models, also may appear in patches because of the variation in intensity with time. Studies of handling qualities by experienced pilots while instrument flying in a moving-base simulator have shown that a constant level of turbulence intensity is unrealistic whereas varying the intensity as a model for turbulent patch encounters more closely resembles the situation found in nature (cf. Jacobson and Joshi[4]).

Hoblit,[5] in his text dealing with the effect of gust inputs on aircraft response and loads, describes the many sources of atmospheric turbulence in degrees of severity as severe turbulence that is usually storm related such as by a thunderstorm, less severe turbulence arising due to the presence of cumulus clouds, and clear air turbulence, usually much less severe. Sources include wind shear, jet streams, wind over and between mountains, and convective thermals as encountered during afternoon flight in desert country.

An overview of the turbulence problem is shown in Fig. 9.1 in a block diagram format to assist in visualizing the actual process and the steps that will be used in the subsequent modeling. The product of atmospheric turbulence, the velocity field, will be considered as frozen in space momentarily while the aircraft transits the region. This velocity field, the input to the airframe, will be modeled by a deterministic form in Sec. 9.2 and by a random (or spectral) form in Sec. 9.4. The interaction of the aircraft system's aerodynamics with the velocity field will result in forces and moments being developed, which in turn will cause a dynamic response by the airframe and structure. The coupling between the aerodynamics and the airframe response can, especially when structural response modes are considered, require the use of unsteady aerodynamics in the modeling. However, the

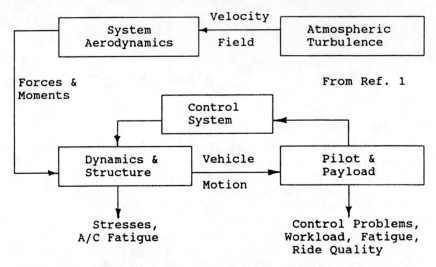

Fig. 9.1 Overview of the turbulence process.

concepts presented in this chapter are introductory in nature and will be oriented toward a rigid airframe using quasi-steady aerodynamics. Readers interested in investigating the effects of the elastic structural modes on the aircraft's turbulence response are referred to the literature (e.g., Hoblit[5]). Finally, as can be seen in Fig. 9.1, the vehicle motion impacts on the pilot, crew, and payload (i.e., passengers, freight, ordnance, etc.) by placing demands on the maintenance of flight control with subsequent effects on fatigue and ride quality.

Before the early 1930s, gust loads were not considered in the design, and instead the maneuver envelope (cf. Fig. 3.6) defined the airframe design loadings. The maneuvering envelope provides the boundary of load factor n_z vs airspeed (EAS) corresponding to the limits allowed from pilot induced control inputs. As aircraft speeds increased, the influence on airframe loadings of gust loadings, as well as pilot inputs, became evident. Early research indicated that aircraft response in atmospheric turbulence could be realistically described by a one-minus-cosine gust input with the amplitude of the input scaled to match the response found in flight tests. Static gust loadings are still determined by the one-minus-cosine vertical gust velocity shape with the aircraft motion constrained to the plunge mode only, although alternate methods based on the spectral descriptions of turbulence also are used. The concept, in accord with *Federal Aviation Regulations*, Part 25[6] (FAR 25), will be described in Sec. 9.2. This approach may be used to define a gust envelope, as shown in Fig. 9.2, and along with the maneuver envelope of Fig. 3.6, provides a basis for establishing design loads.

Straight lines of constant gust velocity (i.e., equivalent airspeed, EAS) vs n_z (radiating from $n_z = 1$ at $V_e = 0$) representing design amplitudes for the gust input are shown by the dashed lines in Fig. 9.2. The appropriate intersections of these lines with the corresponding design airspeeds establish the static gust load factors at the respective corners of the gust envelope. Considerations relating to the determination of the load factor for an assumed gust velocity will be presented in subsequent sections. Except for the $(C_L)_{max}$ boundary, straight line fairing between

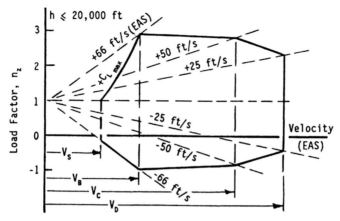

Fig. 9.2 Sketch of the flaps-up gust envelope.

the adjacent corners is used to complete the gust envelope. For altitudes from sea level to 20,000 ft, the magnitudes of the symmetrical gusts during level flight are defined according to FAR 25 as V_B (design rough-air speed), $w_g = \pm 66$ ft/s (EAS); V_C (design cruise speed), $w_g = \pm 50$ ft/s (EAS); and V_D (design dive speed), $w_g = \pm 25$ ft/s (EAS).

The design vertical gust velocities (i.e., w_g [EAS]) are assumed to vary linearly with altitude above 20,000 ft and decrease to ± 38, ± 25, and ± 12.5 ft/s at V_B, V_C, and V_D, respectively, at an altitude of 50,000 ft (e.g., when $h = 35,000$ ft, the design gust value $w_g = \pm 37.5$ ft/s for the design cruise speed V_C).

An introduction to random processes will be presented in Sec. 9.3 in support of the assumption that atmospheric turbulence can be modeled by an appropriate power spectral density. The transfer function approach will be applied in Sec. 9.4 to determine the aircraft gust response. The result will show the power spectrum of the load factor whereas the area under the power spectrum curve will represent the mean square of the load factor. An alternate approach will be used in Sec. 9.5 to show that the Lyapunov equation can directly yield the mean square of the load factor. This latter approach has proven very useful in support of studies to identify the impact of feedback laws to reduce the gust response of elastic airframes especially when they are operating at low altitudes.

9.2 Discrete Gust Response Solutions

9.2.1 Sharp-Edged Gust

The aircraft response when entering an idealized sharp-edged gust, as shown in Fig. 9.3, will prove useful in illustrating key features of the gust problem. Assumptions to make the problem tractable will include the following.

1) The sharp-edged gust is two dimensional; i.e., the gust does not vary in the spanwise direction.

2) The dynamic response is in the vertical (plunging) degree-of-freedom mode only. Consequently, aircraft pitch motion will be neglected.

3) Quasi-steady aerodynamics will be used. This implies that unsteady aerodynamic influences will be neglected, including both Wagner's function, the airfoil

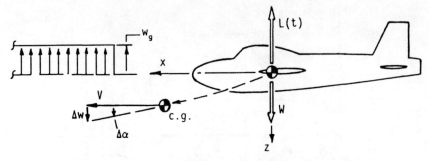

Fig. 9.3 Sketch of an aircraft entering a sharp-edged gust.

lift due to $\alpha = \alpha(t)$ resulting from airfoil vertical motion, and Kussner's function, the lift (buildup) due to an airfoil's entry into a sharp-edged gust normal to the flight path.

It is apparent from Fig. 9.3 that

$$W - L(t) = \frac{W}{g} \frac{d\Delta w}{dt} \quad (9.1)$$

where

L = aerodynamic forces, $L_0 + \Delta L(t)$
Δw = aircraft vertical velocity, + is down
w_g = vertical gust velocity, + is shown up
W = aircraft weight
g = gravitational constant

In level steady flight prior to the gust encounter, Eq. (9.1) becomes

$$W - L_0 = 0$$

While in unsteady flight, Eq. (9.1) takes the form of

$$W - L_0 - \Delta L(t) = \frac{W}{g} \frac{d\Delta w}{dt}$$

which simplifies to

$$\Delta L(t) = -(W/g)\Delta \dot{w} \quad (9.2)$$

The lift due to aerodynamic influences can be expressed in terms of the quasi-steady perturbation terms as

$$\Delta L = QS[C_{L_\alpha}\Delta\alpha + C_{L_{\dot\alpha}}(c/2V)\Delta\dot\alpha] \quad (9.3)$$

Continuing, assume that $C_{L_{\dot\alpha}} \cong 0$ and recognize that the net α perturbation $\Delta\alpha$ is due to both the aircraft motion and the external gust velocity. Forward velocity is assumed to remain unchanged,

$$\Delta\alpha = \frac{\Delta w}{V} - \frac{\Delta w_g}{V} \quad (9.4)$$

Substitution of Eqs. (9.3) and (9.4) into Eq. (9.2) simplifies to a linear, first-order differential equation, i.e.,

$$\lambda \Delta \dot{w} + \Delta w = \Delta w_g(t) \tag{9.5}$$

where the time constant (in seconds) is

$$\lambda = \frac{(W/S)}{C_{L_\alpha}} \frac{2}{\rho V g}$$

The assumption of a sharp-edged gust (cf. Fig. 9.3) encountered at $t = 0$ implies that

$$\Delta w_g(t) = -w_g 1(t)$$

where $1(t)$ is the unit step function.

Solution of Eq. (9.5) due to a step function type of input yields a relation for the aircraft vertical velocity at the c.g. as

$$\Delta w(t) = -w_g(1 - e^{-t/\lambda})1(t) \tag{9.6}$$

The vertical acceleration of the aircraft c.g., from Eq. (9.6), is

$$\Delta \dot{w} = \frac{d\Delta w}{dt} = -\frac{w_g}{\lambda} e^{-t/\lambda} \quad \text{for } t > 0$$

It may be seen that the maximum vertical acceleration of the aircraft c.g. due to its encounter of the idealized sharp-edged gust occurs at $t = 0$, i.e.,

$$(\Delta \dot{w})_{\max} = \Delta \dot{w}(0) = -w_g/\lambda$$

which can be expressed in load factor terms as

$$(\Delta n_z)_{\max} = -\frac{(\Delta \dot{w})_{\max}}{g} = \frac{w_g}{\lambda g}$$

with further simplification to

$$(\Delta n_z)_{\max} = C_{L_\alpha} \frac{\rho V}{2} \frac{w_g}{(W/S)} \tag{9.7}$$

Facts to be observed from Eq. (9.7) relative to aircraft vertical acceleration during a gust encounter include the following.

1) Load factor is directly proportional to lift-curve slope.
2) Load factor is inversely proportional to wing loading, W/S.
3) For a given gust velocity w_g, load factor varies linearly with velocity, as shown on the gust envelope sketch (cf. Fig. 9.2).

Although these observations were made based on the use of simplifying assumptions, the features of aircraft gust response will behave similarly even when more exact models are employed.

The load factor result of Eq. (9.7) is based on quasi-steady aerodynamics. The lift buildup due to $\alpha(t)$ motion by the airfoil and the transient lift due to the entry of the airfoil into a sharp-edged vertical gust do not occur instantaneously as is implied by the quasi-steady assumption. The Wagner and Kussner functions from the theory of unsteady aerodynamics in incompressible flow may be used to account for the finite time required by an airfoil to develop circulation when encountering a temporal change in α.

Example 9.1

Estimate the maximum static gust load factor based on entry into a sharp-edged gust for the Lockheed Jetstar at an airspeed approximating the design cruise V_C. Assume quasi-steady aerodynamics are applied to a pure plunging response for the rigid airframe. From Appendix B.6, find that $M = 0.75$, $h = 20,000$ ft, $V = 778$ fps (TAS), $\rho = 0.001267$ slug-ft^2, $Q = 383.0$ psf, $W = 38,204$ lb, $m = 1187.4$ slugs, $S = 542.5$ ft^2, $c = 10.93$ ft, and $Z_\alpha = -1033.9$ ft-s^{-2}. Also from Fig. 9.2, $w_g = 50$ fps (EAS) = 68.5 (TAS) for V_C.

1) From Z_α, estimate $C_{L_\alpha} = 5.88$ rad^{-1} based on

$$(C_{L_\alpha} + C_D) = -\frac{m}{QS} Z_\alpha$$

2) Maximum load factor under assumptions of Eq. (9.7) is

$$(\Delta n_z)_{max} = C_{L_\alpha} \frac{\rho V}{2} \frac{w_g}{(W/S)} = 2.818 \text{ g} \quad \text{for } w_g = 68.5 \text{ fps (TAS)}$$

Details on the use of the Wagner and Kussner functions from unsteady aerodynamic theory are described by Fung[7] for a sharp-edged gust encounter. For Example 9.1, a gust alleviation factor that reflects the time delay in lift buildup would provide approximately a 15% reduction in $(\Delta n_z)_{max}$.

9.2.2 One-Minus-Cosine Gust

The idealized sharp-edged gust is a very severe type of a velocity profile that seldom occurs in nature. Instead, a discrete gust may be modeled more practically by a ramp input that reaches a peak value in a distance known as the gradient distance. The one-minus-cosine profile, as specified by FAR 25,[6] is more frequently used in determining gust-induced load factors rather than a ramp rising to a steady peak gust value. The evolutionary background for gust modeling appears in the articles by Houbolt[3] and Etkin[1] along with the text by Hoblit.[5]

The one-minus-cosine discrete gust, shown in Fig. 9.4, will be considered as frozen in space while the aircraft transits through it with a constant velocity. The vertical velocity is defined by

$$\Delta w_g = -\frac{w_g}{2}\left(1 - \cos\frac{\pi x}{d}\right) \tag{9.8}$$

where

d = gradient distance (cf. FAR 25), $12.5c$
c = wing chord length
w_g = magnitude of vertical gust velocity

ATMOSPHERIC TURBULENCE 297

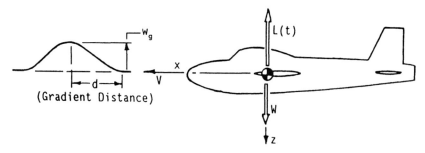

Fig. 9.4 One-minus-cosine discrete gust.

For pure plunging motion, Eq. (9.5) is valid when used in conjunction with the discrete gust of Eq. (9.8), i.e.,

$$\lambda \Delta \dot{w} + \Delta w = \Delta w_g = -\frac{w_g}{2}\left(1 - \cos\frac{\pi x}{d}\right)$$

Assume that the aircraft (cf. Fig. 9.4) encounters the discrete gust at time $t = 0$. Thereafter, the distance traveled into the gust corresponds to $x = Vt$. The vertical response equation, under the quasi-steady aerodynamic assumption, becomes

$$\lambda \Delta \dot{w} + \Delta w = f(t) = -(w_g/2)(1 - \cos \omega t) \quad \text{for } 0 \leq t \leq 2\pi/\omega \quad (9.9)$$

where

λ = time constant of Eq. (9.5), s
$\omega = \pi V/d$, rad/s

A closed-form solution of Eq. (9.9) can be obtained using the Duhamel superposition integral, which is similar to the convolution integral described in Sec. 5.2. Consider the forcing function $f(t)$ as shown in Fig. 9.5 in the form of a tiered layering of small step functions. The Duhamel integral allows one to express the response at a later time t by

$$\Delta w(t) = \int_0^t h(t - \tau) f'(\tau) d\tau \quad (9.10)$$

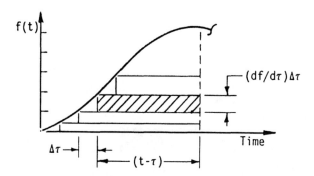

Fig. 9.5 Sketch of convolution by indicial responses.

where

$f'(\tau) = df/d\tau$
$h(t - \tau)$ = response at time t due to unit step input at time τ

Consequently, the solution to Eq. (9.9) may be expressed in integral form as

$$\Delta w(t) = -\frac{w_g \omega}{2} \int_0^t \left[1 - e^{-(t-\tau)/\lambda}\right] \sin \omega \tau \, d\tau \qquad \text{for } 0 \leq t \leq 2\pi/\omega \quad (9.11)$$

which may be integrated to give a solution of

$$\Delta w(t) = \frac{-w_g}{2}\left\{(1 - \cos \omega t) - \frac{1}{1 + (\omega\lambda)^{-2}}\left(e^{-t/\lambda} + \frac{1}{\omega\lambda}\sin \omega t - \cos \omega t\right)\right\} \quad (9.12)$$

This solution is valid during the time while the one-minus-cosine gust is in effect. The maximum normal acceleration will occur in this time frame. Note that Eq. (9.12) satisfies the boundary condition that $\Delta w(0) = 0$. The load factor for the aircraft constrained to the plunging mode becomes

$$\Delta n_z(t) = \frac{w_g}{2g}\left[\omega \sin \omega t + \frac{1}{1 + (\omega\lambda)^{-2}}\left(\frac{1}{\lambda}e^{-t/\lambda} - \frac{1}{\lambda}\cos \omega t - \omega \sin \omega t\right)\right] \quad (9.13)$$

The maximum load factor will occur near to the time for the peak gust value. A numerical procedure is convenient for finding the load factor response when using Eq. (9.13).

Example 9.2

Estimate the load factor response for the Lockheed Jetstar of Example 9.1 due to a one-minus-cosine gust encounter. Assume $w_g = 50$ fps at $V_C = 778$ fps and $h = 20,000$ ft.

1) Determine constants used in the analysis,

$$\omega = \frac{\pi V}{d} = \frac{\pi(778 \text{ fps})}{(12.5)(10.93 \text{ ft})} = 17.890 \text{ rad/s}$$

$$\lambda = \frac{(W/S)}{C_{L_\alpha}}\frac{2}{\rho V g} = 0.753 \text{ s}$$

where T is the period of the gust profile equal to $2\pi/\omega = 0.351$ s.

2) Substituting the constants into Eq. (9.13) provides a numerical relation, which may be solved using MATLAB, i.e.,

```
w=17.890; % Frequency, omega
wg=68.5.; g=32.174; K=0.5*wg/g ;
for i=1:101,
```

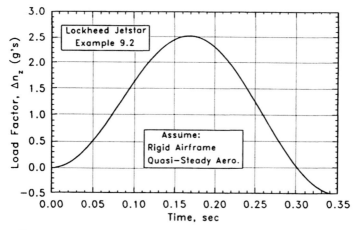

Fig. 9.6 Typical load factor response to a one-minus-cosine gust.

```
t(i)=(0.351/100.)*(i-1);
N(i)=0.09750*sin(w*t(i));
N(i)=N(i) + 1.3168*(exp(-1.324*t(i)) - cos(w*t(i)));
n(i)=K*N(i) ;
end
plot(t,n)
```

The load factor response is shown in Fig. 9.6. The maximum load factor $(\Delta n_z)_{max}$ is found from the time-history response to be 2.525 g occurring at $t = 0.168$ s.

If the pure plunging aircraft response to the vertical gust input of Eq. (9.9) were to include the influence of unsteady aerodynamics, the differential equation would be modified using a form of the convolution integral combined with the following functions, which account for the time lag in lift buildup. Consider, $K_W(t)$ is Wagner's function for lift lag due to $\alpha(t)$ and $K_K(t)$ is Kussner's function for lift lag due to $(\Delta w_g/V)1(t)$. The behavior of these functions is illustrated in Fig. 9.7 (cf. Fung[7] for a discussion of K_W and K_K). The governing differential equation becomes an integral-differential equation when the unsteady influences are included, i.e.,

$$\lambda \Delta \dot{w} + \int_0^t K_W(t-\tau)\Delta \dot{w}(\tau)d\tau = \int_0^t K_K(t-\tau)\Delta \dot{w}_g(\tau)d\tau \qquad (9.14)$$

Details relating to Eq. (9.14) and its solution by numerical methods are described by Hoblit.[5] Although the K_W and K_K functions of Fig. 9.7 are valid only for $M = 0$, their usage has been found to provide a good approximation for gust modeling in subsonic flight.[5] The maximum aircraft load factor response to the one-minus-cosine gust input has traditionally been modeled by applying an empirical gust alleviation factor to the quasi-steady response for a sharp-edged gust (in

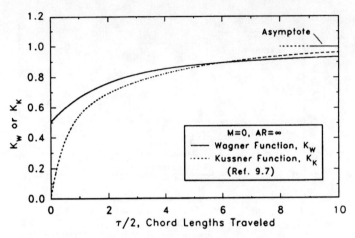

Fig. 9.7 Sketch of Wagner and Kussner functions.

accord with the estimation procedure cited in FAR 25[6]), i.e.,

$$(\Delta n_z)_{max} = K_g (\Delta n_z)'_{max} = K_g \left[C_{L_\alpha} \frac{\rho V}{2} \frac{w_g}{(W/S)} \right] \quad (9.15)$$

where

$(\Delta n_z)'_{max}$ = sharp-edged gust response, Eq. (9.7)
μ = airplane mass ratio, $[2(W/g)/\rho S c C_{L_\alpha}]$
K_g = gust alleviation factor, $[0.88\mu/(5.3 + \mu)]$

Example 9.3

Estimate the load factor response for the Lockheed Jetstar of Example 9.1 due to a one-minus-cosine gust encounter when the influence of unsteady aerodynamics is considered. Assume $w_g = 68.5$ fps at $V_C = 778$ fps, and $h = 20,000$ ft.

1) Determine the dimensionless mass parameter μ from

$$\mu = (2m)/(\rho S c C_{L_\alpha}) = 53.76$$

2) Find the gust alleviation factor,

$$K_g = (0.88\mu)/(5.3 + \mu) = 0.801$$

3) Estimate the maximum load factor by Eq. (9.15) using K_g in conjunction with $(\Delta n_z)'_{max}$ found in Example 9.1 for the idealized sharp-edged gust, i.e.,

$$(\Delta n_z)_{max} = K_g (\Delta n_z)'_{max} = (0.801)(2.818) = 2.257 \text{ g}$$

Comparison of the load factor estimate obtained from Example 9.2 (i.e., based on quasi-steady aerodynamics) with the estimate from Example 9.3 (i.e., based on $M = 0$ unsteady aerodynamics) indicates that the quasi-steady aerodynamic basis provides an estimate that is conservative by approximately 10%.

$$\frac{[(\Delta n_z)_{max}]_{Ex.9.3}}{[(\Delta n_z)_{max}]_{Ex.9.2}} = \frac{2.257}{2.525} = 0.89$$

Let us consider the aircraft responding with the short-period approximation as developed in Sec. 6.3.1 modified to reflect gust inputs of $\alpha_g(t)$ and $q_g(t)$ in place of control inputs. The state equation can be written for the short-period response of the rigid airframe based on quasi-steady aerodynamics along with the assumption that $Z_{\dot{\alpha}}$ and Z_q are negligible, i.e.,

$$\begin{Bmatrix} \dot{\alpha} \\ \dot{q} \end{Bmatrix} = \begin{bmatrix} Z_\alpha/V & 1 \\ M'_\alpha & M'_q \end{bmatrix} \begin{Bmatrix} \alpha \\ q \end{Bmatrix} + \begin{Bmatrix} Z_\alpha/V \\ M'_\alpha \end{Bmatrix} \alpha_g + \begin{Bmatrix} 0 \\ M_q \end{Bmatrix} q_g \qquad (9.16)$$

with normal acceleration output given by

$$\Delta n_z = -Z_\alpha(\alpha + \alpha_g)/(32.174)$$

Although the pure plunging response to a one-minus-cosine gust was the historical basis for establishing an empirical relation to predict aircraft load factors in accord with flight-test measurements, the use of Eq. (9.16) will provide an insight as to the import of increasing the airframe degrees of freedom when representing the airframe dynamics. Consideration of the one-minus-cosine vertical gust input being used to define a time-history of q_g will be left as a problem.

Example 9.4

Estimate the load factor response for the Lockheed Jetstar of Example 9.1 due to a one-minus-cosine gust encounter when the quasi-steady response dynamics are modeled by the short-period approximation. Assume $w_g = 68.5$ fps at $V_C = 778$ fps, and $h = 20{,}000$ ft.

1) The aircraft plant $[A]$ may be obtained from Appendix B.6, i.e.,

$$[A] = \begin{bmatrix} -1.329 & 1.0 \\ -12.126 & -1.537 \end{bmatrix}$$

$$B = [-1.329 \quad -12.126]^T$$

2) Obtain the time-history response for load factor during $0 \le t \le 0.35$ s as in Example 9.2 using MATLAB in conjunction with Eq. (9.16).

```
% Input problem constants in addition to [A] and {B}
Zalph = -1033.9; wg = 68.5; g =32.174;
c = 10.93; d = 12.5*c; V = 778.;
w = pi*V/d; % Frequency, omega
for i=1:101,
t(i)=(0.351/100.)*(i-1);
ALPHg(i)=(0.5*wg/V)*(1. - cos(w*t(i)) );
```

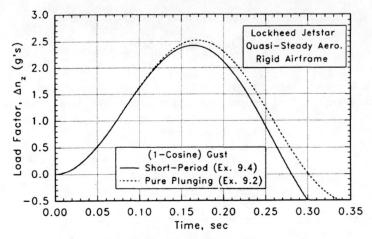

Fig. 9.8 Effects on gust response of degree of freedom.

```
end
C=[-Zalph/g 0.]; D=-Zalph/g;
[nz,X]=lsim(A,B,C,D,ALPHg',t');
plot(t,nz)
```

The maximum load factor, shown in Fig. 9.8, was found to be $(\Delta n_z)_{max} = 2.423 g$ occurring at $t = 0.165$ s.

Comparison is shown in Fig. 9.8 of the load factor time histories for both the pure plunging and the two-degree-of-freedom responses. The addition of pitch angle rotation to the response model results in the maximum load factor decreasing by about 0.1 g after the startup transient has occurred. Although the examples considered applied to the Lockheed Jetstar, the trends may be considered as being typical for most aircraft.

9.3 Random Processes

Atmospheric turbulence, although modeled approximately in Sec. 9.2 by discrete vertical gust considerations, is in reality a random process. Analysis of its effects involves the mathematics of generalized harmonic theory combined with power spectrum methods. The purpose of this section will be to introduce a few principles of random processes in order to facilitate estimation of the root-mean-square gust response of an aircraft.

Definition: A random function is characterized by the fact that knowledge of the past does not allow one to predict the future.

There are many references available dealing with the subject of random processes and their effects on system dynamics. Classical views have been presented by Tsien,[8] Etkin,[9] and Fung.[7] Modern control theory views are presented by Friedland[10] and McLean.[11]

ATMOSPHERIC TURBULENCE

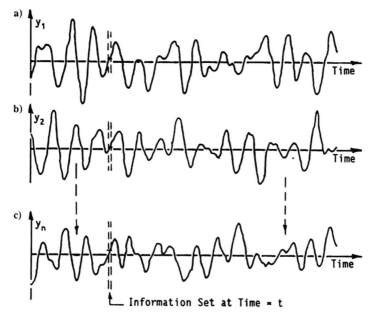

Fig. 9.9 Ensemble of random functions.

9.3.1 Ensemble of Random Functions

Let $y(t)$ correspond to a random function of time such as a velocity component of turbulent flow at a spatial location, the lift acting on an airfoil passing through a gust field, etc. Next consider n identical processes generating the nondeterministic information $y(t)$ as shown in Fig. 9.9. The collection of information forms an ensemble or set, which at time t has values $y_i(t)$ for $i = 1$ to n, with related statistical properties such as a mean, mean square, etc., i.e., $y_1(t), y_2, \ldots, y_n(t)$ is the set of information at time t.

The set of information at any time t may range in value from $-\infty$ to $+\infty$ although the set may be bounded by physical limitations. The value of any member of the set is not deterministic and, therefore, is described statistically by the probability of its occurrence. Properties of the probability P are that probability P is a positive number; P will range in value from 0 to 1, i.e., $0 \leq P \leq 1$; $P = 1$ if the event is certain; and $P = 0$ if the event cannot occur.

The probability is expressed by Eq. (9.17) as a functional for a value x where $F(x)$ is a monotonically increasing function that satisfies the relations of $F(x = -\infty) = 0$ and $F(x = +\infty) = 1$ as shown in Fig. 9.10,

$$P(y \leq x) = F(x) \qquad (9.17)$$

The derivative of $F(x)$ is defined as the probability distribution function (PDF) [$=p(x)$], by the relation

$$F'(x) = \frac{dF}{dx} = p(x)$$

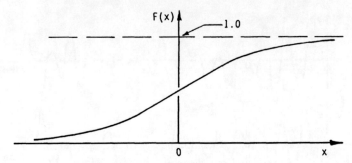

Fig. 9.10 Sketch of the probability function.

From the monotonic property of $F(x)$, it is evident that $p(x) \geq 0$. The probability of an occurrence with value x is given by

$$F(x) = \int_{-\infty}^{x} p(z)\,dz \qquad (9.18)$$

and the probability that an event will occur between $x = A$ and $x = B$ is

$$F(B) - F(A) = \int_{A}^{B} p(z)\,dz \qquad (9.19)$$

The first and second moments of the probability distribution function have a practical meaning to the dynamics of a system under random excitation.

Mean value.   The first moment of the PDF is known as the average or mean value and is expressed by

$$E(x) = \int_{-\infty}^{+\infty} z p(z)\,dz = m = \bar{x} \qquad (9.20)$$

Definition: $E(x) = m$ is the expectation for the mean value of x.

Mean square.   The second moment of the PDF provides an estimate for the mean square value, i.e.,

$$E(x^2) = \int_{-\infty}^{+\infty} z^2 p(z)\,dz = \overline{x^2} \qquad (9.21)$$

Definition: $E(x^2) = \overline{x^2}$ is the expectation for the mean square of x.
The root mean square, a frequent value of interest, is

$$\text{rms} = [E(x^2)]^{\frac{1}{2}}$$

The variation of the mean square about the mean value, also a statistical term of interest, is known as the variance of x and is given by

$$\sigma^2 = \int_{-\infty}^{+\infty} (z-m)^2 p(z)\,dz$$

Expanding the variance definition yields

$$\sigma^2 = \int_{-\infty}^{+\infty} (z^2 - 2mz + m^2)^2 p(z)\,dz = E(x^2) - 2mE(x) + m^2$$

But it is recognized from Eq. (9.20) that $m = E(x)$; hence,

$$\sigma^2 = E\{[x - E(x)]^2\} = E(x^2) - E^2(x) \tag{9.22}$$

Definition: σ is the standard deviation.

It should be noted in the proceeding discussion that $E(x) = \bar{x}$ and $E(x^2) = \overline{x^2}$ are interchangeable notations. In general, the expectation notation will be used, where appropriate, in the following discussions.

A common type of probability function that occurs frequently in nature is the normal (or Gaussian) distribution defined as

$$F(x) = \frac{1}{\sqrt{2\pi}} \int_{-\infty}^{x} e^{-z^2/2}\,dz \tag{9.23}$$

whereas the PDF of the normal distribution, expressed as

$$p(x) = \frac{1}{\sqrt{2\pi}} e^{-x^2/2} \tag{9.24}$$

may be recognized as the bell-shaped curve, as shown in Fig. 9.11. The normal distribution, as given by Eq. (9.23), corresponds to $E(x) = 0$ and $\sigma = 1$. The probability of an occurrence using the Gaussian (normal) distribution is a well-known relation expressed in many mathematical tables by the erf(x), with values

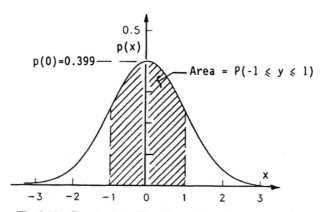

Fig. 9.11   Sketch of the Gaussian probability distribution.

Table 9.1 Error function

x	erf(x)
1	0.683
2	0.955
3	0.997

as typically shown in Table 9.1, i.e.,

$$P(-x \leq y \leq +x) = \text{erf}(x) = \frac{1}{\sqrt{2\pi}} \int_{-x}^{+x} e^{-z^2/2} dz$$

Table 9.1 makes the well-known statement that a random process with a Gaussian probability distribution has a 99.7% probability that the event is less than or equal to 3σ.

A more general expression for a Gaussian probability distribution for a random process with $E(x) = m$ and deviation of σ is

$$p(x) = \frac{1}{\sigma\sqrt{2\pi}} \exp\left[-\frac{(x-m)^2}{2\sigma^2}\right] \tag{9.25}$$

The statistical properties described by Eqs. (9.17–9.25) are based on the concept of probability functions for a set of observations at a prescribed time t. Alternatively, statistical averages at time t may be made for the n members of the set as follows:

$$E(x) \cong \overline{x(t)}_n = \frac{1}{n} \sum_{i=1}^{n} x_i(t)$$

and

$$E(x^2) \cong \overline{x^2(t)}_n = \frac{1}{n} \sum_{i=1}^{n} x_i^2(t)$$

These expectations will hold approximately for n large and be exact in the limit as $n \to \infty$, e.g.,

$$E(x) = \lim_{n \to \infty} \frac{1}{n} \sum_{i=1}^{n} x_i(t)$$

An important property of random functions is the assumption that the statistical properties be independent of the time considered. This property leads to the concept of a stationary random function.

Definition: If $E(x)$ and σ are independent of time t, then $x(t)$ is a stationary function in a statistical sense.

Instead of averaging statistically over a large number of concurrent observations, it is convenient instead to average over one observation with respect to time. This concept assumes that the functional is ergodic.

Definition: A process is ergodic when the time average equals the ensemble average.

These two assumptions (i.e., stationarity and ergodicity) when applied to random processes imply with a probability of one that

$$\lim_{n\to\infty} \frac{1}{n} \sum_{i=1}^{n} x_i(t) = \lim_{n\to\infty} \frac{1}{n\Delta t} \sum_{i=1}^{n} x(i\Delta t)\Delta t \qquad (9.26)$$

Equation (9.26) may expressed in integral form as

$$E(x) = \lim_{T\to\infty} \frac{1}{2T} \int_{-T}^{T} x(t)\,dt \qquad (9.27)$$

along with a parallel relation for the mean square expectation,

$$E(x^2) = \lim_{T\to\infty} \frac{1}{2T} \int_{-T}^{T} x^2(t)\,dt \qquad (9.28)$$

9.3.2 Power Spectrum Concepts

The power spectrum represents a frequency viewpoint for describing the square (hence, power) of a random variable that is originally considered in the time domain.

The original time-varying random signal or function $x(t)$, shown in Fig. 9.12a, is processed (or filtered) through a unit rectangular filter, shown in Fig. 9.12b, to yield a truncated signal $x_T(t)$ that is zero when $|t| > T$ as shown in Fig. 9.12c. The signal or function shown in Fig. 9.12c is absolutely integrable because T is

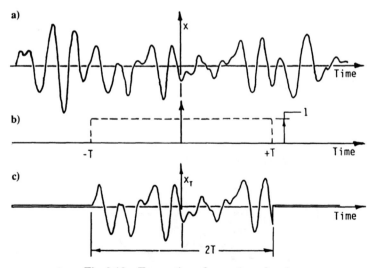

Fig. 9.12   Truncation of a random signal.

finite and the function is assumed to be of bounded variation. Hence,

$$\int_{-\infty}^{\infty} |x_T| \, dt = \int_{-T}^{T} |x_T| \, dt = \text{exists}!$$

Consequently, a Fourier transform of $x_T(t)$ exists and may be expressed as

$$x_T(t) = \frac{1}{\sqrt{2\pi}} \int_{-\infty}^{\infty} X_T(\omega) e^{+i\omega t} \, d\omega$$

$$X_T(\omega) = \frac{1}{\sqrt{2\pi}} \int_{-T}^{+T} x_T(t) e^{-i\omega t} \, dt$$
(9.29)

Since $X_T(\omega)$ in Eq. (9.29) is a complex quantity whereas $x_T(t)$ is a real quantity, it follows that

$$X_T^*(-\omega) = X_T(+\omega)$$

where $X_T^*(-\omega)$ is the complex conjugate of $X_T(-\omega)$.

Another useful relation in the development of the power spectral density is Parseval's theorem, which can be described in the preceding notation as

$$\int_{-T}^{T} x_T^2(t) \, dt = \int_{-\infty}^{\infty} |X_T(\omega)|^2 \, d\omega$$
(9.30)

Example 9.5

Verify Parseval's theorem in the form of Eq. (9.30).

1) Apply the Fourier transform definition of Eq. (9.29) to the left-hand side of Eq. (9.30), i.e.,

$$\int_{-T}^{T} x_T^2(t) \, dt = \int_{-T}^{T} x_T(t) \left\{ \frac{1}{\sqrt{2\pi}} \int_{-\infty}^{\infty} X_T(\omega) e^{+i\omega t} \, d\omega \right\} dt$$

2) Then let $\hat{\omega} = -\omega$ and $d\hat{\omega} = -d\omega$. Also note that $X_T(-\hat{\omega}) = X_T^*(+\omega)$, which allows the Fourier transform in the preceding relation to be restated as

$$\int_{-\infty}^{\infty} X_T(\omega) e^{+i\omega t} \, d\omega = \int_{-\infty}^{\infty} X_T^*(\hat{\omega}) e^{-i\hat{\omega} t} \, d\hat{\omega}$$

3) Next, reverse the order of integration in the double integral,

$$\int_{-T}^{T} x_T^2(t) \, dt = \int_{-\infty}^{\infty} X_T^*(\hat{\omega}) \left\{ \frac{1}{\sqrt{2\pi}} \int_{-T}^{T} x_T(t) e^{-i\hat{\omega} t} \, dt \right\} d\hat{\omega}$$

$$= \int_{-\infty}^{\infty} X_T^*(\hat{\omega}) X_T(\hat{\omega}) \, d\hat{\omega}$$

Hence, verifying that

$$\int_{-T}^{T} x_T^2(t)\,dt = \int_{-\infty}^{\infty} |X_T(\omega)|^2\,d\omega \qquad \text{QED}$$

Power spectral density. The development of the power spectral density follows from applying Parseval's theorem to Eq. (9.28) to obtain an alternate form for the mean square that involves a frequency-dependent functional, i.e.,

$$E(x^2) = \lim_{T \to \infty} \int_{-\infty}^{\infty} \frac{|X_T(\omega)|^2}{2T}\,d\omega = \lim_{T \to \infty} \int_{0}^{\infty} \frac{|X_T(\omega)|^2}{T}\,d\omega$$

The limiting action on the integrand in the preceding expression leads to the definition of the power spectral density, i.e.,

$$\lim_{T \to \infty} \frac{|X_T(\omega)|^2}{T} = \phi(\omega)$$

Therefore, the expectation for the mean square may be described statistically in terms of its frequency content by

$$E(x^2) = \int_{0}^{\infty} \phi(\omega)\,d\omega \qquad (9.31)$$

Properties of $\phi(\omega)$ to note include that 1) it is a symmetric function, $\phi(\omega) = \phi(-\omega)$; 2) the contribution to $E(x^2)$ in range from (ω) to $(\omega + d\omega)$ is $\phi(\omega)d\omega$; and 3) the total contribution to $E(x^2)$ from frequencies in the range of $\omega = 0$ to ω_0 is

$$\int_{0}^{\omega_0} \phi(\omega)\,d\omega$$

Autocorrelation function. The second moment of a stationary random process with a time lag τ is known as the autocorrelation function and is expressed as

$$R_{xx}(\tau) = E[x(t)x(t+\tau)] = \lim_{T \to \infty} \frac{1}{2T} \int_{-T}^{T} x(t)x(t+\tau)\,dt \qquad (9.32)$$

Inasmuch as the time origin has no influence when determining a statistical property for a stationary random process, it follows that

$$R_{xx}(\tau) = \lim_{T \to \infty} \frac{1}{2T} \int_{-T}^{T} x(t-\tau)x(t)\,dt \qquad (9.33)$$

By comparing Eqs. (9.32) and (9.33), the following may be concluded.
1) It is a symmetric function,

$$R_{xx}(\tau) = R_{xx}(-\tau)$$

2) The autocorrelation function with zero time lag is the mean square,
$$R_{xx}(0) = E(x^2)$$
3) And for a random process,
$$\lim_{\tau \to \infty} R_{xx}(\tau) = 0$$

A Fourier transform of the autocorrelation function will exist based on the following reasonable assumptions.

1) The absolute integral
$$\int_0^\infty |R_{xx}(\tau)| \, d\tau = \text{exists}$$

2) $R_{xx}(\tau)$ is of bounded variation.

The Fourier transform of $R_{xx}(\tau)$ takes the form
$$f(\omega) = \frac{1}{\sqrt{2\pi}} \int_{-\infty}^\infty R_{xx}(\tau) e^{-i\omega\tau} \, d\tau$$

But because $R_{xx}(\tau) = R_{xx}(-\tau)$, the Fourier transform of $f(\omega)$ may be restated as
$$f(\omega) = \frac{1}{\sqrt{2\pi}} \int_0^\infty R_{xx}(\tau) [e^{+i\omega\tau} + e^{-i\omega\tau}] \, d\tau$$

which becomes the Fourier cosine transform, i.e.,
$$f(\omega) = \sqrt{\frac{2}{\pi}} \int_0^\infty R_{xx}(\tau) \cos(\omega\tau) \, d\tau$$

which has an inverse of
$$R_{xx}(\tau) = \sqrt{\frac{2}{\pi}} \int_0^\infty f(\omega) \cos(\omega\tau) \, d\omega$$

Because $R_{xx}(\tau) = E(x^2)$ when $\tau = 0$, it is apparent that
$$E(x^2) = \sqrt{\frac{2}{\pi}} \int_0^\infty f(\omega) \, d\omega$$

Comparison with Eq. (9.31) shows that $f(\omega)$ differs from $\phi(\omega)$ only by a constant, i.e.,
$$\phi(\omega) = \sqrt{(2/\pi)} f(\omega)$$

Finally, it can be stated (i.e., the Weiner–Kintchine relation[8]) that the autocorrelation has the interesting property of being the Fourier cosine transform of the power spectral density, i.e.,
$$R_{xx}(\tau) = \int_0^\infty \phi(\omega) \cos(\omega\tau) \, d\omega \tag{9.34}$$

and
$$\phi(\omega) = \frac{2}{\pi} \int_0^\infty R_{xx}(\tau) \cos(\omega\tau) \, d\tau$$

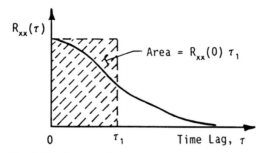

Fig. 9.13 Sketch of the autocorrelation function.

Correlation length. In general, the area under the curve of $R_{xx}(\tau)$ vs time lag τ will have a finite value for a random-natured functional. From Fig. 9.13, it can be seen that the area under the autocorrelation curve can be considered in the form

$$\int_0^\infty R_{xx}(\tau)\,d\tau = R_{xx}(0)\tau_1$$

where τ_1 is the equivalent time lag.

The lag time τ_1 can be considered as related to a scale of turbulence in a moving airstream by the expression that

$$\tau_1 = L/V \qquad (9.35)$$

where L is the scale of turbulence.

If one considered two spatial points in a moving airstream that were aligned with the freestream velocity, then on an approximate statistical basis, it could be assumed that any flow measurements farther apart than the distance L (shown in Fig. 9.14) would be uncorrelated. The scale of turbulence can be interpreted as a statistical measure of the size of turbulent eddies in the airflow.

1) L small represents a small grain of turbulence such as found in wind-tunnel flow (on the order of 1 in. or less).

2) L large represents a large grain of turbulence such as in the wake of a mountain or in a storm encounter (on the order of 1000–2000 ft).

Isotropic turbulence. Wind-tunnel tests using a hot-wire anemometer system to measure rapid variations of tunnel turbulent velocity have shown that (cf. Fung[7] and Tsien[8]) approximate autocorrelations for the $u(t)$ and $w(t)$ velocity components as

$$R_{uu}(\tau) = R_{uu}(0)e^{-V\tau/L_1} \qquad (9.36)$$

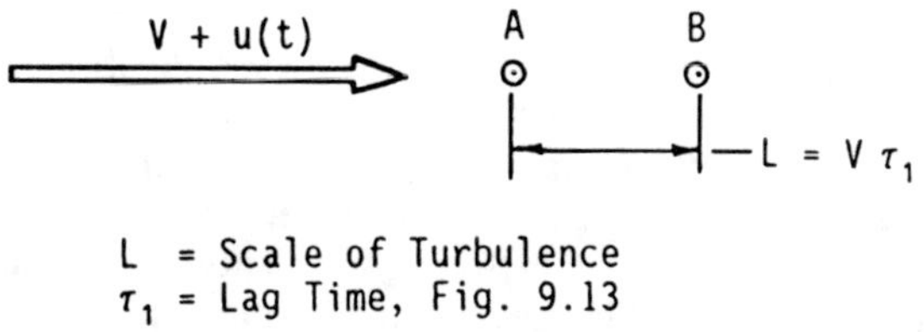

L = Scale of Turbulence
τ_1 = Lag Time, Fig. 9.13

Fig. 9.14 Scale of turbulence.

and

$$R_{ww}(\tau) = R_{ww}(0)e^{-V\tau/L_2}[1 - (V\tau/2L_2)] \quad (9.37)$$

where

L_1 = scale of turbulence in the $u(t)$ direction
L_2 = scale of turbulence in the $w(t)$ direction

A Fourier cosine transform of $R_{uu}(\tau)$ [Eq. (9.36)] yields

$$\phi_u(\omega) = \frac{2}{\pi}\int_0^\infty R_{uu}(\tau)\cos(\omega\tau)\,\mathrm{d}\tau = \phi_u(0)\frac{1}{1+\Omega_1^2} \quad (9.38)$$

where

$$\phi_u(0) = \frac{2}{\pi}\frac{L_1}{V}R_{uu}(0) = \frac{2}{\pi}\frac{L_1}{V}\sigma_u^2$$

and the dimensionless frequency is $\Omega_1 = (\omega L_1/V)$.

A similar transform applied to the vertical turbulence autocorrelation, Eq. (9.37), gives

$$\phi_w(\omega) = \phi_w(0)\frac{1+3\Omega_2^2}{(1+\Omega_2^2)^2} \quad (9.39)$$

where

$$\phi_w(0) = \frac{1}{\pi}\frac{L_2}{V}R_{ww}(0) = \frac{1}{\pi}\frac{L_2}{V}\sigma_w^2$$

$$\Omega_2 = (\omega L_2/V)$$

The turbulence relations in the lateral z direction are generally assumed as the same as in the vertical y direction on the basis of having isotropic flow.

Relations, Eqs. (9.38) and (9.39), are known as the Dryden turbulence model. They are considered valid for use in describing isotropic turbulence as found in natural phenomena, providing the scale of turbulence is appropriate.

9.3.3 Random Process Summary

The development of random processes described in Sec. 9.3 has included the following items: probability concepts; PDF; moments of the PDF, where the first moment is the mean and the second moment is mean square; variance, σ^2; standard deviation, σ; Gaussian (normal) probability distribution, including the bell-shaped curve and $[F(x) - F(-x)] = \mathrm{erf}(x)$; statistical averages; stationary function; ergodic hypothesis; power spectral density; autocorrelation function; correlation length, i.e., scale of turbulence; and Dryden turbulence model.

In the following sections, the power spectral density of turbulence will play a significant role in determining the expected load factors by aircraft encountering atmospheric gusts. However, an appreciation of the background for the $\phi(\omega)$ concept should provide the reader with a stronger foundation for its usage.

An alternate functional to describe atmospheric turbulence is the von Kármán model. Although more accurate than the Dryden model at low frequencies, the von Kármán model does not readily lend itself to numerical analyses of aircraft response to gust inputs. Consequently, the techniques to be described in the following sections will favor the usage of the Dryden model.

9.4 Random Gust Response

9.4.1 System Response to Random Excitation

The system response to a random input will be first illustrated for a second-order harmonic oscillator that is excited by a white noise signal. Consider the stable, linear second-order harmonic oscillator with constant coefficients in conjunction with a random-natured forcing function $x(t)$, i.e.,

$$\ddot{y} + 2\zeta\omega_n \dot{y} + \omega_n^2 y(t) = x(t) \tag{9.40}$$

Assume that the random forcing function is of bounded variation and absolutely integrable. With this understanding, a Fourier transform can be applied to both sides of Eq. (9.40). Then if $Y(\omega) = \mathcal{F}[y(t)]$ and $X(\omega) = \mathcal{F}[x(t)]$ represent Fourier transforms in accord with Eq. (9.29), Eq. (9.40) can be viewed in the transform domain as

$$Z(\omega)Y(\omega) = X(\omega)$$

or

$$Y(\omega) = [Z(\omega)]^{-1} X(\omega) = G(\omega)X(\omega) \tag{9.41}$$

where

$Z(\omega)$ = complex-natured impedance, $[(\omega_n^2 - \omega^2) + 2i\zeta\omega_n\omega]$
$G(\omega)$ = complex-natured transfer function, $[Z(\omega)]^{-1}$

It will be noted that the results of Eq. (9.41) parallel Laplace transform techniques for solving a differential equation; however, in the preceding development, the function $x(t)$ was not restricted to starting at $t = 0$. The transfer function could alternatively have been obtained by letting $s = i\omega$ in $G(s)$ where s is the Laplace transform function.

Next, apply the inverse Fourier transform to $Y(\omega)$ to obtain

$$y(t) = \frac{1}{\sqrt{2\pi}} \int_{-\infty}^{\infty} Y(\omega)e^{+i\omega t} d\omega = \frac{1}{\sqrt{2\pi}} \int_{-\infty}^{\infty} G(\omega)X(\omega)e^{+i\omega t} d\omega \tag{9.42}$$

Although Eq. (9.42) was developed using a linear second-order system to act as a filter between the random input and the output, the complex impedance or its inverse, the transfer function, could equally represent more general systems or relations.

Let us consider a truncated form, y_T, in order to develop the mean square expectation of the output, i.e.,

$$E(y^2) = \lim_{T\to\infty} \frac{1}{2T} \int_{-T}^{T} y_T^2 \, dt = \lim_{T\to\infty} \frac{1}{2T} \int_{-\infty}^{\infty} |Y_T(\omega)|^2 \, d\omega$$

where Parseval's theorem, Eq. (9.30), has been used in order to obtain a result in the frequency domain. But by applying Eq. (9.41) to the truncated function and recognizing symmetry conditions, it can be shown that

$$E(y^2) = \lim_{T\to\infty} \frac{1}{2T} \int_{-\infty}^{\infty} |G(\omega)|^2 |X_T|^2 \, d\omega = \lim_{T\to\infty} \frac{1}{T} \int_{0}^{\infty} |G(\omega)|^2 |X_T|^2 \, d\omega$$

The limit expression in the integrand can be interpreted in the form of a power spectral density similar to the technique used in developing Eq. (9.31), i.e.,

$$E(y^2) = \int_{0}^{\infty} |G(\omega)|^2 \phi(\omega) \, d\omega \tag{9.43}$$

where the input is

$$E(x^2) = \int_{0}^{\infty} \phi(\omega) \, d\omega \tag{9.31}$$

If the input were a Gaussian process, it should be noted that the filter action does not alter the output's probability distribution; i.e., the output also is Gaussian.

9.4.2 Aircraft Response to a Turbulent Gust

The aircraft normal load factor, in response to a turbulent vertical gust, may be found by a series application of the Dryden vertical gust model's transfer function (squared) to the aircraft transfer function (squared) of normal load factor to a vertical gust input. This statement is illustrated by the block diagram shown in Fig. 9.15 in accord with the principles implied by Eq. (9.44). The expectation of the normal load factor response is obtained by integrating the power spectral density, also shown in Fig. 9.15.

The Dryden vertical gust model, Eq. (9.39), may be expressed in a transfer function format by recognizing that

$$\phi_{w_g}(\omega) = |G_{w_g}(\omega)|^2 \sigma_w^2 \tag{9.44}$$

where the complex natured transfer function is

$$G_{w_g}(\omega) = G_{w_g}(s)\big|_{s=i\omega}$$

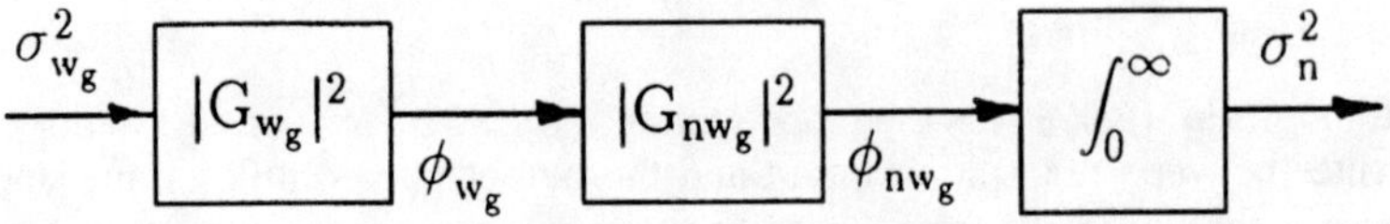

Fig. 9.15 Load factor response to a vertical gust input.

It is conveneint, for numerical analysis purposes, to consider the Dryden vertical gust transfer function [Eq. (9.39)] in terms of the Laplace transform variable, i.e.,

$$G_{w_g}(s) = K \frac{(s+\theta)}{(s+\lambda)^2} \quad (9.45)$$

where

$$K = (3V/\pi L_2)^{\frac{1}{2}}, \; s^{-1/2}$$
$$\theta = V/\sqrt{3}L_2, \; s^{-1}$$
$$\lambda = V/L_2, \; s^{-1}$$

and L_2 is the scale of vertical turbulence gust, in feet.

The airframe longitudinal response is based on the short-period approximation [Eq. (9.16)] where it is noted that $\alpha = w/V$ and $\alpha_g = w_g/V$. In state-variable format,

$$\dot{x} = Ax + Bw_g$$
$$a_n = Cx + Dw_g$$

where $[A]$, $\{B\}$, $[C]$, and D are in accord with Eq. (9.16) whereas $\{x\} = [w \; q]^T$. The transfer function of $G_{nw_g}(s)$ becomes

$$G_{nw_g}(s) = C[sI - A]^{-1}B + D$$

which yields the appropriate power spectral density as

$$\phi_{nw_g}(\omega) = \left|G_{nw_g}(s)\right|^2_{s=i\omega}$$

and, finally, gives the output power spectral density as

$$\phi_n(\omega) = \phi_{w_g}(\omega)\phi_{nw_g}(\omega) \quad (9.46)$$

The expected value for normal load factor is obtained from the integration of the output power spectral density, i.e.,

$$E(n_z^2) = \int_0^\infty \phi_n(\omega)\,d\omega \quad (9.47)$$

Example 9.6

Use the transfer function approach to determine the expected value of normal acceleration when the Lockheed Jetstar ($M = 0.75$ at $h = 20,000$ ft) encounters a turbulent vertical gust with a $\sigma_w = 20$ ft/s. Represent the vertical gust by the Dryden model using a scale of turbulence $L_2 = 1750$ ft.

1) Turbulence constants for the Dryden vertical gust model [Eq. (9.45)] are

$$K = (3V/\pi L_2)^{\frac{1}{2}} = 0.6516 \text{ s}^{-1/2}$$
$$\theta = V/\sqrt{3}L_2 = 0.2567 \text{ s}^{-1}$$
$$\lambda = V/L_2 = 0.4446 \text{ s}^{-1}$$

which establishes the vertical gust transfer function as

$$G_{w_g}(s) = (0.6516)\frac{(s+0.2567)}{(s^2+0.8892s+0.1977)}$$

2) Aircraft matrices are

$$[A] = \begin{bmatrix} -1.329 & 778.0 \\ -0.0156 & -1.537 \end{bmatrix}$$

$$\{B\} = [-1.329 \quad -0.0156]^T$$

$$[C] = [-1.329 \quad 0.0]$$

$$D = -1.329$$

$$\{x\} = [w \quad q]^T$$

3) Obtain the transfer functions followed by a numerical analysis of Eqs. (9.44–9.47). Constants from steps 1 and 2 are assumed to have been entered and available using MATLAB.

```
W=0.0:0.10:40.0; % Set frequency range, rad/sec
% Establish Dryden vertical gust model
NUMGUST=K*[ 1.0  0.2567 ]; % Polynomial in powers of s
DENGUST=[ 1.0  0.8892  0.1977 ];
[MAGGUST,PHASE]=bode(NUMGUST,DENGUST,W);
XIN2=MAGGUST.*MAGGUST; % |G(w)|^2 for input
% Find |G(w)|^2 for aircraft load factor response
[NUMNW,DENNW]=ss2tf(A,B,C,D,1);
disp(NUMNW)
        -1.3290  -2.0247  0.0; % Polynomial in powers of s
disp(DENNW)
         1.0000   2.8660  14.1795 ;
[MAGNW,PHASE]=bode(NUMNW,DENNW,W);
NWG2=MAGNW.*MAGNW;
% Find output P.S.D. response
YOUT=XIN2.*NWG2;
% Trapezoidal rule for integrations
```

```
N=401; SIGMA=0.0; NZ2=0.0; DELW=0.10;
    for i=1:(N-1),
    SIGMA=SIGMA + 0.5*DELW*( XIN2(i)+XIN2(i+1) );
    NZ2=NZ2 + 0.5*DELW*( YOUT(i)+YOUT(i+1) );
    end
disp(SIGMA)
    0.9894 ; % $\sigma_W^2$, ft$^2$-sec$^{-2}$
disp(NZ2)
    0.4435; % $\sigma_n^2$, ft$^2$-sec$^{-4}$
```

4) From the MATLAB listing of step 3, find that $\sigma_n = 0.6660$ ft-s^{-2} due to gust $\sigma_w = 0.995$ ft-s^{-1}. The final acceleration response due to $\sigma_w = 20$ ft-s^{-1} is $\sigma_n = 13.390$ ft-s^{-2} (0.416 g).

The numerical estimations for σ_w (normalized input) and σ_n (output) in Example 9.5 were determined by using a trapezoidal integral approximation for a finite frequency range of $0 < \omega < 40.0$. The normalized gust transfer function, Eq. (9.44), would yield $\sigma_w = 1.0$ ft/s if ω_{max} were infinite; however, the frequency truncation resulted in $\sigma_w = 0.995$, which corresponds to a 0.5% error. The three-sigma value for aircraft normal load factor is estimated as being 1.25 g with a 99.7% probability of not exceeding this value when encountering a turbulent vertical gust with a variance σ_w of 20 ft/s.

A graphical representation of the concepts used in Example 9.6 is shown in Fig. 9.16. Figure 9.16a is a spectral representation for the Dryden vertical gust model when normalized to a unit area. The aircraft normal load factor transfer function due to a vertical gust input is shown in Fig. 9.16b with a peak response value occurring in the neighborhood of the short-period frequency. Figure 9.16c, the product of $|G_{w_g}(\omega)|^2 |G_{nw}(\omega)|^2$, represents the frequency distribution of the aircraft normal acceleration.

The frequency response distributions shown in Figs. 9.16b and 9.16c correspond to the behavior of a rigid airframe using quasi-steady aerodynamics. The peak response in Fig. 9.16b, in the neighborhood of $\omega = 4$ rad/s, can be translated for the Example 9.6 case into a dimensionless frequency (as used in aeroelasticity analyses) of $k = \omega c/2V \cong 0.03$. Application of the Sears function for a sinusoidal vertical gust input (cf. Fung[7]) will show that the unsteady aerodynamic lift input would be approximately 8% less than the quasi-steady estimate for $k \cong 0.03$.

If airframe elastic modes such as wing bending or torsion were included in the analysis, it would be necessary to consider their coupling with the unsteady aerodynamics. Their spectral influences would normally be expected to occur at frequencies near to or above the upper frequency limit shown in Fig. 9.16. Also, because structural modes are frequently lightly damped, their presence would be evident by the appearance of local spikes in the frequency response plot.

Aircraft gust responses, when both unsteady aerodynamics and elastic structural influences are considered, are viewed as beyond the scope of this chapter. Further comments on these influences are discussed by Hoblit.[5]

The aircraft load factor response from turbulent gusts can be broadened by recognizing that a turbulent vertical gust has both w_g and q_g gust inputs. The Dryden model for vertical gust velocity w_g has been defined in a power spectral

318 INTRODUCTION TO AIRCRAFT FLIGHT DYNAMICS

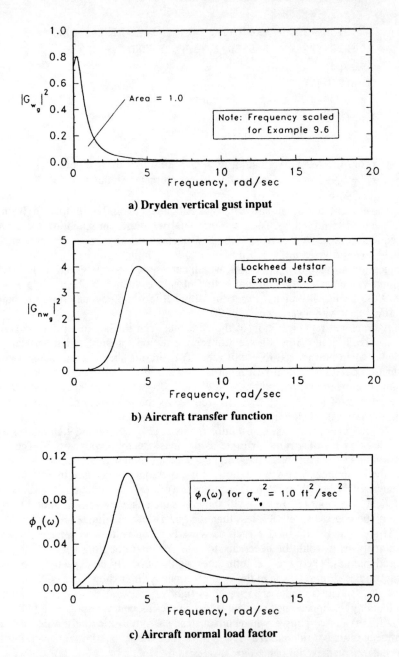

a) Dryden vertical gust input

b) Aircraft transfer function

c) Aircraft normal load factor

Fig. 9.16 Aircraft spectral response resulting from a vertical gust.

density form by Eq. (9.39) and in an equivalent transfer function form by Eq. (9.45). The q_g gust input may be deduced from the spectra of the relevant w_g slope at the aircraft origin in accord with principles described by Etkin[1] and the military flying qualities specifications.[12] The relation that

$$q_g = \frac{\partial w_g}{\partial x} \qquad (9.48)$$

translates for the Dryden gust model to[12]

$$\phi_{q_g}(\Omega) = \frac{(\Omega/L_2)^2}{1 + (4\Omega b/\pi L_2)^2} \phi_{w_g}(\Omega) \qquad (9.49)$$

where

b = wing span
L_2 = scale of vertical turbulence gust
Ω = dimensionless frequency, $(\omega L_2/V)$

In a transfer function form, Eq. (9.49) becomes

$$\phi_{q_g}(\omega) = |G_{q_g}(\omega)|^2 \sigma_w^2 \qquad (9.50)$$

where the complex-natured transfer function is

$$G_{q_g}(\omega) = G_{q_g}(s)\big|_{s=i\omega} \qquad (9.51)$$

The transfer function form, Eq. (9.51), may be expressed in terms of the Laplace transform variable in a form similar to Eq. (9.44), i.e.,

$$G_{q_g}(s) = \frac{\pi}{4b} \frac{s}{(s+\gamma)} G_{w_g}(s) \qquad (9.52)$$

where $\gamma = \pi V/4b$, s^{-1}.

It should be noted in Eq. (9.49) that the frequency-dependent conversion factor between $\phi_{w_g}(\Omega)$ and $\phi_{q_g}(\Omega)$ contains a $(1/L_2)^2$ term in the numerator in order to preserve dimensional consistency between $\sigma_{w_g}^2$ and $\sigma_{q_g}^2$.

Application of Eq. (9.49) to the Lockheed Jetstar case of Example 9.6 resulted in $\sigma_n = 0.083\ g$ due to the turbulent gust pitch input that corresponded to having $\sigma_w = 20$ ft/s. The combined σ_n for the Example 9.6 situation, which includes the concurrent response due to both w_g and q_g, is estimated as being $\sigma_n = 0.424\ g$, i.e.,

$$\sigma_n = \left[(n_z)_{w_g}^2 + (n_z)_{q_g}^2\right]^{\frac{1}{2}} = 0.424\ g$$

9.5 Lyapunov Equation Usage for Gust Responses

9.5.1 Background

In Sec. 9.4, the frequency variation of the power spectral density function provided information concerning the spectral content of the gust input and the load factor output. In addition to providing a perspective of the frequency content, clues were also given as to whether the rigid-airframe assumption should be modified to include the influence of structural dynamic responses. As was pointed out, the variance of both the input and the output required a numerical integration of the respective spectral distributions. However, it was noted that not having the upper integration limit at infinity, as stated by Eq. (9.31), resulted in a small error when estimating the variance.

An alternate approach to determine the respective variances without a numerical integration is the purpose of this section. State-space techniques, based on material already presented, will be used to obtain the covariance matrix directly.

9.5.2 Direct Solution for the Covariance Matrix

Consider an nth order, LTI system that is excited by white noise with a zero mean. The initial state vector relation is given by

$$\dot{x} = Ax(t) + Bw(t) \tag{9.53}$$

where $w(t)$ is the white noise input with standard deviation σ. White noise has the property of being the most uncorrelated random function. This implies that the autocorrelation function involves the Dirac delta function, i.e.,

$$E[w(t)w^T(\tau)] = \sigma^2 \delta(t - \tau) \tag{9.54}$$

It is useful to note that the following product of two vectors will result in a square matrix:

$$xx^T = \underbrace{\{x\}}_{(n\times 1)} \underbrace{[x]}_{(1\times n)} = \underbrace{[xx^T]}_{(n\times n)}$$

When the vector x is viewed as the random response to a white noise input, as posed by Eq. (9.53), the response is given as expectations in accord with principles described in Sec. 9.3 and notated as $E[xx^T]$, that is, an $(n \times n)$ matrix containing expected values of xx^T. $E[xx^T]$ is defined as the covariance matrix with the diagonal elements representing the variances because the output means are zero, which is in accord with the zero mean white noise input, i.e.,

$$E(x_1^2) = \sigma_1^2, \qquad E(x_2^2) = \sigma_2^2, \qquad \ldots, \qquad E(x_n^2) = \sigma_n^2$$

If Eq. (9.53) were postmultiplied by x^T, one would find that

$$\dot{x}x^T = Axx^T + Bwx^T \tag{9.55}$$

The transpose of Eq. (9.55) may also be obtained as

$$x\dot{x}^T = xx^T A^T + xw^T B^T \tag{9.56}$$

Equations (9.55) and (9.56) may be put into covariance matrix format by applying the time limiting operations as described in Sec. 9.3. The resulting relations are

$$E[\dot{x}x^T] = AE[xx^T] + BE[wx^T] \tag{9.57}$$

and

$$E[x\dot{x}^T] = E[xx^T]A^T + E[xw^T]B^T \qquad (9.58)$$

It is appropriate to consider a solution for the response $x(t)$ due to a zero mean, white noise input. The solution to the state-space equation may be recognized from Eq. (5.61) where now the transition matrix is expressed as $\phi(t)$, i.e.,

$$x(t) = \phi(t)x_0 + \int_0^t \phi(t-\tau)Bw(\tau)d\tau \qquad (9.59)$$

$E[xw^T]$ is obtained by postmultiplying Eq. (9.59) by the transpose of the noise input, $w(t)^T$, and finding the expected values. The result of applying the time limiting operations is

$$E[x(t)w(t)^T] = \phi(t)E[x_0 w(t)^T] + \int_0^t \phi(t-\tau)BE[w(\tau)w(t)^T]d\tau \qquad (9.60)$$

It will be recognized in Eq. (9.60) that

$$E[x_0 w(t)^T] = [0]$$

because the random forcing function has no correlation with the initial conditions. Consequently, Eq. (9.60) simplifies to

$$E[x(t)w(t)'] = \int_0^t \phi(t-\tau)BE[w(\tau)w(t)^T]d\tau \qquad (9.61)$$

As stated earlier by Eq. (9.54), the autocorrelation of white noise can be expressed in terms of the Dirac delta function due to its being the most uncorrelated random function. The Dirac delta function property, as given by Eqs. (5.4) and (5.5) in Sec. 5, allows the following integral to be written involving the autocorrelation of white noise:

$$\int_{-t}^{t} E[w(\tau)w(t)^T]d\tau = \lim_{t \to 0} \int_{-t}^{t} \sigma^2 \delta(t-\tau)d\tau = \sigma^2$$

Consequently,

$$\int_0^t E[w(\tau)w(t)^T]d\tau = \frac{1}{2}\sigma^2$$

and

$$\lim_{t \to 0} \int_0^t \phi(t-\tau)B\frac{\sigma^2}{2}\delta(t-\tau)d\tau = \phi(0)B\frac{\sigma^2}{2} = B\frac{\sigma^2}{2} \qquad (9.62)$$

because the transition matrix at $t = 0$ is

$$\phi(0) = I = \text{unit identity matrix}$$

Applying the simplifying outcome of Eq. (9.62) to Eq. (9.61), we find that

$$E[x(t)w(t)^T] = B\sigma^2/2 \qquad (9.63)$$

and

$$E[w(t)x(t)^T] = \sigma^2 B^T/2$$

Continuing in the development, the time derivative of $E[xx^T]$ may be obtained by application of the chain rule, i.e.,

$$\frac{d}{dt}E[xx^T] = E[\dot{x}x^T] + E[x\dot{x}^T]$$

Applying the results of Eqs. (9.57) and (9.58), the time derivative becomes

$$\frac{d}{dt}E[xx^T] = AE[xx^T] + BE[wx^T] + E[xx^T]A^T + E[xw^T]B^T$$

and simplifies further using the results of Eq. (9.63),

$$\frac{d}{dt}E[xx^T] = AE[xx^T] + E[xx^T]A^T + B\sigma^2 B^T$$

However, it should be recognized that the expected values contained in the covariance matrix will attain a constant set of values in the limit for a stationary process. Therefore, the time derivative of the covariance matrix must be zero, which provides the relation that

$$AE[xx^T] + E[xx^T]A^T + B\sigma^2 B^T = [0] \qquad (9.64)$$

Equation (9.64) is the same as Lyapunov's equation described in Sec. 5.7 except that here the matrix quantities apply to random-natured functions and the solution to the equation yields the covariance matrix. In Sec. 5.7, Lyapunov's equation was employed to confirm asymptotic stability without solving for the system dynamics. It is uncertain whether Lyapunov anticipated the Eq. (9.64) usage of his equation when he developed the relations at the beginning of the 20th century.

The interested reader may find further development of this section's subject in many references including the exposition by Bryson and Ho.[13] The solution of Lyapunov's equation may be found using numerical software such as MATLAB. The two examples that follow will illustrate the principles involved in using the equation to solve problems where white noise input is provided to a transfer function that represents the spectral form of a power spectral density.

Consider the expectation in the form of Eq. (9.31), i.e.,

$$E(x^2) = \int_0^\infty \phi(\omega)d\omega \qquad (9.31)$$

where the power spectral density is given by

$$\phi(\omega) = 1/(1+\omega^2) \qquad (9.65)$$

Solution for Eq. (9.65) where $\phi(\omega)$ resembles the Dryden model for a turbulent horizontal gust is

$$\int_0^\infty \frac{1}{1+\omega^2}d\omega = \tan^{-1}\omega\Big|_0^\infty = \frac{\pi}{2} \tag{9.66}$$

The transfer function form of the preceding $\phi(\omega)$ is

$$G(s) = 1/(1+s) \tag{9.67}$$

because

$$\phi(\omega) = |G(\omega)|^2_{s=i\omega}$$

Example 9.7

Find the variance of a first-order system when excited by white noise with $\sigma^2 = 1.0$. When noise excitation is provided to the assumed transfer function, the equivalent differential equation for Eq. (9.67) is

$$\dot{x}_1 = -x_1(t) + n(t)$$

which translates into equivalent state-space relations as

$$A = -1 \quad \text{and} \quad B = +1$$

```
% Solve Lyapunov's Eq. using the MATLAB lyap function
% Input noise is assumed as unity
   Q = lyap(A,B*B');
   disp(Q)
   0.5000;  % (1/pi)*Expectation
```

It will be noted in Example 9.7 that the expectation (or variance) differed from the predicted result [Eq. (9.66)] by a pi factor. This scaling difference arises from the choice of the constant in the definition relating expectation to the power spectral density. In many references (e.g., Gelb[14]) the relation is given as

$$E(x^2) = \frac{1}{2\pi}\int_{-\infty}^\infty \phi(\omega)d\omega = \frac{1}{\pi}\int_0^\infty \phi(\omega)d\omega \tag{9.68}$$

where the symmetry property of the power spectral density function has been used to alter the limits of integration. MATLAB considers the expectation in the form of Eq. (9.68), which results in the pi factor being introduced when converting $\phi(\omega)$ into an equivalent state-space form.

The definition for the expectation (or variance) [Eq. (9.31)] used in this text is consistent with the definitions of the Dryden gust model found in the literature when considering atmospheric turbulence.[7,12]

The Dryden turbulence model for vertical gusts has been put into transfer function format [Eq. (9.45)] and used as an input when making an estimate of the aircraft load factor spectral response (cf. Example 9.6). The transfer function, when considered in the time domain, involves a second-order differential equation, i.e.,

$$G_{w_g}(s) = K\frac{(s+\theta)}{(s+\lambda)^2} \qquad (9.44)$$

which corresponds to

$$\ddot{w}_g + 2\lambda \dot{w}_g + \lambda^2 w_g(t) = K[\dot{n} + \theta n(t)] \qquad (9.69)$$

where

w_g = vertical gust velocity output
n = white noise input

Equation (9.69) can be converted to state-space format by several methods. The following approach is in accord with Ogata[15] and requires the inclusion of an auxiliary state variable due to the presence of the $\dot{n}$ term in the input.

Let

$$y_1 = w_g - \beta_0 n$$
$$y_2 = \dot{w}_g - \beta_0 \dot{n} - \beta_1 n = \dot{y}_1 - \beta_1 n$$

Taking the time derivative of y_2 and including the statement of Eq. (9.69) yields

$$\dot{y}_2 = \ddot{w}_g - \beta_0 \ddot{n} - \beta_1 \dot{n} = (-\lambda^2 w_g - 2\lambda \dot{w}_g + K\dot{n} + K\theta n) - \beta_0 \ddot{n} - \beta_1 \dot{n}$$

Consideration of these three relations leads to

$$\beta_0 = 0, \qquad \beta_1 = K$$
$$w_g = y_1, \qquad \dot{w}_g = y_2 + Kn$$

which provides the two state equations

$$\dot{y}_1 = y_2 + Kn$$
$$\dot{y}_2 = -\lambda^2 y_1 - 2\lambda y_2 + K(\theta - 2\lambda)n$$

Or, in matrix notation,

$$\begin{Bmatrix} \dot{y}_1 \\ \dot{y}_2 \end{Bmatrix} = \begin{bmatrix} 0 & 1 \\ -\lambda^2 & -2\lambda \end{bmatrix} \begin{Bmatrix} y_1 \\ y_2 \end{Bmatrix} + \begin{Bmatrix} K \\ K(\theta - 2\lambda) \end{Bmatrix} n \qquad (9.70)$$

It should be noted that y_1 corresponds to the vertical gust velocity w_g whereas y_2 does not have a direct physical meaning. If Eq. (9.69) were of third order, an additional state variable, y_3, would be required by an extension of the steps shown here.

Example 9.6 indicated that when the vertical gust transfer function was excited by white noise with unit magnitude, the output variance of w_g should also be of unit magnitude. The finite range of the frequencies considered resulted in a slight

ATMOSPHERIC TURBULENCE

error. The use of Lyapunov's equation removes the integration limit difficulty, but requires a conversion of the transfer function into state-variable format.

Example 9.8

Find the variance for the output of the Dryden vertical gust when excited by white noise with input $\sigma^2 = 1.0$. Use the Lockheed Jetstar condition of Example 9.6.

From Example 9.6, appropriate constants are

$$K = 0.6516 \text{ s}^{-1/2}$$

$$\theta = 0.2567 \text{ s}^{-1}$$

$$\lambda = 0.4446 \text{ s}^{-1}$$

The plant and control matrices, per Eq. (9.70), are

$$A = \begin{bmatrix} 0 & 1 \\ -0.1977 & -0.8892 \end{bmatrix}$$

$$B = \begin{bmatrix} 0.6516 & -0.4121 \end{bmatrix}^T$$

```
% Apply Lyapunov's Eq. to the Dryden Vertical Gust Model
% Input noise assumed as unity
   Q = lyap(A,B*B'); disp(Q)
              0.3184    -0.2123
             -0.2123     0.1427
% pi*Q(1,1) = 1.00 ... Q.E.D.
```

In Example 9.8, the $Q(1,1)$ term corresponded to the variance produced by the Dryden vertical gust model when excited by white noise having $\sigma^2 = 1.0$. As explained following Example 9.7, the covariance output using the MATLAB software differs from the notation used in this text by the pi constant. The $Q(2,2)$ term corresponds to the variance for the y_2 variable. As would be expected, the covariance matrix is symmetrical about the lead diagonal.

This second example, in addition to providing confirmation of the scaling factor, sets the stage for the following material in Sec. 9.5.3 relating to the load factor response of an aircraft when experiencing vertical turbulence.

9.5.3 Aircraft Response to Vertical Turbulence

The results of Sec. 9.5.2 concerning the use of Lyapunov's equation to obtain the variance of the Dryden vertical gust model as output due to white noise excitation may be extended to the variance estimation of aircraft load factor by appropriate partitioning of the state-space relations.

The y_1 and y_2 components of Eq. (9.70) for the Dryden vertical gust model in Sec. 9.5.2 are recognized as

$$Y = \begin{Bmatrix} y_1 \\ y_2 \end{Bmatrix} = \begin{Bmatrix} w_g \\ y_2 \end{Bmatrix}$$

and Eq. (9.70) may be expressed symbolically as

$$\dot{Y} = FY + Gn(t) \qquad (9.71)$$

where F and G are the corresponding matrices of Eq. (9.70).

A suitable form of the short-period response of an aircraft may be obtained from a variation of Eq. (9.16) with α replaced by w/V. In terms of $w_g (= y_1)$ and y_2, the short-period equation becomes

$$\begin{Bmatrix} \dot{w} \\ \dot{q} \end{Bmatrix} = \begin{bmatrix} Z_\alpha/V & V \\ M'_\alpha/V & M'_q \end{bmatrix} \begin{Bmatrix} w \\ q \end{Bmatrix} + \begin{bmatrix} Z_\alpha/V & 0 \\ M'_\alpha/V & 0 \end{bmatrix} \begin{Bmatrix} y_1 \\ y_2 \end{Bmatrix}$$

and may be expressed symbolically as

$$\dot{X} = AX + BY \qquad (9.72)$$

Equations (9.71) and (9.72) are coupled by virtue of the Y state vector and may be combined by partitioning as follows:

$$\dot{Z} = \begin{Bmatrix} \dot{X} \\ \hline \dot{Y} \end{Bmatrix} = \begin{bmatrix} A & | & B \\ \hline 0 & | & F \end{bmatrix} \begin{Bmatrix} X \\ \hline Y \end{Bmatrix} + \begin{Bmatrix} 0 \\ \hline G \end{Bmatrix} n(t)$$

or in symbolic format

$$\dot{Z} = HZ + Mn(t) \qquad (9.73)$$

Application of Lyapunov's equation to find the covariance matrix applicable to Eq. (9.73) will result in a 4×4 matrix because the partitioned vector Z is of order 4×1.

The normal acceleration response for the aircraft is

$$a_n = (Z_\alpha/V)w + (Z_\alpha/V)w_g$$

which results in

$$a_n = CZ + Dn \qquad (9.74)$$

where

$$C = [(Z_\alpha/V) \quad 0 \quad (Z_\alpha/V) \quad 0] \quad \text{and} \quad D = 0$$

When the output from Lyapunov's equation is the covariance matrix Q, i.e.,

$$Q = E[ZZ^T]$$

then the variance of the normal acceleration is given by

$$\sigma_n^2 = CQC^T \tag{9.75}$$

Example 9.9

Find the variance of the aircraft load factor due to an encounter with a vertical gust based on the Dryden turbulence model. Use the Lockheed Jetstar condition of Example 9.6 where the deviation of the vertical gust was assumed as $\sigma_{w_g} = 20$ ft/s.

Combining information from previous related examples, one has

$$A = \begin{bmatrix} -1.329 & 778.0 \\ -0.0156 & -1.537 \end{bmatrix}$$

$$B = \begin{bmatrix} -1.329 & 0.0 \\ -0.0156 & 0.0 \end{bmatrix}$$

$$F = \begin{bmatrix} 0.0 & 1.0 \\ -0.1977 & -0.8892 \end{bmatrix}$$

$$G = \begin{Bmatrix} 0.6515 \\ -0.4121 \end{Bmatrix}$$

$$C = [-1.329 \quad 0.0 \quad -1.329 \quad 0.0]$$

$$D = 0.0$$

```
% Apply Lyapunov's Eq. to the Aircraft Response Model
% Input White Noise assumed as Unity
    H = [A,B;zeros(A),F];
    M = [zeros(G);G];
    Q = lyap(H,M*M'); disp(Q)
        0.3458    0.0001   -0.2904    0.1959
        0.0001    0.0000   -0.0002    0.0001
       -0.2904   -0.0002    0.3184   -0.2123
        0.1959    0.0001   -0.2123    0.1427
    AN2 = (pi)*C*Q*C'; disp(AN2)
        0.4624; % Variance of Load Factor
```

The $Q(3,3)$ term in the example corresponds to $(1/\pi)E(\sigma_{w_g}^2)$ and agrees with $Q(1,1)$ found in Example 9.8. The AN2 result translates to a $\sigma_n = 0.680$ ft-s^{-2} due to the unit input of $\sigma_{w_g} = 1.0$ ft-s^{-1}. Scaling to $\sigma_{w_g} = 20.0$ ft-s^{-1} provides a $\sigma_n = 13.600$ ft-s^{-2} ($= 0.423$ g). Comparison with $\sigma_n = 0.416$ g as found in Example 9.6 indicates that the finite frequency limit on the integration of the spectral distribution resulted in approximately a 2% error in the estimation of airframe load factor response when performing a numerical integration. It may be concluded that the agreement between the two approaches for obtaining an estimate for the output deviation of aircraft load factor is well within engineering accuracy.

Obtaining the aircraft load factor response during turbulent gust encounters using the Lyapunov equation has advantages for aircraft design studies that are aimed at improving the handling qualities and ride comfort. Stability augmentation concepts for gust alleviation may be readily introduced into the governing equations when they are in state-space format using partitioning techniques as illustrated in this section when the gust and airframe response relations were combined. Further discussions with respect to handling quality improvements may be found in an article by Swaim et al.[16]

9.5.4 Comments on Design Values for Turbulent Gusts

The first portions of this chapter considered the aircraft as responding to discrete gust inputs, many of which assumed a one-minus-cosine form. The stated design magnitudes for discrete gust encounters in Sec. 9.1 are in agreement with the governmental regulations.[6,12] Latter parts of this chapter introduced random variable concepts along with the corresponding system responses to these type of inputs. This latter view is based on an assumption that considering the gust environment by a continuous, stochastic-natured turbulence model is more realistic than considering it as a collection of individual gusts having simple geometric shapes. Of course, it is recognized from a historical perspective that modeling atmospheric turbulence by discrete gust inputs during the first-half of the 20th century was reasonable inasmuch as the aircraft designed under this aegis were safe. However, more recent aircraft designs differ in their gust response characteristics and, consequently, a more representative type of gust encounter definition has been introduced. Appendix G of FAR 25[6] reflects an amendment (1990) to treat the design criteria for continuous gusts.

Continuous gusts can be represented from either design envelope or mission profile viewpoints. Support of the former representation is consistent with the goal of Chapter 9, which is to introduce the reader to various techniques for modeling airframe dynamic responses subject to using quasi-steady aerodynamics on a rigid airframe.

The selection of a standard deviation for the vertical gust of Example 9.6 of $\sigma_w = 20$ fps was arbitrary; however, it is representative of a value near to design values. The FAR 25[6] indicates design vertical gust velocities, U_σ, to use in conjunction with the design envelope at the V_C gust envelope corner speed. If one considers the design random gust velocity as a peak value, as described by Hoblit,[5] then it is reasonable to assume that $U_\sigma \cong 3\sigma$.

FAR 25 states the following.

At design velocity V_C (EAS): $U_\sigma = 85$ ft/s true gust velocity for $0 < h \leq$ 30,000 ft and linearly decreases to 30 ft/s at $h = 80,000$ ft.

At design velocity V_B (EAS): $U_\sigma = 1.32(U_\sigma)V_C$.

At design velocity V_C (EAS): $U_\sigma = 0.50(U_\sigma)V_C$.

At speeds between V_B and V_C, and between V_C and V_D, U_σ is obtained by linear interpolation. The ratios used to adjust the design vertical gust velocities between V_B, V_C, and V_D are in accord with the corresponding ratios used when considering discrete gusts such as the one-minus-cosine type.

ATMOSPHERIC TURBULENCE 329

These values for U_σ are approximate and do not reflect details such as offered by considering an aircraft's mission profile that can lead to concerns such as probability of exceedance during an aircraft's lifetime. Also, gust design criteria are in an evolutional state and refinements in the guidelines will continue to appear. As stated by Hoblit,[5] "Design levels are based on the strength of past satisfactory airplanes."

References

[1] Etkin, B., "Turbulent Wind and Its Effect on Flight," 1980 Wright Brothers Lecture, *Journal of Aircraft*, Vol. 18, No. 5, 1981, pp. 327–345.

[2] Hunsaker, J. C., and Wilson, E. B., "Report on Behavior of Aeroplanes in Gusts," NACA TR. 1, 1915.

[3] Houbolt, J. C., "Atmospheric Turbulence," *AIAA Journal*, Vol. 11, No. 4, 1973, pp. 421–437.

[4] Jacobson, I. D., and Joshi, D. S., "Handling Qualities of Aircraft in the Presence of Simulated Turbulence," *Journal of Aircraft*, Vol. 15, No. 4, 1978, pp. 254–256.

[5] Hoblit, F. M., *Gust Loads on Aircraft: Concepts and Applications*, AIAA Education Series, AIAA, Washington, DC, 1988, Chaps. 1–3, Appendix A, p. 77.

[6] Anon., "Part 25, Airworthiness Standards: Transport Category Airplanes," *Federal Aviation Regulations*, U.S. Government Printing Office, Washington, DC, Feb. 1991.

[7] Fung, Y. C., *An Introduction to the Theory of Aeroelasticity*, Wiley, New York, 1955 (reissued Dover, New York, 1969), pp. 283–305, 343–350, 411.

[8] Tsien, H. S., *Engineering Cybernetics*, McGraw–Hill, New York, 1954, Chap. 9.

[9] Etkin, B., *Dynamics of Atmospheric Flight*, Wiley, New York, 1972, Chap. 13.

[10] Friedland, B., *Control System Design*, McGraw–Hill, New York, 1986, Chap. 10.

[11] McLean, D., *Automatic Flight Control Systems*, Prentice–Hall, Englewood Cliffs, NJ, 1990, Chap. 5.

[12] Anon., "Flying Qualities of Piloted Airplanes," MIL SPEC MIL-F-8785C, U.S. Government Printing Office, Washington, DC, Nov. 1980.

[13] Bryson, A. E., Jr., and Ho, Y. C., *Applied Optimal Control*, Ginn and Co., Waltham, MA, 1969, pp. 328–334.

[14] Gelb, A., *Applied Optimal Estimation*, M.I.T. Press, Cambridge, MA, 1974, pp. 40–42.

[15] Ogata, K., *Modern Control Engineering*, 2nd ed., Prentice–Hall, Englewood Cliffs, NJ, 1990, pp. 314–315.

[16] Swaim, R. L., Schmidt, D. K., Roberts, P. A., and Hinsdale, A. J., "An Analytical Method for Ride Quality of Flexible Airplanes," *AIAA Journal*, Vol. 15, No. 1, 1977, pp. 4–7.

Problems

9.1. Reconsider Example 9.4 to include an estimate for the load factor response due to the influence of q_g that is induced by w_g. Compare the normal load factor response between w_g and $(w_g + q_g)$ in order to identify the importance of the induced q_g term on gust response. Hint: Assume, as the aircraft travels along a distance $x = Vt$ after encountering the $(1 - \cos x)$ vertical gust, that the induced pitch rate corresponds to

$$q_g = \frac{\partial w_g}{\partial x}$$

9.2. Show that, in the limit, erf(x) tends to 1.0 when x tends to ∞ as implied by Table 9.1. Hint: Multiply erf(x) by erf(y) and change coordinates to R and θ such that dx dy becomes Rdrdθ.

9.3. Compare the time histories of a second-order system output, Eq. (9.40), with the excitation of a random, white noiselike input. Assume that the system is defined by $\omega_n = 1.25$ rad/s, mass $m = 1$, and $\zeta = 0.1$. Compare the mean and variance of input and output.

Optional: Repeat the problem with $\zeta = 0.05$. Did you find approximate agreement with the thoretical behavior, i.e., that the output variance (mean square) varied inversely with damping? Hint: Select a time span of approximately 20 times the natural period of the system; i.e., in MATLAB notation, let $t = 0. : 0.10 : 100.0$. The random, white noiselike input can be generated as a vector using the *rand* function available in MATLAB. Use the same input vector for both system cases.

9.4. Verify using the Fourier cosine transfer:

a) The power spectral density, Eq. (9.38), based on the Dryden model of the autocorrelation function, Eq. (9.36), that describes the horizontal turbulent gust.

Remark: This power spectral density has the form of a first-order Markov process.[10]

b) The power spectral density, Eq. (9.39), based on the Dryden model of the autocorrelation function, Eq. (9.37), describing the vertical turbulent gust.

9.5. Apply the Dryden model for the turbulence spectral profile of q_g to the Lockheed Jetstar of Example 9.6 in order to find the load factor contribution from the induced pitch rate component of the vertical gust. As in Example 9.6, assume that $\sigma_w = 20$ ft/s.

9.6. Repeat Problem 9.5 using the Lyapunov equation approach to find the load factor contribution from q_g.

Optional: Compare the result with that obtained in Problem 9.5. Hint: Equation (9.52), which describes $G_{q_g}(s)$, will involve a third-order differential equation in the time domain combined with a noise input that includes an $\dot{n}$ term. The state-space representation will involve a state vector Y having three components with y_1 corresponding to q_g.

Appendix A
Atmospheric Table

The atmospheric table (Table A.1) corresponds to the U.S. Standard Atmosphere, 1976. The altitude (Alt.) is the geometric height for the standard atmosphere. The gas constant for air $R = 1716.50$ ft^2/s^2 °R, the gravitational constant $g = 32.174$ ft/s^2, and $\rho_0 = 0.0023769$ lb-s^2/ft^4.

Table A.1 Standard atmosphere properties

Alt., ft	T, °R	p, lb/ft^2	ρ/ρ_0	ρ, lb-s^2/ft^4	a, ft/s
0	518.67	2116.2	1.00000	0.0023769	1116.44
1,000	515.10	2040.8	0.97107	0.0023082	1112.60
2,000	511.54	1967.7	0.94278	0.0022409	1108.76
3,000	507.97	1896.7	0.91513	0.0021751	1104.89
4,000	504.41	1827.7	0.88811	0.0021109	1100.98
5,000	500.84	1760.9	0.86170	0.0020482	1097.08
6,000	497.28	1696.0	0.83590	0.0019869	1093.18
7,000	493.71	1633.1	0.81070	0.0019270	1089.27
8,000	490.15	1572.0	0.78609	0.0018684	1085.33
9,000	486.59	1512.9	0.76206	0.0018113	1081.36
10,000	483.02	1455.6	0.73859	0.0017555	1077.40
11,000	479.46	1400.0	0.71568	0.0017011	1073.43
12,000	475.88	1346.2	0.69333	0.0016480	1069.42
13,000	472.34	1294.1	0.67151	0.0015961	1065.42
14,000	468.78	1243.6	0.65022	0.0015455	1061.38
15,000	465.22	1194.8	0.62946	0.0014962	1057.35
16,000	461.66	1147.5	0.60921	0.0014480	1053.31
17,000	458.09	1101.7	0.58946	0.0014011	1049.25
18,000	454.53	1057.5	0.57021	0.0013553	1045.14
19,000	450.97	1014.7	0.55144	0.0013107	1041.04
20,000	447.42	973.3	0.53317	0.0012673	1036.94
21,000	443.86	933.2	0.51534	0.0012249	1032.81
22,000	440.30	894.6	0.49798	0.0011837	1028.64
23,000	436.74	857.2	0.48108	0.0011435	1024.48
24,000	433.18	821.2	0.46462	0.0011044	1020.31
25,000	429.62	786.3	0.44859	0.0010663	1016.11

(*Cont.*)

Table A.1 (Continued)

Alt., ft	T, °R	p, lb/ft^2	ρ/ρ_0	ρ, lb-s^2/ft^4	a, ft/s
26,000	426.07	752.7	0.43300	0.0010292	1011.88
27,000	422.51	720.1	0.41782	0.0009931	1007.64
28,000	418.95	688.9	0.40305	0.0009580	1003.41
29,000	415.40	658.8	0.38869	0.0009239	999.15
30,000	411.84	629.7	0.37473	0.0008907	994.85
31,000	408.28	601.6	0.36115	0.0008584	990.55
32,000	404.73	574.6	0.34795	0.0008270	986.22
33,000	401.17	548.5	0.33513	0.0007966	981.89
34,000	397.62	523.5	0.32267	0.0007670	977.53
35,000	394.06	499.3	0.31058	0.0007382	973.13
36,000	390.51	476.1	0.29883	0.0007103	968.73
37,000	389.97	453.9	0.28525	0.0006780	968.08
38,000	389.97	432.6	0.27191	0.0006463	968.08
39,000	389.97	412.4	0.25920	0.0006161	968.08
40,000	389.97	393.1	0.24708	0.0005873	968.08
42,000	389.97	357.2	0.22452	0.0005336	968.08
44,000	389.97	324.6	0.20402	0.0004849	968.08
46,000	389.97	295.0	0.18540	0.0004407	968.08
48,000	389.97	268.1	0.16848	0.0004005	968.08
50,000	389.97	243.6	0.15311	0.0003639	968.08
52,000	389.97	221.4	0.13914	0.0003307	968.08
54,000	389.97	201.2	0.12645	0.0003006	968.08
56,000	389.97	182.8	0.11492	0.0002731	968.08
58,000	389.97	166.2	0.10444	0.0002482	968.08
60,000	389.97	151.0	0.09492	0.0002256	968.08
65,000	389.97	118.9	0.07475	0.0001777	968.08
70,000	392.25	93.7	0.05857	0.0001392	970.90
75,000	394.97	74.0	0.04591	0.0001091	974.28
80,000	397.69	58.5	0.03606	0.0000857	977.62
85,000	400.42	46.3	0.02837	0.0000674	980.94
90,000	403.14	36.8	0.02236	0.0000531	984.28
95,000	405.85	29.2	0.01765	0.0000419	987.60
100,000	408.57	23.3	0.01396	0.0000332	990.91

Appendix B
Aircraft Stability Derivatives

1) Geometric data and dimensional stability derivatives are provided for the following seven aircraft: Tables B.1a–B.1d, U.S. Navy A-4D attack aircraft; Tables B.2a–B.2d, U.S. Navy A-7A attack aircraft; Tables B.3a–B.3d, U.S. Air Force F-4C fighter aircraft; Tables B.4a–B.4f, McDonnell–Douglas DC-8 jet transport; Tables B.5a–B.5f, Convair CV-880M jet transport; Tables B.6a–B.6e, Lockheed Jetstar utility jet transport; and Tables B.7a–B.7d, North American Navion general aviation aircraft. The information was obtained from reports by Heffley and Jewell[1] and Teper.[2]

2) Dimensionless stability derivatives are provided for a few aircraft in the transport category in support of material and exercises in Chapter 3 dealing with aircraft static stability and control.

3) Center-of-gravity listings refer to the physical position of the c.g. on the reference chord; e.g., "c.g. at 0.25c" implies that $(x/c)_{c.g.} = -0.25$ where it is understood that x is in body-axis coordinates and $x = 0$ corresponds to the forward edge of the reference chord.

4) The derivative information is all oriented with respect to aircraft stability axes, which are defined as that particular body-axis orientation where the x coordinate system is coincident with the projection of the total steady-state velocity vector on the aircraft's plane of symmetry.

5) Tabulated angle-of-attack values correspond to the angles between the x coordinates of the stability and aircraft reference body axes. Body reference axes are defined by the airframe manufacturer and are established during the design process. The x body reference axis is usually described as parallel to a waterline (WL). Note that all angle related derivatives use radian measure.

6) The roll (L) and yaw (N) moment dimensional stability derivatives are presented in two forms, primed and unprimed. The primed form is obtained by correcting the unprimed form for the influence of the cross product of inertia term I_{xz}, as described in Chapter 4 and shown next for an example case of derivatives with respect to sideslip angle β.

$$L'_\beta = \left[L_\beta + N_\beta \frac{I_{xz}}{I_x} \right] G$$

$$N'_\beta = \left[N_\beta + L_\beta \frac{I_{xz}}{I_z} \right] G$$

(B.1)

where

$$G = \frac{1}{1 - \left(I_{xz}^2 / I_x I_z\right)}$$

7) The longitudinal characteristic equation is expressed as

$$(s^2 + 2\zeta_{sp}\omega_{sp}s + \omega_{sp}^2)(s^2 + 2\zeta_p\omega_p s + \omega_p^2) = 0$$

or, when the phugoid mode has real roots, as

$$(s^2 + 2\zeta_{sp}\omega_{sp}s + \omega_{sp}^2)(s + 1/T_{p1})(s + 1/T_{p2}) = 0$$

The lateral-directional characteristic equation is expressed as

$$(s^2 + 2\zeta_d\omega_d s + \omega_d^2)(s + 1/T_R)(s + 1/T_S) = 0$$

The subscript notation refers to the following modes:

sp = short period
p = phugoid
d = Dutch roll
R = roll
S = spiral

B.1 U.S. Navy A-4D Attack Aircraft

The data given in Tables B.1a–B.1d are for the normal cruise configuration, clean airplane, where $W = 17{,}578$ lb, mass $m = 546$ slugs, $S = 260$ ft², $b = 27.5$ ft, $\bar{c} = 10.8$ ft, and the c.g. is at $0.25\bar{c}$. Figure B.1 is a sketch of the aircraft.

Note that the absence of derivatives implies negligible values. The stability derivatives are relative to the stability axes. All angles are in radian measure.

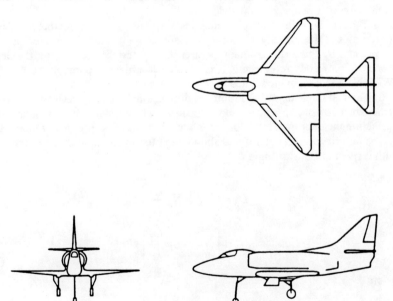

Fig. B.1 Sketch of A-4D aircraft.

AIRCRAFT STABILITY DERIVATIVES

Table B.1a Geometric data

Condition	1	2	3	4	5	6
h, ft	0(S/L)	15,000	15,000	15,000	35,000	35,000
M	0.4	0.4	0.6	0.9	0.6	0.9
V, ft/s	447	423	634	952	584	876
Q, lb/ft^2	237	134	301	677	126	283
I_x, slug-ft^2	8,020	8,200	8,010	8,060	8,190	8,010
I_y, slug-ft^2	25,900	25,900	25,900	25,900	25,900	25,900
I_z, slug-ft^2	29,270	29,090	29,280	29,230	29,100	29,280
I_{xz}, slug-ft^2	−441	−1,989	41	1,042	−1,952	227
α_{trim}, deg	4.7	8.9	3.4	0.7	8.8	2.9

Table B.1b Longitudinal dimensional derivatives

Condition	1	2	3	4	5	6
h, ft	0(S/L)	15,000	15,000	15,000	35,000	35,000
M	0.4	0.4	0.6	0.9	0.6	0.9
X_u, s^{-1}	−0.0160	−0.0148	−0.0129	−0.0635	−0.0128	−0.0353
X_α, ft/s^2	−1.742	−12.46	−3.721	−45.28	−20.61	−47.04
Z_u, s^{-1}	−0.156	−0.160	−0.104	−0.135	−0.114	−0.120
Z_α, ft/s^2	−398.5	−219.6	−518.9	−1404.5	−218.3	−586.7
M_u, ft^{-1}-s^{-1}	0.0004	0.0005	0.0004	−0.0095	0.0004	−0.0050
M_α, s^{-2}	−10.233	−5.639	−12.97	−35.96	−5.402	−14.99
$M_{\dot\alpha}$, s^{-1}	−0.342	−0.204	−0.353	−0.858	−0.160	−0.389
M_q, s^{-1}	−1.151	−0.670	−1.071	−1.934	−0.484	−0.876
X_{δ_e}, ft/s^2	4.14	2.55	4.02	−20.98	2.69	−6.05
Z_{δ_e}, ft/s^2	−42.56	−22.94	−57.02	−102.98	−23.73	−42.90
M_{δ_e}, s^{-2}	−13.73	−7.40	−19.46	−33.81	−8.10	−14.80

Table B.1c Lateral-directional dimensional derivatives condition

Condition	1	2	3	4	5	6
h, ft	0(S/L)	15,000	15,000	15,000	35,000	35,000
M	0.4	0.4	0.6	0.9	0.6	0.9
Y_β, ft/s^2	−110.94	−62.42	−144.6	−345.2	−60.38	−139.8
L_β, s^{-2}	−27.67	−14.01	−35.00	−87.19	−14.24	−40.32
L_p, s^{-1}	−1.732	−0.988	−1.516	−2.492	−0.671	−1.134
L_r, s^{-1}	0.933	0.607	0.874	1.346	0.464	0.672
N_β, s^{-2}	15.16	8.223	18.78	46.43	7.864	19.65
N_p, s^{-1}	0.040	0.000	0.040	0.125	−0.004	0.041
N_r, s^{-1}	−0.639	−0.401	−0.566	−0.958	−0.291	−0.428
Y_{δ_r}, ft/s^2	19.65	10.83	25.09	52.24	10.46	21.78

(Cont.)

Table B.1c (Continued)

Condition	1	2	3	4	5	6
L_{δ_r}, s^{-2}	7.305	2.802	9.961	24.05	2.739	8.568
N_{δ_r}, s^{-2}	−6.732	−3.651	−8.397	−17.41	−3.517	−7.241
Y_{δ_a}, ft/s^2	−2.599	−0.795	−2.409	−5.291	−0.478	−2.420
L_{δ_a}, s^{-2}	17.27	8.757	21.27	37.48	7.998	16.88
N_{δ_a}, s^{-2}	0.334	−0.246	0.479	1.462	−0.139	0.414
L'_β, s^{-2}	−28.53	−16.27	−34.90	−81.56	−16.38	−39.77
L'_p, s^{-1}	−1.736	−1.004	−1.516	−2.488	−0.681	−1.134
L'_r, s^{-1}	0.968	0.717	0.872	1.227	0.542	0.660
N'_β, s^{-2}	15.59	9.336	18.73	43.53	8.963	19.35
N'_p, s^{-1}	0.066	0.069	0.038	0.036	0.042	0.032
N'_r, s^{-1}	−0.653	−0.450	−0.565	−0.914	−0.327	−0.423
L'_{δ_r}, s^{-2}	7.682	3.750	9.918	21.90	3.635	8.365
N'_{δ_r}, s^{-2}	−6.848	−3.907	−8.383	−16.63	−3.760	−7.176
L'_{δ_a}, s^{-2}	17.26	8.965	21.27	37.84	8.162	16.90
N'_{δ_a}, s^{-2}	0.073	−0.859	0.508	2.811	−0.686	0.545

Table B.1d Eigenvalue summary

Condition	1	2	3	4	5	6
h, ft	0(S/L)	15,000	15,000	15,000	35,000	35,000
M	0.4	0.4	0.6	0.9	0.6	0.9
Longitudinal						
ζ_{sp}	0.356	0.285	0.301	0.344	0.215	0.248
ω_{sp}, s^{-1}	3.356	2.443	3.720	6.237	2.360	3.954
$\zeta_p(1/T_{p_1})$	0.072	0.067	0.087	(0.111)	0.091	(0.057)
$\omega_p(1/T_{p_2})$	0.107	0.113	0.076	(−0.066)	0.084	(−0.047)
Lateral-directional						
ζ_d	0.112	0.095	0.089	0.097	0.068	0.065
ω_d, s^{-1}	3.961	3.059	4.340	6.618	2.996	4.404
$1/T_R$, s^{-1}	1.744	1.015	1.535	2.480	0.702	1.138
$1/T_s$, s^{-1}	0.009	0.005	0.006	0.007	0.004	0.007

B.2 U.S. Navy A-7A Attack Aircraft

The data given in Tables B.2a–B.2d are for the normal cruise configuration, clean airplane, where $W = 21,889$ lb, mass $m = 680$ slugs, $S = 375$ ft^2, $b = 38.7$ ft, $c = 10.8$ ft, and the c.g. is at $0.30c$. Longitudinal control is provided by the all-moving horizontal tail. Figure B.2 is a sketch of the aircraft.

Note that the absence of derivatives implies negligible values. The stability derivatives are relative to the stability axes. All angles are in radian measure.

AIRCRAFT STABILITY DERIVATIVES 337

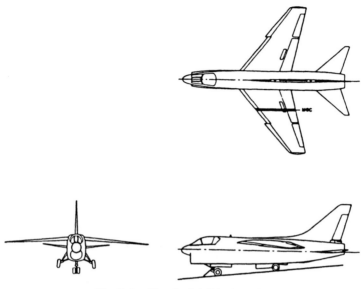

Fig. B.2   Sketch of A-7A aircraft.

Table B.2a   Geometric data

Condition	1	2	3	4	5	6
h, ft	0(S/L)	0(S/L)	15,000	15,000	35,000	35,000
M	0.6	0.9	0.6	0.9	0.6	0.9
V, ft/s	670	1,005	634	952	584	876
Q, lb/ft^2	534	1,200	301	677	126	283
I_x, slug-ft^2	13,480	13,490	13,490	13,480	13,790	13,480
I_y, slug-ft^2	58,970	58,970	58,970	58,970	58,970	58,970
I_z, slug-ft^2	67,720	67,700	67,700	67,710	67,400	67,710
I_{xz}, slug-ft^2	193	950	−848	572	−4145	−659
α_{trim}, deg	2.9	2.1	4.0	2.5	7.5	3.8

Table B.2b Longitudinal dimensional derivatives

Condition	1	2	3	4	5	6
h, ft	0(S/L)	0(S/L)	15,000	15,000	35,000	35,000
M	0.6	0.9	0.6	0.9	0.6	0.9
X_u, s^{-1}	−0.0156	−0.0760	−0.0098	−0.0464	−0.0093	−0.0230
X_α, ft/s^2	−26.58	−93.94	−16.96	−53.90	−33.35	−29.74
Z_u, s^{-1}	−0.124	−0.107	−0.109	−0.091	−0.112	−0.091
Z_α, ft/s^2	−1284.0	−3413.6	−733.6	−2015.2	−316.1	−881.3
M_u, ft^{-1}-s^{-1}	0.0002	−0.0003	0.0001	−0.0032	0.0002	−0.0026
M_α, s^{-2}	−15.57	−40.41	−9.096	−27.68	−4.183	−13.02
$M_{\dot\alpha}$, s^{-1}	−0.207	−0.372	−0.134	−0.267	−0.065	−0.143
M_q, s^{-1}	−1.110	−1.570	−0.696	−1.070	−0.330	−0.539
X_{δ_s}, ft/s^2	−0.02	−0.06	−0.01	0.01	0.01	0.01
Z_{δ_s}, ft/s^2	−165.2	−318.2	−99.84	−209.2	−43.57	−99.62
M_{δ_s}, s^{-2}	−30.60	−58.60	−18.90	−41.70	−8.19	−20.20

Table B.2c Lateral-directional dimensional derivatives

Condition	1	2	3	4	5	6
h, ft	0(S/L)	0(S/L)	15,000	15,000	35,000	35,000
M	0.6	0.9	0.6	0.9	0.6	0.9
Y_β, ft/s^2	−210.4	−516.5	−118.6	−295.0	−49.46	−127.0
L_β, s^{-2}	−44.57	−98.77	−28.59	−66.05	−13.60	−29.89
L_p, s^{-1}	−4.401	−9.699	−2.665	−6.153	−1.295	−2.956
L_r, s^{-1}	1.341	1.791	0.979	1.112	0.639	0.705
N_β, s^{-2}	8.126	22.15	4.787	13.62	2.416	6.444
N_p, s^{-1}	0.022	0.116	−0.001	0.071	−0.020	0.026
N_r, s^{-1}	−0.968	−1.614	−0.591	−1.022	−0.288	−0.490
Y_{δ_r}, ft/s^2	51.52	62.91	34.07	52.34	15.59	30.39
L_{δ_r}, s^{-2}	11.08	13.60	6.487	11.20	1.854	5.927
N_{δ_r}, s^{-2}	−9.207	−11.76	−5.948	−9.371	−2.754	−5.473
Y_{δ_a}, ft/s^2	−7.034	−8.611	−4.155	−6.576	−1.559	−3.740
L_{δ_a}, s^{-2}	28.46	25.20	17.66	24.12	7.859	14.24
N_{δ_a}, s^{-2}	0.559	0.281	0.360	0.383	0.098	0.205
L'_β, s^{-2}	−44.45	−97.30	−28.91	−65.49	−14.59	−30.22
L'_p, s^{-1}	−4.401	−9.700	−2.667	−6.152	−1.313	−2.959
L'_r, s^{-1}	1.327	1.679	1.017	1.069	0.739	0.729
N'_β, s^{-2}	8.000	20.78	5.149	13.07	3.313	6.738
N'_p, s^{-1}	0.009	−0.020	0.033	0.019	0.061	0.054
N'_r, s^{-1}	−0.964	−1.590	−0.604	−1.013	−0.334	−0.496
L'_{δ_r}, s^{-2}	10.95	12.78	6.866	10.81	2.732	6.197
N'_{δ_r}, s^{-2}	−9.176	−11.58	−6.034	−9.280	−2.922	−5.533
L'_{δ_a}, s^{-2}	28.47	25.24	17.65	24.15	7.977	14.24
N'_{δ_a}, s^{-2}	0.641	0.636	0.139	0.587	−0.393	0.067

Table B.2d   Eigenvalue summary

Condition	1	2	3	4	5	6
h, ft	0(S/L)	0(S/L)	15,000	15,000	35,000	35,000
M	0.6	0.9	0.6	0.9	0.6	0.9
Longitudinal						
ζ_{sp}	0.385	0.395	0.316	0.316	0.224	0.231
ω_{sp}, s^{-1}	4.206	6.763	3.146	5.475	2.085	3.685
$\zeta_p(1/T_{p_1})$	0.097	0.776	0.061	(0.088)	0.064	(0.062)
$\omega_p(1/T_{p_2})$	0.077	0.048	0.074	(−0.050)	0.082	(−0.050)
Lateral-directional						
ζ_d	0.202	0.218	0.155	0.175	0.116	0.128
ω_d, s^{-1}	2.906	4.684	2.295	3.664	1.805	2.582
$1/T_R$, s^{-1}	4.465	9.745	2.702	6.171	1.282	2.919
$1/T_s$, s^{-1}	0.041	0.018	0.044	0.021	0.032	0.019

B.3 U.S. Air Force F-4C Fighter Aircraft

The data given in Tables B.3a–B.3d are for normal cruise configuration, clean airplane, where $W = 38,924$ lb, mass $m = 1210$ slugs, $S = 530$ ft^2, $b = 38.67$ ft, $\bar{c} = 16.0$ ft, c.g. is at $0.289\bar{c}$, and WL = 27.65 in. Longitudinal control is provided by the all-moving horizontal tail. Lateral control is provided by combined ailerons and wing spoilers. Figure B.3 is a sketch of the aircraft.

Note that the absence of derivatives implies negligible values. The stability derivatives are relative to the stability axes. All angles are in radian measure.

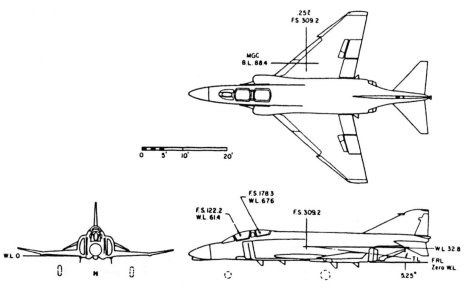

Fig. B.3   Sketch of F-4C aircraft.

Table B.3a Geometric data

Condition	1	2	3	4	5	6
h, ft	0(S/L)	15,000	35,000	35,000	35,000	45,000
M	0.8	0.9	0.6	0.9	1.2	1.5
V, ft/s	893	952	584	876	1,167	1,452
Q, lb/ft^2	948	677	126	283	503	489
I_x, slug-ft^2	24,980	24,970	27,360	25,040	24,970	25,040
I_y, slug-ft^2	122,190	122,190	122,190	122,190	122,190	122,190
I_z, slug-ft^2	139,790	139,800	137,410	139,730	139,800	139,730
I_{xz}, slug-ft^2	1,576	1,175	−16432	−3033	−1030	−3033
α_{trim}, deg	0.3	0.5	9.4	2.6	1.6	2.6

Table B.3b Longitudinal dimensional derivatives

Condition	1	2	3	4	5	6
h, ft	0(S/L)	15,000	35,000	35,000	35,000	45,000
M	0.8	0.9	0.6	0.9	1.2	1.5
X_u, s^{-1}	−0.0162	−0.0215	−0.0176	−0.0123	−0.0136	−0.0072
X_α, ft/s^2	−0.820	−5.93	−24.25	−7.43	−16.53	−29.98
Z_u, s^{-1}	−0.073	−0.144	−0.116	−0.113	−0.010	−0.012
Z_α, ft/s^2	−1374.9	−1103.2	−162.2	−475.4	−848.3	−716.7
$Z_{\dot\alpha}$, ft/s	−2.42	−2.00	−0.591	−1.01	−1.24	−0.519
Z_q, ft/s	−8.20	−6.00	−1.82	−2.89	−4.09	−2.24
M_u, ft^{-1}-s^{-1}	−0.0017	−0.0044	−0.0001	−0.0028	0.0022	0.0025
M_α, s^{-2}	−17.76	−17.00	−1.927	−7.877	−29.03	−28.94
$M_{\dot\alpha}$, s^{-1}	−0.592	−0.457	−0.144	−0.234	−0.288	−0.122
M_q, s^{-1}	−1.360	−0.993	−0.307	−0.487	−0.746	−0.488
X_{δ_s}, ft/s^2	0.00	0.00	−0.01	0.00	−0.01	0.00
Z_{δ_s}, ft/s^2	−141.0	−107.0	−20.98	−49.65	−90.44	−70.67
M_{δ_s}, s^{-2}	−32.30	−25.00	−4.90	−11.40	−20.70	−16.00

Table B.3c Lateral-directional dimensional derivatives

Condition	1	2	3	4	5	6
h, ft	0(S/L)	15,000	35,000	35,000	35,000	45,000
M	0.8	0.9	0.6	0.9	1.2	1.5
Y_β, ft/s^2	−299.2	−204.7	−33.05	−80.68	−176.2	−171.3
L_β, s^{-2}	−27.21	−27.85	−8.252	−17.35	−13.23	−9.977
L_p, s^{-1}	−3.034	−2.264	−0.687	−1.223	−1.372	−0.983
L_r, s^{-1}	0.876	0.678	0.281	0.409	0.329	0.319
N_β, s^{-2}	16.03	11.97	2.155	5.403	12.59	10.18
N_p, s^{-1}	0.009	0.008	−0.006	−0.032	−0.021	−0.008

(*Cont.*)

AIRCRAFT STABILITY DERIVATIVES 341

Table B.3c (Continued)

Condition	1	2	3	4	5	6
N_r, s^{-1}	−0.753	−0.541	−0.149	−0.246	−0.403	−0.317
Y_{δ_r}, ft/s^2	39.47	26.75	6.600	12.44	15.40	14.35
L_{δ_r}, s^{-2}	7.774	5.286	−0.346	1.503	2.765	1.600
N_{δ_r}, s^{-2}	−7.920	−5.666	−1.403	−2.656	−3.251	−2.076
Y_{δ_a}, ft/s^2	−6.644	−4.751	−0.882	−1.989	−3.524	−2.890
L_{δ_a}, s^{-2}	22.15	17.48	4.243	9.967	10.93	6.798
N_{δ_a}, s^{-2}	0.556	0.447	−0.124	−0.042	0.443	0.215
L'_β, s^{-2}	−26.22	−27.30	−10.29	−18.06	−13.75	−11.24
L'_p, s^{-1}	−3.036	−2.265	−0.736	−1.222	−1.371	−0.985
L'_r, s^{-1}	0.829	0.653	0.400	0.440	0.345	0.359
N'_β, s^{-2}	15.74	11.74	3.385	5.795	12.69	10.42
N'_p, s^{-1}	−0.025	−0.011	0.082	−0.006	−0.011	0.014
N'_r, s^{-1}	−0.743	−0.536	−0.197	−0.256	−0.406	−0.325
L'_{δ_r}, s^{-2}	7.279	5.021	0.536	1.830	2.900	1.856
N'_{δ_r}, s^{-2}	−7.838	−5.624	−1.467	−2.696	−3.272	−2.116
L'_{δ_a}, s^{-2}	22.20	17.51	4.651	10.00	10.91	6.790
N'_{δ_a}, s^{-2}	0.807	0.594	−0.680	−0.259	0.362	0.068

Table B.3d Eigenvalue summary

Condition	1	2	3	4	5	6
h, ft	0(S/L)	15,000	35,000	35,000	35,000	45,000
M	0.8	0.9	0.6	0.9	1.2	1.5
Longitudinal						
ζ_{sp}	0.393	0.308	0.259	0.224	0.162	0.102
ω_{sp}, s^{-1}	4.433	4.246	1.413	2.850	5.425	5.396
$\zeta_p(1/T_{p_1})$	(0.051)	(0.074)	0.083	(0.044)	0.191	0.155
$\omega_p(1/T_{p_2})$	(−0.037)	(−0.061)	0.076	(−0.045)	0.045	0.040
Lateral-directional						
ζ_d	0.125	0.097	0.088	0.049	0.073	0.067
ω_d, s^{-1}	4.017	3.460	1.823	2.432	3.576	3.234
$1/T_R$, s^{-1}	3.102	2.336	0.650	1.322	1.407	0.994
$1/T_s$, s^{-1}	0.005	0.008	0.017	0.010	0.002	0.000

B.4 McDonnell–Douglas DC-8 Jet Transport

The data given in Tables B.4a–B.4f are for the transport where $S = 2600.0$ ft^2, $b = 142.3$ ft, $\bar{c} = 23.0$ ft, and the c.g. is at $0.15\bar{c}$. Here condition 1 is the power approach, condition 2 is holding, condition 3 is cruise, and condition 4 = V_{NE}. Figure B.4 is a sketch of the aircraft.

Note that the absence of derivatives implies negligible values. The stability derivatives are relative to the stability axes. All angles are expressed in radian measure.

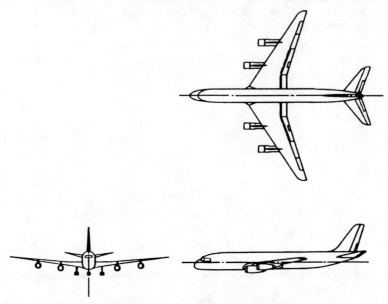

Fig. B.4 Sketch of McDonnell–Douglas DC-8 jet transport.

Table B.4a Geometric data

Condition	1	2	3	4
h, ft	0(S/L)	15,000	33,000	33,000
M	0.218	0.443	0.84	0.88
V, ft/s	243	468	825	868
Q, lb/ft^2	70.4	164	271	300
W, lb	190,000	190,000	230,000	230,000
I_x, slug-ft^2	3.09E+06	3.11E+06	3.77E+06	3.77E+06
I_y, slug-ft^2	2.94E+06	2.94E+06	3.56E+06	3.56E+06
I_z, slug-ft^2	5.58E+06	5.88E+06	7.13E+06	7.13E+06
I_{xz}, slug-ft^2	28,000	−64500	45,000	53,700
δ_{flap}, deg	35	0.0	0.0	0.0

Table B.4b Longitudinal dimensional derivatives

Condition	1	2	3	4
h, ft	0(S/L)	15,000	33,000	33,000
M	0.218	0.443	0.84	0.88
X_u, s^{-1}	−0.0291	−0.0071	−0.0140	−0.0463
X_α, ft/s^2	15.32	15.03	3.544	−22.36
Z_u, s^{-1}	−0.251	−0.133	−0.074	0.062
Z_α, ft/s^2	−152.8	−354.0	−664.3	−746.9
M_u, ft^{-1}-s^{-1}	0.0000	0.0000	−0.0008	−0.0025
M_α, s^{-2}	−2.118	−5.010	−9.149	−12.003
$M_{\dot\alpha}$, s^{-1}	−0.260	−0.337	−0.420	−0.449
M_q, s^{-1}	−0.792	−0.991	−0.924	−1.008
X_{δ_e}, ft/s^2	0.0	0.0	0.0	0.0
Z_{δ_e}, ft/s^2	−10.17	−23.70	−34.69	−39.06
M_{δ_e}, s^{-2}	−1.351	−3.241	−4.589	−5.120

Table B.4c Lateral-directional dimensional derivatives

Condition	1	2	3	4
h, ft	0(S/L)	15,000	33,000	33,000
M	0.218	0.443	0.84	0.88
Y_β, ft/s^2	−27.05	−47.19	−71.73	−81.28
L_β, s^{-2}	−1.334	−2.684	−4.449	−5.111
L_p, s^{-1}	−0.949	−1.234	−1.184	−1.299
L_r, s^{-1}	0.611	0.391	0.337	0.352
N_β, s^{-2}	0.762	1.272	2.175	2.497
N_p, s^{-1}	−0.119	−0.048	−0.013	−0.008
N_r, s^{-1}	−0.268	−0.253	−0.231	−0.254
Y_{δ_r}, ft/s^2	5.782	13.47	18.38	20.35
L_{δ_r}, s^{-2}	0.185	0.375	0.561	0.639
N_{δ_r}, s^{-2}	−0.389	−0.861	−1.173	−1.298
Y_{δ_a}, ft/s^2	0.0	0.0	0.0	0.0
L_{δ_a}, s^{-2}	0.725	1.622	2.120	2.329
N_{δ_a}, s^{-2}	0.050	0.036	0.052	0.062
L'_β, s^{-2}	−1.327	−2.711	−4.424	−5.076
L'_p, s^{-1}	−0.950	−1.233	−1.184	−1.299
L'_r, s^{-1}	0.609	0.397	0.335	0.349
N'_β, s^{-2}	0.756	1.302	2.148	2.459
N'_p, s^{-1}	−0.124	−0.035	−0.021	−0.017
N'_r, s^{-1}	−0.264	−0.257	−0.228	−0.251
L'_{δ_r}, s^{-2}	0.181	0.393	0.547	0.621
N'_{δ_r}, s^{-2}	−0.388	−0.866	−1.169	−1.294
L'_{δ_a}, s^{-2}	0.726	1.622	2.120	2.330
N'_{δ_a}, s^{-2}	0.053	0.018	0.065	0.080

Table B.4d Eigenvalue summary

Condition	1	2	3	4
h, ft	0(S/L)	15,000	33,000	33,000
M	0.218	0.443	0.84	0.88
Longitudinal				
ζ_{sp}	0.522	0.435	0.342	0.324
ω_{sp}, s^{-1}	1.619	2.400	3.146	3.592
$\zeta_p(1/T_{p_1})$	0.060	0.031	0.244	(0.107)
$\omega_p(1/T_{p_2})$	0.164	0.089	0.024	(−0.070)
Lateral-directional				
ζ_d	0.109	0.106	0.079	0.086
ω_d, s^{-1}	0.996	1.198	1.498	1.598
$1/T_R$, s^{-1}	1.121	1.330	1.258	1.365
$1/T_s$, s^{-1}	−0.013	0.007	0.004	0.004

Table B.4e Longitudinal dimensionless derivatives

Condition	1	2	3	4
h, ft	0(S/L)	15,000	33,000	33,000
M	0.218	0.443	0.84	0.88
C_D	0.1095	0.0224	0.0188	0.0276
C_{D_α}	0.487	0.212	0.272	0.486
C_{D_M}	0.0202	0.0021	0.1005	0.365
C_L	1.038	0.445	0.326	0.295
C_{L_α}	4.810	4.876	6.744	6.899
C_{L_M}	0.020	0.048	0.000	−1.200
C_{m_α}	−1.478	−1.501	−2.017	−2.413
C_{m_M}	−0.006	−0.02	−0.17	−0.50
$C_{m_{\dot\alpha}}$	−3.84	−4.10	−6.62	−6.83
C_{m_q}	−11.70	−12.05	−14.60	−15.20
Elevator				
C_{D_δ}	0.0	0.0	0.0	0.0
C_{L_δ}	0.328	0.328	0.352	0.358
C_{m_δ}	−0.943	−0.971	−1.008	−1.016

Table B.4f Lateral-directional derivatives

Condition	1	2	3	4
h, ft	0(S/L)	15,000	33,000	33,000
M	0.218	0.443	0.84	0.88
C_{Y_β}	−0.8727	−0.6532	−0.7277	−0.7449
C_{ℓ_β}	−0.1582	−0.1375	−0.1673	−0.1736
C_{ℓ_p}	−0.385	−0.416	−0.516	−0.538
C_{ℓ_r}	0.248	0.132	0.147	0.146
C_{n_β}	0.1633	0.1232	0.1547	0.1604
C_{n_p}	−0.087	−0.031	−0.011	−0.006
C_{n_r}	−0.196	−0.161	−0.190	−0.199
Rudder				
C_{Y_δ}	0.1865	0.1865	0.1865	0.1865
C_{ℓ_δ}	0.0219	0.0192	0.0211	0.0217
C_{n_δ}	−0.0834	−0.0834	−0.0834	−0.0834
Aileron				
C_{Y_δ}	0.0	0.0	0.0	0.0
C_{ℓ_δ}	0.0860	0.0831	0.0797	0.0791
C_{n_δ}	0.0106	0.0035	0.0037	0.0040

B.5 Convair CV-880M Jet Transport

The data given in Tables B.5a–B.5f are for the transport where $S = 2000.0$ ft^2, $b = 120.0$ ft, and $\bar{c} = 18.94$ ft. Here condition 1 is landing, gear down, $\delta_{\text{flap}} = 50$ deg. Condition 2 is power approach, gear up, $\delta_{\text{flap}} = 35$ deg. For conditions 1 and 2, $W = 126,000$ lb, and the c.g. is at $0.195\bar{c}$. Conditions 3–5 are gear up, flaps up, $W = 155,000$ lb, $m = 4818$ slugs, with the c.g. at $0.25\bar{c}$. Figure B.5 is a sketch of the aircraft.

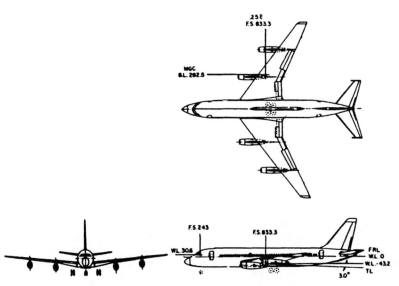

Fig. B.5 Sketch of Convair CV-880M jet transport.

Note that the absence of derivatives implies negligible values. The stability derivatives are relative to the stability axes. All angles are expressed in radian measure. The lateral control consists of hydraulically actuated wing spoilers combined with spring-tab deflected ailerons. Maximum spoiler deflection (60 deg) is possible up to $V_e = 200$ kn. At speeds greater than 200 KEAS, maximum spoiler deflection is hinge-moment limited approximately inversely in proportion to the Q ratio.

Table B.5a Geometric data

Condition	1	2	3	4	5
h, ft	0(S/L)	0(S/L)	23,000	35,000	35,000
M	0.203	0.249	0.86	0.80	0.86
V, ft/s	227	278	881	779	837
Q, lb/ft^2	61.0	91.8	444	224	259
I_x, slug-ft^2	1.170E+6	1.164E+6	1.516E+6	1.527E+6	1.523E+6
I_y, slug-ft^2	2.450E+6	2.450E+6	2.510E+6	2.510E+6	2.510E+6
I_z, slug-ft^2	3.570E+6	3.576E+5	4.094E+6	4.083E+6	4.087E+6
I_{xz}, slug-ft^2	−2.202E+5	−1.833E+5	−1.264E+5	−2.093E+5	−1.820E+5
α_{trim}, deg	5.20	4.32	2.80	4.65	4.04

Table B.5b Longitudinal dimensional derivatives

Condition	1	2	3	4	5
h, ft	0(S/L)	0(S/L)	23,000	35,000	35,000
M	0.203	0.249	0.86	0.80	0.86
X_u, s^{-1}	−0.0423	−0.0270	−0.0080	−0.0057	−0.0059
X_α, ft/s^2	18.70	19.39	19.34	18.24	18.39
Z_u, s^{-1}	−0.300	−0.241	−0.077	−0.089	−0.082
Z_α, ft/s^2	−149.8	−216.1	−816.5	−448.4	−528.3
$Z_{\dot\alpha}$, ft/s	−3.46	−4.26	−4.93	−3.06	−3.49
Z_q, ft/s	−10.2	−12.3	−12.6	−8.39	−9.19
M_u, ft^{-1}-s^{-1}	−0.0001	−0.0001	0.0000[a]	0.0000[a]	0.0000[a]
M_α, s^{-2}	−0.359	−1.283	−3.827	−2.193	−2.885
$M_{\dot\alpha}$, s^{-1}	−0.165	−0.200	−0.335	−0.185	−0.203
M_q, s^{-1}	−0.481	−0.585	−0.850	−0.493	−0.530
X_{δ_e}, ft/s^2	0.0	0.0	0.0	0.0	0.0
Z_{δ_e}, ft/s^2	−4.970	−7.150	−20.63	−13.55	−15.44
M_{δ_e}, s^{-2}	−0.443	−0.647	−2.340	−1.490	−1.650

[a]High-speed values of $M_u = 0$ because tuck-under compensator is included in basic control system.

Table B.5c Lateral-directional dimensional derivatives

Condition	1	2	3	4	5
h, ft	0(S/L)	0(S/L)	23,000	35,000	35,000
M	0.203	0.249	0.86	0.80	0.86
Y_β, ft/s^2	−31.50	−41.14	−149.8	−75.44	−90.39
L_β, s^{-2}	−2.984	−3.725	−10.15	−6.218	−7.291
L_p, s^{-1}	−1.305	−1.560	−1.166	−0.846	−0.859
L_r, s^{-1}	1.022	0.811	0.421	0.414	0.426
N_β, s^{-2}	0.593	0.860	3.172	1.700	2.022
N_p, s^{-1}	−0.094	−0.065	−0.005	−0.011	−0.006
N_r, s^{-1}	−0.236	−0.246	−0.335	−0.167	−0.180
Y_{δ_r}, ft/s^2	6.947	10.10	25.69	17.09	18.08
L_{δ_r}, s^{-2}	0.259	0.428	1.286	0.658	0.786
N_{δ_r}, s^{-2}	−0.408	−0.590	−0.885	−0.994	−0.978
$Y_{\delta_{sp}}$, ft/s^2	−2.430	−1.477	−3.685	−1.913	−2.178
$L_{\delta_{sp}}$, s^{-2}	1.198	1.444	2.902	1.635	1.976
$N_{\delta_{sp}}$, s^{-2}	0.080	0.044	0.193	0.097	0.090
L'_β, s^{-2}	−3.132	−3.891	−10.44	−6.497	−7.573
L'_p, s^{-1}	−1.302	−1.562	−1.169	−0.850	−0.862
L'_r, s^{-1}	1.079	0.857	0.450	0.441	0.450
N'_β, s^{-2}	0.786	1.059	3.494	2.033	2.359
N'_p, s^{-1}	−0.014	0.015	0.031	0.033	0.033
N'_r, s^{-1}	−0.303	−0.290	−0.349	−0.190	−0.200
L'_{δ_r}, s^{-2}	0.340	0.525	1.363	0.799	0.908
N'_{δ_r}, s^{-2}	−0.429	−0.617	−0.927	−1.035	−1.018
$L'_{\delta_{sp}}$, s^{-2}	1.197	1.449	2.893	1.633	1.976
$N'_{\delta_{sp}}$, s^{-2}	0.007	−0.030	0.103	0.014	0.002

Table B.5d Eigenvalue summary

Condition	1	2	3	4	5
h, ft	0(S/L)	0(S/L)	23,000	35,000	35,000
M	0.203	0.249	0.86	0.80	0.86
Longitudinal					
ζ_{sp}	0.794	0.600	0.493	0.400	0.381
ω_{sp}, s^{-1}	0.821	1.291	2.130	1.563	1.782
$\zeta_p(1/T_{p_1})$	0.085	0.059	0.077	0.041	0.049
$\omega_p(1/T_{p_2})$	0.149	0.145	0.049	0.057	0.053
Lateral-directional					
ζ_d	0.118	0.136	0.133	0.091	0.094
ω_d, s^{-1}	1.021	1.113	1.879	1.430	1.539
$1/T_R$, s^{-1}	1.493	1.686	1.171	0.869	0.873
$1/T_s$, s^{-1}	0.009	0.012	0.018	0.008	0.008

Table B.5e Longitudinal dimensionless derivatives

Condition	1	2	3	4	5
h, ft	0(S/L)	0(S/L)	23,000	35,000	35,000
M	0.203	0.249	0.86	0.80	0.86
C_D	0.154	0.080	0.019	0.024	0.023
C_{D_u}	0.43	0.27	0.07	0.15	0.13
C_L	1.029	0.686	0.175	0.346	0.300
C_{L_α}	4.66	4.53	4.41	4.80	4.90
C_{m_α}	−0.381	−0.904	−0.571	−0.650	−0.740
$C_{m_{\dot\alpha}}$	−4.14	−4.13	−4.64	−4.48	−4.58
C_{m_q}	−12.20	−12.10	−11.81	−12.00	−12.01
Elevator					
C_{D_δ}	0.0	0.0	0.0	0.0	0.0
C_{L_δ}	0.160	0.153	0.112	0.146	0.144
C_{m_δ}	−0.470	−0.456	−0.349	−0.441	−0.423

Table B.5f Lateral-directional dimensionless derivatives[a]

Condition	1	2	3	4	5
h, ft	0(S/L)	0(S/L)	23,000	35,000	35,000
M	0.203	0.249	0.86	0.80	0.86
C_{Y_β}	−1.011	−0.878	−0.813	−0.812	−0.842
C_{ℓ_β}	−0.239	−0.197	−0.144	−0.177	−0.179
C_{ℓ_p}	−0.394	−0.382	−0.244	−0.312	−0.294
C_{ℓ_r}	0.308	0.200	0.088	0.153	0.146
C_{n_β}	0.145	0.140	0.122	0.129	0.133
C_{n_p}	−0.087	−0.049	−0.003	−0.011	−0.005
C_{n_r}	−0.218	−0.185	−0.189	−0.165	−0.165
Rudder					
C_{Y_δ}	0.223	0.216	0.139	0.184	0.169
C_{ℓ_δ}	0.021	0.023	0.018	0.019	0.019
C_{n_δ}	−0.099	−0.096	−0.053	−0.076	−0.064
Lateral					
C_{Y_δ}	−0.078	−0.032	−0.020	−0.021	−0.020
C_{ℓ_δ}	0.0958	0.0763	0.0413	0.0465	0.0485
C_{n_δ}	0.020	0.007	0.007	0.007	0.006

[a]Lateral control derivatives combine effects of wing spoilers with ailerons and are expressed in terms of spoiler deflection.

B.6 Lockheed Jetstar Utility Jet Transport

The data given in Tables B.6a–B.6d are for the transport where $S = 542.5\,\text{ft}^2$, $b = 53.75\,\text{ft}$, $\bar{c} = 10.93\,\text{ft}$, the c.g. is at $0.25\bar{c}$, and WL $= 94.2$. Condition 1 is power approach, gear down, flaps at 40%, $W = 23,904\,\text{lb}$, $m = 743.0$ slugs. Conditions 2–6 are normal cruise, clean configuration, $W = 38,204\,\text{lb}$, and $m = 1187.4$ slugs. Table B.6e is for the power approach condition. Figure B.6 is a sketch of the aircraft.

Note that the absence of derivatives implies negligible values. The stability derivatives are relative to the stability axes. All angles are in radian measure.

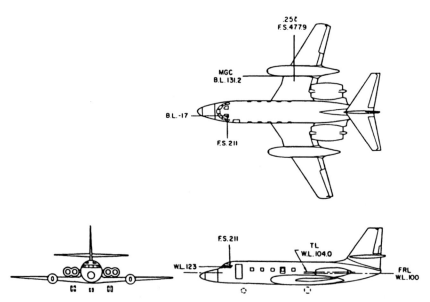

Fig. B.6 Sketch of Lockheed jet transport.

Table B.6a Geometric data

Condition	1	2	3	4	5	6
h, ft	0(S/L)	0(S/L)	20,000	20,000	40,000	40,000
M	0.2	0.4	0.55	0.75	0.65	0.8
V, ft/s	223	447	570	778	629	774
Q, lb/ft^2	59.3	237	206	383	116	176
I_x, slug-ft^2	42,550	118,680	118,760	118,580	119,410	118,710
I_y, slug-ft^2	126,110	135,880	135,880	135,880	135,880	135,880
I_z, slug-ft^2	159,830	243,620	243,540	243,720	242,890	243,590
I_{xz}, slug-ft^2	−7924	−3668	−4758	−613	−10178	−4104
α_{trim}, deg	6.5	4.0	4.5	2.6	7.0	4.2

Table B.6b   Longitudinal dimensional derivatives

Condition	1	2	3	4	5	6
h, ft	0(S/L)	0(S/L)	20,000	20,000	40,000	40,000
M	0.2	0.4	0.55	0.75	0.65	0.8
X_u, s^{-1}	−0.0369	−0.0121	−0.0086	−0.0162	−0.0076	−0.0042
X_α, ft/s^2	−0.642	14.35	13.20	7.188	−4.774	−16.98
Z_u, s^{-1}	−0.300	−0.154	−0.120	−0.085	−0.104	−0.085
Z_α, ft/s^2	−221.0	−552.9	−501.5	−1033.9	−295.2	−511.8
M_u, ft-s^{-1}	0.0002	−0.0005	0.0000	−0.0012	0.0001	−0.0048
M_α, s^{-2}	−2.232	−6.522	−6.120	−12.72	−3.561	−5.629
$M_{\dot\alpha}$, s^{-1}	−0.205	−0.380	−0.276	−0.447	−0.150	−0.217
M_q, s^{-1}	−0.546	−1.030	−0.724	−1.090	−0.380	−0.506
X_{δ_e}, ft/s^2	0.010	−0.001	0.009	0.002	−0.004	−0.001
Z_{δ_e}, ft/s^2	−17.31	−43.31	−37.62	−73.58	−21.86	−34.69
M_{δ_e}, s^{-2}	−2.26	−8.38	−7.47	−14.60	−4.27	−6.78

Table B.6c   Lateral-directional dimensional derivatives

Condition	1	2	3	4	5	6
h, ft	0(S/L)	0(S/L)	20,000	20,000	40,000	40,000
M	0.2	0.4	0.55	0.75	0.65	0.8
Y_β, ft/s^2	−31.26	−78.16	−67.87	−129.88	−38.89	−60.49
L_β, s^{-2}	−3.539	−4.914	−4.048	−4.639	−2.516	−1.972
L_p, s^{-1}	−1.810	−1.297	−0.932	−1.341	−0.487	−0.634
L_r, s^{-1}	0.655	0.245	0.176	0.135	0.132	0.089
N_β, s^{-2}	1.598	3.584	3.247	5.836	1.981	2.784
N_p, s^{-1}	−0.151	−0.111	−0.078	−0.091	−0.049	−0.041
N_r, s^{-1}	−0.207	−0.263	−0.180	−0.252	−0.102	−0.120
Y_{δ_r}, ft/s^2	7.592	18.94	16.48	28.85	9.062	12.55
L_{δ_r}, s^{-2}	0.887	1.421	1.208	2.293	0.580	1.079
N_{δ_r}, s^{-2}	−0.715	−1.896	−1.630	−2.762	−0.896	−1.227
Y_{δ_a}, ft/s^2	0.0	0.0	0.0	0.0	0.0	0.0
L_{δ_a}, s^{-2}	2.148	3.118	2.853	5.820	1.662	2.619
N_{δ_a}, s^{-2}	−0.147	−0.249	−0.246	−0.312	−0.220	−0.221
L'_β, s^{-2}	−3.872	−5.027	−4.182	−4.670	−2.695	−2.069
L'_p, s^{-1}	−1.798	−1.294	−0.930	−1.340	−0.484	−0.633
L'_r, s^{-1}	0.700	0.253	0.183	0.136	0.141	0.093
N'_β, s^{-2}	1.790	3.660	3.328	5.848	2.094	2.819
N'_p, s^{-1}	−0.062	−0.092	−0.060	−0.088	−0.029	−0.031
N'_r, s^{-1}	−0.242	−0.267	−0.183	−0.252	−0.107	−0.122
L'_{δ_r}, s^{-2}	1.030	1.480	1.274	2.307	0.658	1.122
N'_{δ_r}, s^{-2}	−0.766	−1.918	−1.655	−2.768	−0.923	−1.246
L'_{δ_a}, s^{-2}	2.195	3.127	2.865	5.821	1.687	2.628
N'_{δ_a}, s^{-2}	−0.256	−0.296	−0.302	−0.327	−0.291	−0.265

Table B.6d Eigenvalue summary

Condition	1	2	3	4	5	6
h, ft	0(S/L)	0(S/L)	20,000	20,000	40,000	40,000
M	0.2	0.4	0.55	0.75	0.65	0.8
Longitudinal						
ζ_{sp}	0.529	0.474	0.362	0.381	0.259	0.289
ω_{sp}, s^{-1}	1.663	2.793	2.599	3.765	1.932	2.455
$\zeta_p(1/T_{p_1})$	0.049	0.060	0.048	(0.030)	0.049	(0.102)
$\omega_p(1/T_{p_2})$	0.193	0.082	0.078	(−0.016)	0.074	(−0.134)
Lateral-directional						
ζ_d	0.083	0.073	0.050	0.069	0.026	0.045
ω_d, s^{-1}	1.452	1.970	1.856	2.446	1.464	1.689
$1/T_R$, s^{-1}	1.950	1.444	1.044	1.419	0.576	0.680
$1/T_s$, s^{-1}	−0.011	0.005	0.003	0.002	−0.000	−0.000

Table B.6e Dimensionless stability derivatives, power approach condition

Longitudinal		Lateral-directional	
C_D	0.095	C_{Y_β}	−0.722
C_{D_α}	0.752	C_{ℓ_β}	−0.087
C_L	0.743	C_{ℓ_p}	−0.37
C_{L_α}	5.00	C_{ℓ_r}	0.13
C_{m_α}	−0.80	C_{n_β}	0.148
$C_{m_{\dot\alpha}}$	−2.96	C_{n_p}	−0.12
C_{m_q}	−8.02	C_{n_r}	−0.16
Elevator		Rudder	
C_{L_δ}	0.40	C_{Y_δ}	0.175
C_{m_δ}	−0.81	C_{ℓ_δ}	0.022
		C_{n_δ}	−0.066
		Aileron	
		C_{ℓ_δ}	0.053
		C_{n_δ}	−0.014

B.7 North American Navion General Aviation Aircraft

The data given in Tables B.7a–B.7d are for the aircraft where $S = 180.0 \text{ ft}^2$, $b = 33.4$ ft, $\bar{c} = 5.7$ ft, and the c.g. is at $0.295\bar{c}$. Condition 1 is the cruise configuration, $W = 2750$ lb, $m = 85.5$ slugs. The stability derivatives are relative to the stability axes. All angles are in radian measure. Figure B.7 is a sketch of the aircraft.

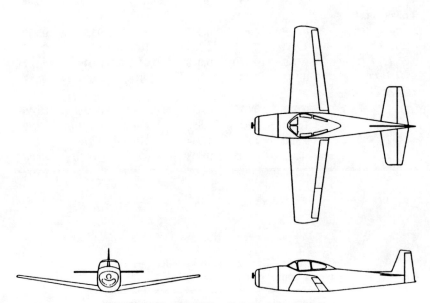

Fig. B.7 Sketch of North American navion.

Table B.7a Geometric data

Condition	1
h, ft	0(S/L)
M	0.158
V, ft/s	176
Q, lb/ft^2	36.8
I_x, slug-ft^2	1048
I_y, slug-ft^2	3000
I_z, slug-ft^2	3530
I_{xz}, slug-ft^2	0
α_{trim}, deg	0.6

Table B.7b Dimensional stability derivatives

Longitudinal		Lateral-directional	
X_u, s^{-1}	-0.0451	Y_β, ft/s^2	43.72
X_α, ft/s^2	6.590	L_β, s^{-2}	-15.62
Z_u, s^{-1}	-0.366	L_p, s^{-1}	-8.215
Z_α, ft/s^2	-348.1	L_r, s^{-1}	2.144
M_α, s^{-2}	-8.598	N_β, s^{-2}	4.395
$M_{\dot\alpha}$, s^{-1}	-0.909	N_p, s^{-1}	-0.342
M_q, s^{-1}	-2.030	N_r, s^{-1}	-0.744
Z_{δ_e}, ft/s^2	-27.52	Y_{δ_r}, ft/s^2	12.17
M_{δ_e}, s^{-2}	-11.19	L_{δ_r}, s^{-2}	2.492
		N_{δ_r}, s^{-2}	-4.495
		L_{δ_a}, s^{-2}	28.34
		N_{δ_a}, s^{-2}	-0.217

Table B.7c Dimensional stability derivatives

Longitudinal		Lateral-directional	
C_D	0.051	C_{Y_β}	-0.564
C_{D_u}	0.330	C_{ℓ_β}	-0.074
C_L	0.415	C_{ℓ_p}	-0.410
C_{L_α}	4.44	C_{ℓ_r}	0.107
C_{m_u}	-0.683	C_{n_β}	0.070
$C_{m_{\dot\alpha}}$	-4.46	C_{n_p}	-0.058
C_{m_q}	-9.96	C_{n_r}	-0.125
Elevator		Rudder	
C_{L_δ}	0.355	C_{Y_δ}	0.157
C_{m_δ}	-0.889	C_{ℓ_δ}	0.012
		C_{n_δ}	-0.072
		Aileron	
		C_{ℓ_δ}	0.134
		C_{n_δ}	-0.003

Table B.7d Eigenvalue summary

Longitudinal		Lateral-directional	
ζ_{sp}	0.710	ζ_d	0.201
ω_{sp}, s^{-1}	3.613	ω_d, s^{-1}	2.358
ζ_p	0.082	$1/T_R, s^{-1}$	8.250
ω_p, s^{-1}	0.210	$1/T_s, s^{-1}$	0.009

References

[1] Heffley, R. K., and Jewell, W. F., "Aircraft Handling Qualities Data," NASA CR-2144, Dec. 1972.

[2] Teper, G. L., "Aircraft Stability and Control Data," Systems Control Technology, Inc., Rept. STI-TR 176-1, Hawthorne, CA, April 1969.

Appendix C
Span Load Program

The relations used to establish the aerodynamic influence coefficient matrix $[A]$, as described in Sec. 2.4, will be expanded using a model suitable for subsonic flow on a straight-tapered wing with sweepback. In addition, a listing will be provided of a computer program (SPANLD.BAS) that can be used on a personal computer to develop span load solutions as an illustration of the procedure used when predicting wing-related stability derivatives. It is important to recognize that the operator symbology denoted by the use of an $[A]$ matrix for a span load solution is very general in order to develop concepts rather than rote techniques.

C.1 Aerodynamic Influence Coefficients

The aerodynamic influence coefficients that populate the $[A]$ matrix are used to solve the span load problem, i.e.,

$$[A]\{\ell/Q\} = \{\alpha\} \tag{C.1}$$

The $\{\alpha\}$ column vector corresponds to requiring that the induced downwash angles (due to the span load distribution) at the control points match the input flow angle distribution. As mentioned in Sec. 2.4, this statement corresponds to satisfying the flow-tangency boundary condition at a prescribed number of wing control points. The span load solution is obtained by premultiplying both sides of Eq. (C.1) by the inverse of $[A]$, i.e.,

$$\{\ell/Q\} = [A]^{-1}\{\alpha\} \tag{C.2}$$

The span load solution can be numerically integrated to obtain: 1) wing lift coefficients and pitching moments for symmetric loadings and 2) rolling moment coefficients for antisymmetric loadings. The selection of the $\{\alpha\}$ input vector will depend on the stability derivative of interest.

The wing will be modeled by either a modified Weissinger approach or a panel method depending on the load detail needed. The wing geometry used in the model (Fig. C.1) shows the right- and left-hand vortices for a location at an arbitrary station j. A control point is located at station i corresponding to the rearward aerodynamic center of the i vortex. The bound vortex portion of the horseshoe vortex is shown as having sweepback, an assumption that allows a more representative model for a swept-back wing. As mentioned in Sec. 2.4, the aerodynamic influence coefficient matrix will be composed of a term from the right-hand wing panel combined with an appropriate complementary term from the left-hand panel depending on whether a symmetric or an antisymmetric span load solution is being determined.

The geometric locations describing the corners of the j vortex and the i control point are determined by the program from the input of wing geometry information, i.e., aspect ratio (AR), taper ratio ($\lambda = c_t/c_r$), wing span, reference sweep

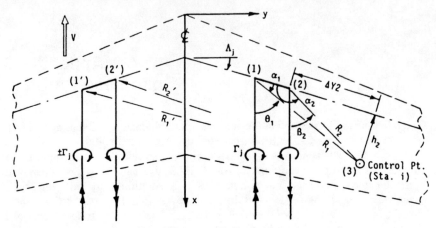

Fig. C.1 Wing vortex geometry.

angle, and desired subsonic Mach number. Specifying the number of spanwise stations with one chordwise station yields a solution corresponding to the modified Weissinger approach. The declaration of the chordwise station number greater than one automatically changes the span load solution to a panel method form (cf. Figs. 2.20 and 2.21).

The Biot–Savart law, Eq. (2.54), which is in accord with Fig. 2.18, is restated for convenience. The induced downwash at a point p from a vortex element is

$$\Delta w_p = \frac{\Gamma(\cos\theta_1 + \cos\theta_2)}{4\pi h} \tag{C.3}$$

The normal distance h between the i control point (shown as 3 in Fig. C.1) and the trailing vortex elements involves lateral distances such as $[Y3(i) - Y2(j)]$. The normal distance h to the swept-back bound vortex element involves a coordinate transformation. The location of the vortex corners (cf. Fig. C.1) are vortex corner 1 at $X1(j), Y1(j)$ and vortex corner 2 at $X2(j), Y2(j)$. The location of the control point is control point 3 at $X3(i), Y3(i)$.

The contribution of the right-hand vortex having vortex strength Γ_j at station j to the downwash angle at control point i is given by

$$\left[\frac{\Delta w(i,j)}{V}\right]_{\text{RH}} = a(i,j)_{\text{RH}} \left(\frac{\ell}{Q}\right)_j$$

where

$$\left(\frac{\ell}{Q}\right)_j = \frac{2\Gamma(j)}{V}$$

and

$$a(i,j)_{\text{RH}} = \frac{1}{8\pi}\left[\frac{(1+\cos\theta_1)}{h_1} + \frac{(\cos\alpha_1 + \cos\alpha_2)}{h_2} - \frac{(1+\cos\beta_2)}{h_3}\right] \tag{C.4}$$

The trailing vortex contributions to Eq. (C.4) involve the following:

$$h_1 = Y3(i) - Y1(j)$$
$$h_3 = Y3(i) - Y2(j)$$
$$R_1 = \{[Y3(i) - Y1(j)]^2 + [X3(i) - X1(j)]^2\}^{1/2}$$
$$R_2 = \{[Y3(i) - Y2(j)]^2 + [X3(i) - X2(j)]^2\}^{1/2}$$

and

$$\cos \theta_1 = [X3(i) - X1(j)]/R_1$$
$$\cos \beta_2 = [X3(i) - X2(j)]/R_2$$

The sweepback angle for the bound vortex portion of the j horseshoe is

$$\Lambda_j = \arctan\left\{\frac{[X2(j) - X1(j)]}{[Y2(j) - Y1(j)]}\right\}$$

The coordinate transformation to establish the normal distance to the bound vortex element h_2 is obtained by applying an orthogonal matrix $[T_\Lambda]$, i.e.,

$$h_2 = [X3(i) - X2(j)]\cos \Lambda(j) - [Y3(i) - Y2(j)]\sin \Lambda(j)$$

A similar rotational transformation to determine the lateral distances in the swept-back coordinate system yields

$$\Delta Y1 = [X3(i) - X1(j)]\sin \Lambda(j) + [Y3(i) - Y1(j)]\cos \Lambda(j)$$
$$\Delta Y2 = [X3(i) - X2(j)]\sin \Lambda(j) + [Y3(i) - Y2(j)]\cos \Lambda(j)$$

which permits evaluation of

$$\cos \alpha_1 = \Delta Y1/R_1 \quad \text{and} \quad \cos \alpha_2 = -\Delta Y2/R_2$$

The lateral distance $\Delta Y2$, which is in the swept-back coordinate system similar to h_2, is shown in Fig. C.1 for illustrative purposes.

The influence coefficient for the left-hand vortex element may be obtained from Eq. (C.4) by the following substitution:

$$X1'(j) = X2(j), \quad Y1'(j) = -Y2(j)$$
$$X2'(j) = X1(j), \quad Y2'(j) = -Y1(j)$$

It should be noted that in the determination of the normal distance h_2 for the j bound vortex, if the bound vortex were aligned with the i control point, then h_2 would be zero and a singularity would occur in Eq. (C.4). In actuality, this situation yields no downwash contribution from the bound vortex element. Consequently, a logic test is necessary in any computational scheme in order to preclude division by zero.

When a subsonic Mach number is prescribed, the X coordinates for the complete set of points ($X1$, $X2$, and $X3$) are altered by the compressibility factor F_c, in keeping with the Prandtl–Glauert–Goethert transformation rule, cf. Eq. (2.58).

The $a(i, j)$ element of the $[A]$ matrix has two forms, a symmetric form, $[A]_s$,

$$a(i, j)_s = a(i, j)_{\text{RH}} + a(i, j)_{\text{LH}}$$

and an antisymmetric form, $[A]_a$,

$$a(i, j)_a = a(i, j)_{\text{RH}} - a(i, j)_{\text{LH}}$$

C.2 Program Listing

A computer program listing, written in Microsoft QuickBASIC®, suitable for use on a personal computer to determine span load distributions is provided. The program, SPANLD.BAS, includes many comments, which should enable the reader to track the logic flow. The program's logic flow is shown in Fig. C.2.

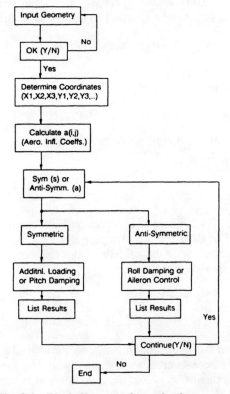

Fig. C.2 Block diagram of span load program.

```
10      ' Program ''SPANLD.BAS''
        '
        ' Solve Span-Load Problem for Straight-Tapered Wing
        ' Vortex Lattice Method uses Swept Bound Vortices
        ' Inputs Consist of:
        ' AR    = Wing Aspect Ratio = B∧2/S
        ' TR    = Wing Taper Ratio = Ct/Cr
        ' SWP25 = 0.25 Chd. Sweep Angle, + is Sweepback
        ' MACH  = Subsonic Mach no. for Compressibility Correctn.
        ' M     = No. of Equal Length Spanwise Stas., R.H. Wing
        ' N     = No. of Equal Chordwise Stas.
        ' ** Comments **
        ' Vortex Lattice Solution OK with MxN panels on R.H. Wing
        ' N = 1 for Elementary (Modified Weissinger) Lifting Line
        '         Theory
        ' Wingspan, B, Set to 1000 Inches (by Default)
100     DIM X1(100), Y1(100), X2(100), X3(100), Y3(100)
        DIM SWP(100), SWPM(100), AILSTA(10), ALPHA(100)
        DIM A1(100,100), A2(100,100), ASYM(100,100),AANT(100,100)
        ' Variables below used in ELU/SLVB Subroutines (S/R's)
        DIM A(100,100) XIN(100,100)
150     ' Input Wing Geometric Information
        CLS
        INPUT " Aspect Ratio, AR ="; AR
        INPUT " Taper Ratio, TR ="; TR
        INPUT " c/4 Sweep (deg.) ="; SWP25D
        INPUT " Mach No. ="; MACH
        ' Determine Prandtl-Glauert Compress. Factor, FC
        IF MACH > .95 THEN GOTO 151
        FC = 1! / SQR(1! - MACH ∧ 2): GOTO 152
151     FC = 1!:  ' Default FC to Incompress. Value
152     ' Continue ... Dummy Statement
        INPUT " No. Spanwise Stas. ="; M
        INPUT "No. Chordwise Stas. ="; N
        '
        ' Echo Check Input Information
        PRINT
        PRINT TAB(10); "Listing of Wing Geometry Input Data"
        GOSUB 1050: ' S/R to List Input Data for Echo Check
        PRINT
        '
        INPUT " Is Wing Data OK, Y/N:"; A$
        IF A$ = "Y" GOTO 160 : IF A$ = "y" GOTO 160
        GOTO 150
        '
160     ' Wing Input Data Echo Checked OK
        PI =4! * ATN(1!): ' Establish Constant
        SWP25 = SWP25D * PI / 180!: ' Convert Sweep to Radians
180     ' Establish No. of R.H. Wing Panels
```

```
         KMAX = M * N
200   ' Determine Initial Wing Geometry
      B = 1000!: ' DEFAULT Value of Wing Span, Inches
      DELB = .5 * B / M: ' Spanwise Spacing, Vortex Elements
      CR = 2! * B /(AR * (1! + TR)): ' Root Chord, Inches
      S = B ∧ 2 / AR: ' Wing Area, Sq. In.
      TANLE = TAN(SWP25) + (.5 * CR * (1! - TR)) / B: 'Tan L.E.
      ' Develop Mean Aero. Chord Information
      MAC = 2! * CR * ((1! + TR) - (TR / (1! + TR))) / 3!
      IF TR = 1! THEN GOTO 210
      YMAC = .5 * B * (1! - (MAC / CR)) / (1! - TR): GOTO 211
210   YMAC = .5 * B
211   XMAC25 = (YMAC * TANLE) + .25 * MAC
      '
      ' Print MAC Location Information
      PRINT
      GOSUB 1060
      '
250   ' Find Coords., Wing Vortex Lattice Corners & Control Pts.
      CONST1 = 1! -TR
      FOR I = 1 TO M
      C1= CR * (1! - (CONST1 * (I -1!) / M))
      ' C1 is Total Wing Chord at Inbd. Vortex
      C2= CR * (1! - (CONST1 * I / M))
      ' C2 is Total Wing Chord at Outbd. Vortex
      C3= CR * (1! - (CONST1 * (I -.5) / M))
      ' C3 is Wing Chord at Control Pt. Sta.
      DELC1 = C1 / N: DELC2 = C2 / N: DELC3 = C3 / N
      '
      FOR J = 1 TO N
      K = (I - 1) * N + J: ' Create Vortex Lattice Numbering
      Y1(K) = (I - 1) * DELB: Y2(K) = Y1(K) + DELB
      Y3(K) = Y1(K) + (.5 * DELB)
      X1(K) = DELC1 * (J - .75) + Y1(K) * TANLE
      X2(K) = DELC2 * (J - .75) + Y2(K) * TANLE
      X3(K) = DELC3 * (J - .25) + Y3(K) * TANLE
      TANSWP = (X2(K) - X1(K)) / (Y2(K) - Y1(K))
      SWP(K) = ATN(TANSWP)
      SWPM(K) = FC * TANSWP
      SWPM(K) = ATN(TANSWPM)
      NEXT J: ' End of Inner Do Loop
      NEXT I: ' End of Outer Do Loop
      '
300   ' Develop Aerodynamic Influence Coeffs.
      ' A1(K1,K2) = R.H. Wing while A2(K1,K2) = L.H. Wing
      FOR K1 = 1 TO KMAX: ' K1 Sta. is at the Control Point
      FOR K2 = 1 TO KMAX: ' K2 Sta. is at the Vortex Station
      NUM1 = X3(K1) - X1(K2): NUM2 = X3(K1) - X2(K2)
      ' Modify X Coords. using Prandtl-Glauert Factor
```

```
        NUM1 = FC * NUM1: NUM2 = FC * NUM2
        DEN1 = Y3(K1) - Y1(K2): DEN2 = Y3(K1) - Y2(K2)
        R1=SQR(NUM1 ∧ 2 + DEN1 ∧ 2)
        R2=SQR(NUM2 ∧ 2 + DEN2 ∧ 2)
        ' Find Trig. Fns. for Orthog. Transformation
        ' on Swept Bound Vortex
        SINSWP = SIN(SWPM(K2)): COSSWP = COS(SWPM(K2))
        H = NUM1 * COSSWP - DEN1 * SINSWP
        Y1ROT = NUM1 * SINSWP + DEN1 * COSSWP
        Y2ROT = NUM2 * SINSWP + DEN2 * COSSWP
        COSTHET1 = Y1ROT / R1: COSTHET2 = -Y2ROT / R2
        ' Logic Check to Avoid Division by Zero
        IF (ABS(H)) <= .001 THEN GOTO 310
        DELWBD = (COSTHET1 + COSTHET2) / H: GOTO 311
310     DELWBD = 0!: ' No Bound Vortex Downwash Option
311   ' Continue ... Dummy Statement Space
        COS1 = NUM1 / R1: COS2 = NUM2 / R2
        DELWLH = (1! + COS1) / DEN1
        DELWRH = (1! + COS2) / DEN2
        A1(K1,K2) = (DELWLH + DELWBD - DELWRH) / (8! * PI)
        NEXT K2
        NEXT K1
        ' Similar Logic for L.H. Wing Panel Aero. Infl. Coeffs.
315     FOR K1 = 1 TO KMAX: ' K1 Sta. is at the Control Point
        FOR K2 = 1 TO KMAX: ' K2 Sta. is at the Vortex Station
        NUM2 = X3(K1) - X1(K2): NUM1 = X3(K1) - X2(K2)
        ' Modify X Coords. using Prandtl-Glauert Factor
        NUM2 = FC * NUM2: NUM1 = FC * NUM1
        DEN2 = Y3(K1) + Y1(K2): DEN1 = Y3(K1) + Y2(K2)
        R1=SQR(NUM1 ∧ 2 + DEN1 ∧ 2)
        R2=SQR(NUM2 ∧ 2 + DEN2 ∧ 2)
        ' Find Trig. Fns. for Orthog. Transformation
        SINSWP = -SIN(SWPM(K2)): COSSWP = COS(SWPM(K2))
        H = NUM1 * COSSWP - DEN1 * SINSWP
        Y1ROT = NUM1 * SINSWP + DEN1 * COSSWP
        Y2ROT = NUM2 * SINSWP + DEN2 * COSSWP
        COSTHET1 = Y1ROT / R1: COSTHET2 = -Y2ROT / R2
        ' Logic Check to Avoid Division by Zero
        IF (ABS(H)) := .001 THEN GOTO 320
        DELWBD = (COSTHET1 + COSTHET2) / H: GOTO 321
320     DELWBD = 0!: ' No Bound Vortex Downwash Option
321   ' Continue ... Dummy Statement Space
        COS1 = NUM1 / R1: COS2 = NUM2 / R2
        DELWLH = (1! + COS1) / DEN1
        DELWRH = (1! + COS2) / DEN2
        A2(K1,K2) = (DELWLH + DELWBD - DELWRH) / (8! * PI)
        ASYM(K1,K2) = A1(K1,K2) + A2(K1,K2)
        AANT(K1,K2) = A1(K1,K2) - A2(K1,K2): NEXT K2
        NEXT K1
```

```
700  ' Select Symmetric or Anti-Symmetric Solution
     PRINT
     INPUT " Sym. or Anti-Sym. Prob., S/A"; P$
     IF P$ = "S" THEN GOTO 710: IF P$ = "s" THEN GOTO 710
     IF P$ = "A" THEN GOTO 710: IF P$ = "a" THEN GOTO 720
     GOTO 1000
710  PRINT "Select Sym. Solution Type: "
     PRINT TAB(5); "1 = Additional Loading"
     PRINT TAB(5); "2 = Wing Pitch Damping"
     INPUT "Sym. Solution Type, 1 or 2"; A$
     IF A$ = "1" THEN GOTO 800
     IF A$ = "2" THEN GOTO 810
     GOTO 1000
720  PRINT "Select Anti-Sym. Solution Type: "
     PRINT TAB(5); "1 = Wing Roll Damping"
     PRINT TAB(5); "2 = Aileron Roll Control"
     INPUT "Anti-Sym. Solution Type, 1 or 2"; A$
     IF A$ = "1" THEN GOTO 900
     IF A$ = "2" THEN GOTO 910
     GOTO 1000
     '
800  ' Find Additional Loading, Alpha = 1.0 Rad.
     FOR K = 1 TO KMAX: ALPHA(K) =1!: NEXT K
     CLS: PRINT TAB(10); "** Additional Load Solution **"
     GOTO 820
     '
810  ' Find Wing Pitch Damping due to qc/2V = 1.0 Rad.
     FOR K = 1 TO KMAX
     ALPHA(K) = (X3(K) - XMAC25) / (.5 * MAC): NEXT K
     CLS: PRINT TAB(10); "** Wing Pitch Damping Solution **"
     GOTO 820
     '
820  ' Load A(I,J) Matrix and XIN(KMAX) Col. Vector
     FOR I = 1 TO KMAX: XIN(I) = ALPHA(I)
     FOR J = 1 TO KMAX: A(I,J) = ASYM(I,J): NEXT J
     NEXT I '
     ' Find Span-Load Solutn. (L/Q) from XIN(KMAX)
     GOSUB 2000
     ' S/R Returns XIN(I) as L/Q Vector
     LIFT = 0!: MOMENT = 0!: FOR I = 1 TO KMAX
     LIFT = LIFT + XIN(I)
     MOMENT = MOMENT + XIN(I) * (XMAC25 - .5 * (1(I) + X2(I))
     NEXT I
     ' Find Wing Lift & Moment Coefficients
     CL = LIFT * 2! * DELB / S
     CM = MOMENT *2! * DELB /(S * MAC)
     NP = .25 - (CM / CL)
     GOSUB 1050: ' Wing Reference Data Listing
```

SPAN LOAD PROGRAM

```
    GOSUB 1060
        PRINT "CL ="; CL; ", CM = "; CM
        PRINT "Neut. Pt. at (% Cmac)"; NP: PRINT
        '
        ' Print Span-Load Distribution Header
        GOSUB 1100
        CAVE = .5 * CR * (1! + TR)
        ' Sum Chordwise Loads for Sectional Loading
        FOR I = 1 TO M: CLC = 0!
        FOR J = 1 TO n: K = (I - 1) * N + J
        CLC = CLC + XIN(K): NEXT J
        ETA = (I - .5) / M
        C3 = CR * (1! - (CONST1 * (I - .5) / M))
        ' C3 = Chord at Control Point
        CLSEC = CLC / C3
        CLC = CLC / CAVE
        GOSUB 1200: ' Print Span-Load Information
        NEXT I
        GOTO 700
        '
900     ' Anti-Symmetric Solution Branch
        ' Find Alpha for Roll Damping due to pB/2V = 1.0 Rad.
        FOR I = 1 TO M
        ALPDAMP = (I - .5) * 2! * DELB / B
        FOR J = 1 TO N
        K = (I - !) * N + J: XIN(K) = ALPDAMP: NEXT J
        NEXT I
        CLS: PRINT TAB(10); "** Wing Roll Damping **"
        GOTO 920
        '
910     ' Use Statements Below for Aileron Control Effectiveness
        ' Reset Aileron Panel Indices, Max. No. = 10
        GOSUB 1300: ' S/R to Display Panel Indices
        FOR I = TO 10: AILSTA(I) = 0!: NEXT I
        INPUT "No. of Aileron Control Panels ="; AILMAX
        FOR I = 1 TO AILMAX
        PRINT "Input Aileron Panel No. for Index ("; I; ")"
        INPUT " Panel No. ="; AILSTA(I): NEXT I
        ' Echo Check on Keyboard Entries
        PRINT " Index"; TAB(12); "Panel"
        FOR I = 1 TO AILMAX
        PRINT USING "#####.#"; I; AILSTA(I); NEXT I
        INPUT "Aileron Stas. OK, (Y/N)"; A$
        IF A$ = "Y" THEN GOTO 915: IF A$ = "y" THEN GOTO 915
        IF A$ = "N" THEN GOTO 910: IF A$ = "n" THEN GOTO 910
        GOTO 1000
915     CLS
        PRINT TAB(10); "** Aileron Control Effectiveness **"
        PRINT "Ail. Panel Nos. =";
```

```
          FOR I = 1 TO 10
          PRINT USING "####."; AILSTA(I); : NEXT I
          ' Set Ail. Panel Control Points
          FOR I = 1 TO KMAX: XIN(I) = 0!: NEXT I
          FOR = 1 TO AILMAX
          J1 = AILSTA(I): XIN(J1) = -1!: NEXT I
          GOTO 920
          '
      920 ' Solve Span Load Problem
          FOR I = 1 TO KMAX: FOR J = 1 TO KMAX
          A(I,J) = AANT(I,J): NEXT J
          NEXT I
          ' Use S/R ELU/SLVB to Solve Anti-Sym. (L/Q)
          GOSUB 2000
          ROLL = 0!: FOR I = 1 TO KMAX
          ROLL = ROLL + XIN(I) * Y3(I): NEXT I
          CROLL = -ROLL * 2! * DELB / (S * B)
          GOSUB 1050: ' Print Wing Reference Data
          GOSUB 1060
          PRINT: PRINT "Roll Moment Deriv. CL ="; CROLL
          '
          GOSUB 1100: ' Print Header
          CAVE = .5 * CR * (1! + TR)
          ' Branch Sums Chordwise Loads for Sectn. Loading
          FOR I = 1 TO M: CLC = 0!
          FOR J = 1 TO N: K = (I - 1) * N + J
          CLC = CLC + XIN(K): NEXT J
          ETA = (I - .5) / M
          C3 = CR * (1! -(CONST1 * (I-.5)/M)): 'Chd. at Control Pt.
          CLSEC = CLC / CAVE
          GOSUB 1200
          NEXT I
          GOTO 700
          '
     1000 END
          '
          ' S/R's are Listed in Following Groupings
          '
     1050 ' S/R for Echo Check of Wing Input Data
          PRINT TAB(8); "Aspect Ratio, AR ="; AR
          PRINT TAB(9); "Taper Ratio, TR = "; TR
          PRINT TAB(8); "c/4 Sweep (deg.) ="; SWP25D
          PRINT TAB(16); "Mach No. ="; MACH
          PRINT TAB(6); "No. Spanwise/Chordwise Stas. =";M;"/";N
          RETURN
          '
     1060 ' S/R to Print MAC Reference Data
          PRINT "Wing Span, In. ="; B; ", Cmac, In. ="; MAC
          PRINT "c/4 Coord, In, X-c/4 =";XMAC25;", Y-c/4 =";YMAC
```

```
        RETURN
        '
1100    ' S/R to Print Span Load Distrib. Header
        PRINT TAB(5);"2Y/B";TAB(14);"Clc/cave";TAB(24);"Cl-sec"
        RETURN
        '
1200    ' S/R to Print Span Load Output
        PRINT USING "#####.####"; ETA; CLC; CLSEC
        RETURN
        '
1300    ' S/R to Show Wing Panel Layout Numbering
        ' to Help in Aileron Control Option
        CLS
        PRINT TAB(20); "** Wing Panel Numbers **"
        PRINT TAB(2); "ETA(Root)"; TAB(25); "** Wing L.E. **";
              TAB(50); "ETA(Tip)"
        FOR I = 1 TO M
        PRINT USING "###.###"; (I - .5) / M;
        NEXT I
        PRINT
        FOR J = 1 TO N: FOR I = 1 TO M
        PRINT USING "######."; J + (I - 1) * N;
        NEXT I
        PRINT
        NEXT J
        PRINT TAB(25); "** WING T.E. **"
        RETURN
        '
2000    ' S/R ELU
        ' Tri-Diagonalizes the Input Matrix A(KMAX,KMAX)
        ' Input Column Vector, XIN(KMAX), is Replaced by
        ' the Output which is Returned to the Main Program
        NM1 = KMAX - 1
        FOR K = 1 TO NM1: KP1 = K + 1
        FOR I = KP1 TO KMAX
        G = -A(I,K) / A(K,K): A(I,K) = G
        FOR J = KP1 TO KMAX
        A(I,J) = A(I,J) + G * A(K,J): NEXT J
        NEXT I
        NEXT K
        GOSUB 2050: ' Use S/R SLVB for Next Step
        RETURN
        '
2050    ' S/R SLVB
        ' Solves the Tri-Diagonalized Matrix [A] Obtained
        ' from ELU by Back Substitution
        NM1 = KMAX - 1: NP1 = KMAX + 1
        FOR K = 1 TO NM1: KP1 = K + 1
        FOR I = KP1 to KMAX
```

```
XIN(I) = XIN(I) + A(I,K) * XIN(K)
NEXT I
NEXT K
XIN(KMAX) = XIN(KMAX) / A(KMAX, KMAX)
FOR K = 2 TO KMAX
I = NP1 - K
J1 = I + 1
FOR J = J1 TO KMAX
XIN(I) = XIN(I) - A(I,J) * XIN(J)
NEXT J
XIN(I) = XIN(I) / A(I,I)
NEXT K
RETURN
```

Appendix D
Linear Algebra Principles

D.1 Linear Algebra Usage

At first glance, linear algebra principles may be viewed as the usage of matrix expressions for the purpose of describing sets of linear equations in a convenient manner. Consider the following:

$$
\begin{aligned}
y_1 &= a_{11}x_1 + a_{12}x_2 + \cdots + a_{1n}x_n \\
y_2 &= a_{21}x_1 + a_{22}x_2 + \cdots + a_{2n}x_n \\
\cdots &= \cdots + \cdots + \cdots + \cdots \\
y_n &= a_{n1}x_1 + a_{n2}x_2 + \cdots + a_{nn}x_n
\end{aligned}
\tag{D.1}
$$

which may be expressed in a summation form as

$$y_i = \sum_{k=1}^{n} a_{ik}x_k \qquad \text{for } i = 1, 2, \ldots, n \tag{D.2}$$

or more compactly in a matrix operator notation form as

$$\{y\} = [A]\{x\} \tag{D.3}$$

A concise expression for the three equivalent relationships [Eqs. (D.1)–(D.3)] is given by the statement that an n-dimensioned vector $\{x\}$ acted on by an $(n \times n)$ matrix operator $[A]$ yields an n-dimensional vector $\{y\}$.

In general, the order of the vector $\{y\}$ may not be the same as the order of the vector $\{x\}$. However, it will be convenient in much of the matrix operations shown in this text to consider linearly independent vectors where the square matrix $[A]$ is of order $(n \times n)$ and has rank of n (cf. Sec. D.9 for the definition of matrix rank). In that case, Eq. (D.3) will represent a linearly independent equation set and the vector $\{x\}$ can be solved by

$$\{x\} = [A]^{-1}\{y\} \tag{D.4}$$

with the following understanding (cf. Ogata[1]).
1) The matrix $[A]$ is of order $(n \times n)$.
2) The determinant of $[A]$ is nonsingular, i.e., $|A| \neq 0$. Hence, the inverse of $[A]$ exists.
3) $[A][A]^{-1} = [A]^{-1}[A] = [I]$.
4) $[I]$ is the unit diagonal matrix that contains ones on the lead diagonal and zeros on the off-diagonal elements.
5) Vectors $\{x\}$ and $\{y\}$ are column vectors of order $n \times 1$.

In addition to the inverse operation, which is valid for a nonsingular square matrix, another matrix operation to consider is the transpose. For convenience, the column vector is sometimes expressed as the transpose of its row vector, i.e.,

$$\{x\} = [x_1 \quad x_2 \quad \cdots \quad x_n]^T$$

The transpose of a square matrix $[A]$, denoted as $[A]^T$, corresponds to a rotation of the matrix about the lead diagonal, i.e., if $[C] = [A]^T$, then the element $c_{ij} = a_{ji}$.

Equation (D.3) represents the output (column) vector due to a matrix operator acting (i.e., multiplying) on a second input (column) vector. The more general situation for matrix multiplication is

$$\underbrace{[A]}_{(m \times n)} \underbrace{[B]}_{(n \times p)} = \underbrace{[C]}_{(m \times p)} \qquad (D.5)$$

and again, by definition, an element of the $[C]$ matrix is

$$c_{ij} = \sum_{k=1}^{n} a_{ik} b_{kj} \qquad (D.6)$$

where

$a_{ik} = i\text{–}k$ element from the ith row of the $[A]$ matrix
$b_{kj} = k\text{–}j$ element from the jth column of the $[B]$ matrix

It should be noted from Eq. (D.6) that matrix multiplication is generally not commutative, i.e.,

$$[A][B] \neq [B][A]$$

Some useful rules applicable to performing linear algebra with matrix products are as follows.

1) If $[C] = [A][B]$, and both $[A]$ and $[B]$ are nonsingular ($n \times n$) matrices, then the inverse of the product is given by

$$[C]^{-1} = [[A] \quad [B]]^{-1} = [B]^{-1}[A]^{-1} \qquad (D.7)$$

Verification of Eq. (D.7) is straightforward. First premultiply the $[C]$ matrix definition by $[A]^{-1}$ to get

$$[A]^{-1}[C] = [A]^{-1}[A][B] = [I][B] = [B]$$

Next, premultiply the preceding expression by $[B]^{-1}$ to obtain

$$[B]^{-1}[A]^{-1}[C] = [B]^{-1}[B] = [I]$$

Therefore,

$$[C]^{-1} = [B]^{-1}[A]^{-1} \qquad \qquad \text{QED}$$

2) If $[C] = [A][B]$, and both $[A]$ and $[B]$ are $(n \times n)$ matrices, then the transpose of the matrix product is

$$[C]^T = [[A]\ [B]]^T = [B]^T[A]^T \tag{D.8}$$

Verification may be shown by noting that the c_{ij} element in $[C]$ is given by

$$c_{ij} = \sum_{k=1}^{n} a_{ik} b_{kj}$$

whereas

$$(c_{ij})^T = c_{ji} = \sum_{k=1}^{n} a_{jk} b_{ki} = \sum_{k=1}^{n} (b_{ik})^T (a_{kj})^T$$

Therefore, we conclude that

$$[C]^T = [B]^T[A]^T \qquad \text{QED}$$

3) An extension of the preceding operation to a sequence of matrix multiplications logically follows. When $[D] = [A][B][C]$, then the transpose of $[D]$ is given by

$$[D]^T = [[A]\ [B]\ [C]]^T = [C]^T[B]^T[A]^T$$

D.2 Linear Algebra and Vector Concepts

The linear algebra formulations, as expressed by Eqs. (D.1–D.3), contained a solution as described by Eq. (D.4) subject to a constraint of the system being linearly independent. These concepts, including the rules on matrix multiplication, applied nicely to problems dealing with airframe static stability and control (cf. Chapter 3) and to wing span load analyses (cf. Chapter 2). The formulations were employed in a short-hand notation to provide a systematic bookkeeping process. The matrix groupings had distinct algebraic properties, which when carefully followed, provided one with results that were easy to visualize.

Consider now a vector that is described in three-dimensional space as shown in Fig. D.1. The initial coordinate system will be described by an E_3 basis consisting of an orthonormal triad of unit vectors e_1, e_2, and e_3 using a customary right-hand rule. In general, the vector x could represent either a force, moment, velocity, or angular rate. As will be shown later, a vector quantity may be expressed in a second orthonormal set of unit vectors by the application of an orthogonal rotation matrix $[T_\theta]$.

The vector shown in Fig. D.1 may be expressed in terms of its components by

$$x = x_1 e_1 + x_2 e_2 + x_3 e_3 \tag{D.9}$$

An alternate notation is to use a column matrix format to describe the components of the vector, i.e.,

$$x = \begin{Bmatrix} x_1 \\ x_2 \\ x_3 \end{Bmatrix}$$

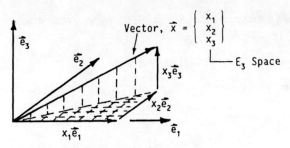

Fig. D.1 Vector in E_3 space.

where it is understood that the three terms in the column matrix of order (3×1) represent scalar values applied to the unit normal vectors e_1, e_2, and e_3, respectively, that are spanning the three space. It is this type of consideration that gives rise to the expression of $\{x\}$ as being a columnar array of numbers representing a vector.

A row matrix may be viewed as the transpose of a column matrix and is denoted by the superscript T, i.e.,

$$x^T = [x_1 \quad x_2 \quad x_3]$$

Similarly, a column vector represents the transpose of the vector transposed because a transpose operation applied twice returns the original matrix form, i.e.,

$$x = (x^T)^T = [x_1 \quad x_2 \quad x_3]^T = \begin{Bmatrix} x_1 \\ x_2 \\ x_3 \end{Bmatrix}$$

D.2.1 Dot Product

A magnitude (squared) of a vector is described as either the inner or dot product, i.e.,

$$|x|^2 = x \cdot x = (x_1 e_1 + x_2 e_2 + x_3 e_3) \cdot (x_1 e_1 + x_2 e_2 + x_3 e_3)$$
$$= (x_1)^2 + (x_2)^2 + (x_3)^2$$

These relations were expanded term by term using the dot-product property of a vector system involving an orthogonal set of unit vectors, namely,

$$e_i \cdot e_j = \delta_{ij}$$

which is the Kronecker delta function, where the Kronecker delta function is defined as

$$\delta_{ij} = \begin{cases} 1 & \text{for } i = j \\ 0 & \text{for } i \neq j \end{cases}$$

The dot product can also be expressed in matrix form as

$$x \cdot x = x^T x = [x_1 \quad x_2 \quad x_3] \begin{Bmatrix} x_1 \\ x_2 \\ x_3 \end{Bmatrix} = \sum_{i=1}^{3} x_i x_i \qquad (D.10)$$

LINEAR ALGEBRA PRINCIPLES

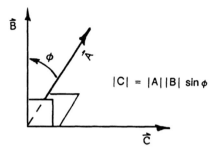

Fig. D.2 Vector cross product of $A \times B$.

D.2.2 Cross Product

Another vector operation that may conveniently be expressed in matrix notation is the cross product. The usual definition of the cross product $A \times B = C$ is that the magnitude of C is given by

$$|C| = |A||B|\sin\phi$$

where ϕ is the angle between A and B, and the direction of C is mutually perpendicular to A and B using a right-hand rule as shown in Fig. D.2.

If we consider vectors A and B as defined in the E_3 vector space using an orthonormal unit vector basis of e_1, e_2, and e_3, then

$$A = a_1 e_1 + a_2 e_2 + a_3 e_3$$
$$B = b_1 e_1 + b_2 e_2 + b_3 e_3$$

and

$$A \times B = (a_2 b_3 - a_3 b_2) e_1 + (a_3 b_1 - a_1 b_3) e_2 + (a_1 b_2 - a_2 b_1) e_3$$

A convenient matrix form that is equivalent to the preceding expression is provided by

$$A \times B = [\hat{A}]\{B\} \tag{D.11}$$

where

$$[\hat{A}] = \begin{bmatrix} 0 & -a_3 & a_2 \\ a_3 & 0 & -a_1 \\ -a_2 & a_1 & 0 \end{bmatrix}$$

$$\{B\} = [b_1 \quad b_2 \quad b_3]^T$$

It will be observed that the A-hat matrix is skew symmetric using the appropriate components from the $\{A\}$ vector. Further discussions on the preceding matrix notation for a cross product may be found in a text by Kane et al.[2] A useful application for Eq. (D.11) will arise when the vector A corresponds to a rotation vector Ω while B represents a vector distance R from a coordinate origin. In the mechanics of a rotating rigid body, the velocity of a point at a vector distance R is given by

$$V = \Omega \times R$$

The preceding vector velocity can be also stated in matrix notation using the format from Eq. (D.11), i.e.,

$$V = [\hat{\omega}]\{R\} \tag{D.12}$$

where the ω-hat matrix is defined by

$$[\hat{\omega}] = \begin{bmatrix} 0 & -\omega_3 & \omega_2 \\ \omega_3 & 0 & -\omega_1 \\ -\omega_2 & \omega_1 & 0 \end{bmatrix}$$

and

$$\{R\} = [r_1 \quad r_2 \quad r_3]^T$$

D.2.3 Coordinate Rotation Matrix

Rotation of a coordinate system alters the components of a vector and involves the use of a matrix transformation relationship. For sake of simplicity, consider a vector in Euclidean two space as shown in Fig. D.3. The original E_2 coordinate system is rotated about the origin (of the vector) by an angle θ, shown as positive for a counterclockwise rotation.

Because the vector is the same physical quantity in either coordinate system, it can be expressed in terms of its components using either the E_2 or F_2 reference spaces. Here

$$V = x_1 e_1 + x_2 e_2$$
$$V = y_1 f_1 + y_2 f_2$$

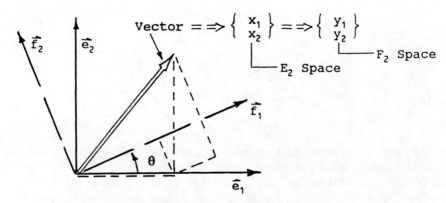

Fig. D.3 Vector expressed in rotated coordinates.

For convenience, define $C_\theta = \cos\theta$ and $S_\theta = \sin\theta$. Using this notation, it can be seen that the F_2 vector space components are

$$y_1 = x_1 C_\theta + x_2 S_\theta$$
$$y_2 = -x_1 S_\theta + x_2 C_\theta$$

When expressed in matrix notation, these relations become

$$\begin{Bmatrix} y_1 \\ y_2 \end{Bmatrix} = \begin{bmatrix} C_\theta & S_\theta \\ -S_\theta & C_\theta \end{bmatrix} \begin{Bmatrix} x_1 \\ x_2 \end{Bmatrix}$$

An equivalent and useful expression in an abbreviated matrix notation is

$$\{y\} = [T_\theta]\{x\} \tag{D.13}$$

where

$$[T_\theta] = \begin{bmatrix} C_\theta & S_\theta \\ -S_\theta & C_\theta \end{bmatrix}$$

If the original coordinates were in the F_2 orientation, then the vector components in the E_2 frame of reference could be obtained by making a clockwise rotation $(-\theta)$, which is equivalent to transposing the $[T_\theta]$ matrix to find $[T_{-\theta}]$. Based on this observation, Eq. (D.13) can be restated to show that

$$\{x\} = [T_\theta]^{-1}\{y\} = [T_{-\theta}]\{y\}$$

from which it can be concluded that

$$[T_\theta]^{-1} = [T_\theta]^T \tag{D.14}$$

The matrix just defined by Eq. (D.13) to make a coordinate system rotation is known as an orthogonal matrix because its inverse equals its transpose and its determinant equals one. A broader definition for an orthogonal matrix (cf. Ogata[1]) is the following.

Definition: A real matrix $[A]$ having the properties that $[A][A]^T = [A]^T[A] = [I]$ and $|A| = \pm 1$ is known as an orthogonal matrix.

D.3 Eigenvalue Problem

Although at first glance the eigenvalue problem appears to have a degree of abstraction that makes its consideration an investment with an uncertain return, it will soon become evident that the concept is an extemely valuable tool for a number of engineering applications including: finding principal stresses and strains in elasticity, finding principal moments of inertia when dealing with engineering problems involving angular momentum concepts, and solving sets of dynamic equations for establishing the motion of physical systems such as airframes.

In developing the concept of the eigenvalue problem, first consider Eq. (D.3) where a matrix operator acted on a vector $\{x\}$ to produce a second vector $\{y\}$. If the second vector $\{y\}$ were the same as the original vector except that its magnitude changed by a constant λ, then one would have the basics of the eigenvalue problem. The eigenvalue problem can be stated as

$$[A]\{x\} = \lambda\{x\} \tag{D.15}$$

where λ is a (real or complex) scalar quantity.

When Eq. (D.15) is satisfied, the scalar quantity λ is labeled as the eigenvalue (or, alternatively, as the characteristic value) and the vector $\{x\}$ is described as the eigenvector (or the mode shape in dynamic related problems). Equation (D.15) may be re-expressed as

$$[\lambda[I] - [A]]\{x\} = \{0\} \tag{D.16}$$

A trivial solution of Eq. (D.16) is that $\{x\} = 0$. A nontrivial solution requires that the characteristic determinant be zero, i.e.,

$$|\lambda[I] - [A]| = 0 \tag{D.17}$$

Expansion of the determinant in Eq. (D.17) will yield the characteristic equation, an nth-order power series in the quantity λ, as

$$\lambda^n + a_{n-1}\lambda^{n-1} + a_{n-2}\lambda^{n-2} + \cdots + a_1\lambda + a_0 = 0 \tag{D.18}$$

The characteristic equation will have n roots, $\lambda_1, \lambda_2, \ldots, \lambda_n$, and can be factored in terms of the roots for an equivalent expression, i.e.,

$$(\lambda - \lambda_1)(\lambda - \lambda_2) \cdots (\lambda - \lambda_n) = 0 \tag{D.19}$$

Some facts to note include the following.

1) If the constant a_0 in Eq. (D.18) were zero, then there would be a zero-valued root.

2) Real constants in Eq. (D.18) require, as can be seen from Eq. (D.19), that if any λ_k were complex (i.e., $\lambda_k = a + ib, i = \sqrt{-1}$), then there also is a complex-conjugate root (shown by the $*$ superscript),

$$\lambda_{k+1} = (\lambda_k)^* = a - ib$$

3) When the order of the characteristic equation is three or higher, modern computer codes prove convenient for finding the roots.

To complete the eigenvalue problem, it is necessary to find the eigenvector $\{x^{(i)}\}$ associated with its corresponding eigenvalue λ_i. This is accomplished by restating Eq. (D.15) in terms of the ith eigenvalue and eigenvector,

$$\lambda_i\{x^{(i)}\} = [A]\{x^{(i)}\} \qquad \text{for } i = 1, 2, \ldots, n \tag{D.20}$$

Each value of i in Eq. (D.20) represents a set of n simultaneous equations with constant coefficients and n unknowns corresponding to the components of the eigenvector associated with λ_i. The magnitude of the eigenvector $\{x^{(i)}\}$ will be arbitrary and can be rescaled or normalized to a unit value without affecting the relative proportions between the components. Multiplying the eigenvector $\{x^{(i)}\}$ by an arbitrary constant in Eq. (D.20) does not alter the meaning of the eigenvalue equation as can be seen,

$$\lambda_i (\text{const})\{x^{(i)}\} = (\text{const})[A]\{x^{(i)}\}$$

Example D.1

Solve the eigenvalue problem for a symmetric, nonsingular (2×2) matrix $[A]$. Symmetry of the $[A]$ matrix will guarantee that the eigenvalues are real (a statement given without proof). Assume

$$[A] = \begin{bmatrix} 1.00 & 1.00 \\ 1.00 & 4.00 \end{bmatrix}$$

1) Form the characteristic equation,

$$|\lambda[I] - [A]| = \left\| \begin{bmatrix} \lambda & 0 \\ 0 & \lambda \end{bmatrix} - \begin{bmatrix} 1.00 & 1.00 \\ 1.00 & 4.00 \end{bmatrix} \right\| = 0$$

which when expanded yields

$$|\lambda[I] - [A]| = \lambda^2 - 5.000\lambda + 3.000 = 0$$

2) Solve the quadratic expression to find the two eigenvalues,

$$\lambda_{1,2} = 2.500 \pm \sqrt{3.250} \quad (0.6972 \quad \text{and} \quad 4.3028)$$

3) Find $\{x^{(1)}\}$ corresponding to $\lambda_1 = 0.6972$ by direct substitution into Eq. (D.20),

$$(0.6972)x_1^{(1)} = x_1^{(1)} + x_2^{(1)}$$

$$(0.6972)x_2^{(1)} = x_1^{(1)} + 4.00x_2^{(1)}$$

The same information as to the relative values of the eigenvector components is contained in either of the two example linear relations. From the first expression, we find

$$x_2^{(1)} = -0.3028x_1^{(1)}$$

The first eigenvector can be expressed as

$$\{x^{(1)}\} = [1.0000 \quad -0.3028]^T$$

Rescaling to find a normalized eigenvector (i.e., $|x^{(1)}| = 1.0$) yields the normalized form

$$\{x^{(1)}\} = [0.9571 \quad -0.2898]^T$$

4) Step 3 may be repeated to find the second eigenvector,

$$\{x^{(2)}\} = [1.0000 \quad 3.3028]^T$$

Rescaling to find a normalized eigenvector (i.e., $|x^{(2)}| = 1.0$) yields the normalized form

$$\{x^{(2)}\} = [0.2898 \quad 0.9571]^T$$

The normalized form of the eigenvectors is not a requirement for solving the eigenvalue problem. However, it frequently does prove convenient to use the normalized form when analyzing linear algebra problems.

D.4 Orthogonality Principles

The concept of orthogonality may be illustrated by using the eigenvectors from Example D.1 to develop the inner product of $\{x^{(1)}\}$ with $\{x^{(2)}\}$, i.e.,

$$\{x^{(1)}\}^T \{x^{(2)}\} = [1.0000 \quad -0.3028] \begin{Bmatrix} 1.0000 \\ 3.3028 \end{Bmatrix} = 1.0000 - 1.0000 = 0$$

Because the inner (or dot) product of the two eigenvectors is zero, it may be concluded that they are orthogonal to each other. As will be shown, this feature resulted because the matrix $[A]$ was real and symmetric and the eigenvalues were distinct.

Using the linear algebra tools developed to this point, it will be shown that the eigenvectors of a symmetric, real matrix with distinct eigenvalues are orthogonal (by pairs). These assumptions imply that $[A] = [A]^T$ and $\lambda_i \neq \lambda_j$. From Eq. (D.20), the eigenvalue problem for the ith and jth eigenvalues can be stated as

$$\lambda_i \{x^{(i)}\} = [A]\{x^{(i)}\} \qquad (D.21)$$

$$\lambda_j \{x^{(j)}\} = [A]\{x^{(j)}\} \qquad (D.22)$$

First, premultiply Eq. (D.21) by $\{x^{(j)}\}^T$ and Eq. (D.22) by $\{x^{(i)}\}^T$ to obtain two quadratic relations,

$$\lambda_i \{x^{(j)}\}^T \{x^{(i)}\} = \{x^{(j)}\}^T [A]\{x^{(i)}\} \qquad (D.23)$$

$$\lambda_j \{x^{(i)}\}^T \{x^{(j)}\} = \{x^{(i)}\}^T [A]\{x^{(j)}\} \qquad (D.24)$$

Using the matrix transposition rules [cf. Eq. (D.8)] developed earlier, transpose both sides of Eq (D.24) to find that

$$\lambda_j \{x^{(j)}\}^T \{x^{(i)}\} = \{x^{(j)}\}^T [A]^T \{x^{(i)}\}$$

Because $[A]^T = [A]$ by assumption, the preceding equation may be restated as

$$\lambda_j \{x^{(j)}\}^T \{x^{(i)}\} = \{x^{(j)}\}^T [A]\{x^{(i)}\} \qquad (D.25)$$

LINEAR ALGEBRA PRINCIPLES

Finally, subtract Eq. (D.25) from Eq. (D.23) to find that

$$(\lambda_i - \lambda_j)\{x^{(j)}\}^T\{x^{(i)}\} = 0$$

leaving the right-hand side as zero. Because distinct eigenvalues were assumed, the preceding relation can only be satisfied by

$$\{x^{(j)}\}^T\{x^{(i)}\} = 0 \qquad \text{for } i \neq j \qquad (D.26)$$

The statement of Eq. (D.26) is the principle of matrix orthogonality. If the eigenvectors had been put in a normalized form such that

$$|x^{(i)}| = 1.0 \qquad \text{for } i = 1, 2, \ldots, n$$

then an nth order modal matrix $[P]$ could have been formed by nesting the normalized eigenvectors in the individual partitions, i.e.,

$$[P] = [\{x^{(1)}\}, \{x^{(2)}\}, \ldots, \{x^{(n)}\}]$$

with the property that

$$[P][P]^T = [P]^T[P] = [I]$$

where $[I]$ is the unit diagonal matrix.

It may be concluded that normalization of the eigenvectors resulted in the modal matrix $[P]$ taking the form of an orthogonal matrix. Consequently, the definition of an orthogonal matrix associated with the rotation of an orthonormal coordinate system (described in Sec. D.2) may be expanded to include a broader class of matrix problems.

Definition: The normalized eigenvectors of a symmetric, real matrix having distinct eigenvalues can be nested by partitioning to form an orthogonal matrix.

D.5 Diagonalization of a Square Matrix

The diagonalization of a real, square nth order matrix $[A]$ will first be shown without any restrictions being placed on its symmetry properties. However, it will be convenient to consider that the eigenvalues are distinct. The eigenvalue problem is restated,

$$\lambda_i\{x^{(i)}\} = [A]\{x^{(i)}\} \qquad \text{for } i = 1, 2, \ldots, n \qquad (D.20)$$

The n eigenvectors can be nested in individual partitions to form a modal matrix $[P]$, i.e.,

$$[P] = [\{x^{(1)}\}, \{x^{(2)}\}, \ldots, \{x^{(n)}\}] \qquad (D.27)$$

Next, define a diagonalized eigenvalue matrix $[D]$ using the same index sequence as used when forming the $[P]$ matrix, i.e.,

$$[D] = \begin{bmatrix} \lambda_1 & 0 & 0 & \cdots & 0 \\ 0 & \lambda_2 & 0 & \cdots & \vdots \\ 0 & 0 & \ddots & 0 & \vdots \\ \vdots & \vdots & 0 & \ddots & 0 \\ 0 & \cdots & \cdots & 0 & \lambda_n \end{bmatrix} \qquad (D.28)$$

The matrix equation sets defined by Eq. (D.20) may be assembled into a single matrix relation using the definitions for the $[P]$ and $[D]$ matrices already given to yield

$$[A][P] = [P][D] \qquad (D.29)$$

In general, if there are distinct eigenvalues and eigenvectors, the matrix $[A]$ can be diagonalized by premultiplying both sides of Eq. (D.29) by $[P]^{-1}$ to obtain

$$[P]^{-1}[A][P] = [D] \qquad (D.30)$$

A transformation that diagonalizes a matrix $[A]$ into a diagonal matrix whose lead diagonal elements correspond to the eigenvalues is called a similarity transformation. It should be noted that the reverse action can take place, i.e.,

$$[A] = [P][D][P]^{-1}$$

This form will prove useful in a subsequent development.

If the matrix $[A]$ were real, symmetric and had distinct eigenvalues, the modal matrix could be formed from the normalized eigenvectors. In that case, the similarity transformation would be simplified by the use of $[P]^T$ for $[P]^{-1}$.

Example D.2

Consider the $[A]$ matrix from Example D.1 and its normalized eigenvectors to illustrate a similarity transformation that converts the $[A]$ matrix into the diagonal eigenvalue matrix,

$$[P]^{-1}[A][P] = [P]^{-1} \begin{bmatrix} 1.00 & 1.00 \\ 1.00 & 4.00 \end{bmatrix} \begin{bmatrix} 0.9571 & 0.2898 \\ -0.2898 & 0.9571 \end{bmatrix}$$

$$= \begin{bmatrix} 0.9571 & -0.2898 \\ 0.2898 & 0.9571 \end{bmatrix} \begin{bmatrix} 0.6673 & 1.2469 \\ -0.2021 & 4.1182 \end{bmatrix}$$

$$= \begin{bmatrix} 0.6972 & 0.0 \\ 0.0 & 4.3028 \end{bmatrix} = [D] \qquad \text{QED}$$

D.6 Eigenvalue Problem for a Symmetric Matrix

It has been stated without proof that a real, symmetric matrix has real eigenvalues. Instead, let us now assume that an unspecified, real, symmetric matrix has a complex eigenvalue and consider the meaning of this assumption on the eigenvalue equation,

$$[A]\{x\} = \lambda\{x\} \qquad (D.15)$$

If the eigenvalue were complex, the eigenvector must also be complex as can be recognized from Eq. (D.15). Next, replace the complex quantities by their complex conjugates,

$$[A]\{x^*\} = \lambda^*\{x^*\} \qquad (D.31)$$

Premultiply Eq. (D.15) by $\{x^*\}^T$ and Eq. (D.31) by $\{x\}^T$ to obtain two quadratic expressions

$$\{x^*\}^T [A]\{x\} = \lambda \{x^*\}^T \{x\}$$
$$\{x\}^T [A]\{x^*\} = \lambda^* \{x\}^T \{x^*\}$$

Finally, subtract the transpose of the second quadratic expression from the first to obtain

$$\{x^*\}^T [[A] - [A^T]]\{x\} = (\lambda - \lambda^*)\{x^*\}^T \{x\}$$

The assumption that the matrix $[A]$ is real, symmetric may be applied to yield

$$(\lambda - \lambda^*) = 0 \qquad (D.32)$$

The statement that the difference between a quantity and its complex conjugate is zero implies that the quantity is real. Demonstrating that the eigenvalues are real when the matrix $[A]$ is real, symmetric also implies that the eigenvectors are real [cf. Eq. (D.15)].

Definition: The eigenvalues and eigenvectors of a real, symmetric matrix are all real.

D.7 Eigenvalue Problem for an Unsymmetric Matrix

In the more general case of a real, square matrix $[A]$ that is not symmetric, the solution of the eigenvalue problem can lead to both complex eigenvalues and eigenvectors. The similarity transformation, as given by Eq. (D.30), is valid even when the eigenvalues and eigenvectors are complex. The only difficulty arises in not being comfortable with the inversion operation for a complex modal matrix. However, modern computer software makes these operations quite feasible. The next example, which is based on the short-period approximation for an aircraft, will provide an illustration of the principle using complex-natured eigenvalues and eigenvectors.

Example D.3

A typical $[A]$ matrix describing the short-period approximation for an aircraft's homogeneous response problem is given by

$$[A] = \begin{bmatrix} -0.805 & 1.000 \\ -8.811 & -1.344 \end{bmatrix}$$

1) Form the characteristic equation from $|\lambda[I] - [A]| = 0$,

$$\begin{vmatrix} (\lambda + 0.805) & -1.000 \\ 8.811 & (\lambda + 1.344) \end{vmatrix} = \lambda^2 + 2.149\lambda + 9.893 = 0$$

2) The roots of the second-order characteristic equation form a complex-conjugate pair, i.e., $\lambda_2 = \lambda_1^*$,

$$\lambda_{1,2} = -1.0745 \pm i\,2.9561$$

where $i = \sqrt{-1}$.

3) To find the first eigenvector, form the eigenvalue equation using the first eigenvector, i.e.,

$$\lambda_1 \{x^{(1)}\} = [A]\{x^{(1)}\}$$

The first equation from the eigenvalue equation is

$$(-1.0745 + i\,2.9561)x_1^{(1)} = -0.805 x_1^{(1)} + x_2^{(1)}$$

or

$$\{x_1^{(1)}\} = \begin{Bmatrix} 1.0000 \\ -0.2695 + i\,2.9561 \end{Bmatrix}$$

The second equation from the eigenvalue equation contains the same information relative to the two components of the first eigenvector. It is easy to show that $\{x_2\} = \{x_1\}^*$.

4) Form the modal matrix $[P]$ from the two eigenvectors, i.e.,

$$[P] = \begin{bmatrix} 1.0000 & 1.0000 \\ (-0.2695 + i\,2.9561) & (-0.2695 - i\,2.9561) \end{bmatrix}$$

5) The inverse of the modal matrix may be found to be

$$[P]^{-1} = \begin{bmatrix} (0.5000 - i\,0.0456) & (0.0 - i\,0.1691) \\ (0.5000 + i\,0.0456) & (0.0 + i\,0.1691) \end{bmatrix}$$

(Try checking it to convince yourself.)

6) The similarity transformation of Eq. (D.30) can be applied to the matrix $[A]$ to yield the complex-natured, diagonal eigenvalue matrix $[D]$.

D.8 Cayley–Hamilton Theorem

A very important property in linear algebra is provided by the Cayley–Hamilton theorem. The theorem may be stated as follows.

Theorem: A nonsingular, square matrix $[A]$ satisfies its characteristic equation in a matrix sense.

To understand the meaning of the theorem, consider the characteristic equation of a nonsingular, square matrix $[A]$ from Eq. (D.18) as

$$\lambda^n + a_{n-1}\lambda^{n-1} + a_{n-2}\lambda^{n-2} + \cdots + a_1\lambda + a_0 = 0 \quad \text{(D.18)}$$

which becomes in terms of the theorem,

$$[A]^n + a_{n-1}[A]^{n-1} + a_{n-2}[A]^{n-2} + \cdots + a_1[A] + a_0[I] = [0] \quad \text{(D.33)}$$

The proof can be demonstrated using three features of linear algebra including: solving for the eigenvalue, performing matrix partitioning, and applying the similarity transformation.

For an illustrative case, consider $[A]$ as a nonsingular, 3×3 matrix. Then from the definition of a characteristic equation [Eq. (D.17)] one finds that

$$\lambda^3 + a_2\lambda^2 + a_1\lambda + a_0 = 0$$

which yields three eigenvalues that satisfy the relation that

$$(\lambda - \lambda_1)(\lambda - \lambda_2)(\lambda - \lambda_3) = 0$$

First, write the characteristic equation for the first eigenvalue as

$$\lambda_1^3 + a_2\lambda_1^2 + a_1\lambda_1 + a_0 = 0$$

Next, develop a row matrix by partitioning in zero terms, i.e.,

$$\lambda_1^3[1\ \ 0\ \ 0] + a_2\lambda_1^2[1\ \ 0\ \ 0] + a_1\lambda_1[1\ \ 0\ \ 0] + a_0[1\ \ 0\ \ 0] = [0\ \ 0\ \ 0]$$

followed by performing similar linear algebra manipulations on the second and third characteristic equations to obtain

$$\lambda_2^3[0\ \ 1\ \ 0] + a_2\lambda_2^2[0\ \ 1\ \ 0] + a_1\lambda_2[0\ \ 1\ \ 0] + a_0[0\ \ 1\ \ 0] = [0\ \ 0\ \ 0]$$

$$\lambda_3^3[0\ \ 0\ \ 1] + a_2\lambda_3^2[0\ \ 0\ \ 1] + a_1\lambda_3[0\ \ 0\ \ 1] + a_0[0\ \ 0\ \ 1] = [0\ \ 0\ \ 0]$$

Continuing with the matrix expansions, the row matrix manipulations can be modified to yield equivalent forms as

$$[\lambda_1^3\ \ 0\ \ 0] + a_2[\lambda_1^2\ \ 0\ \ 0] + a_1[\lambda_1\ \ 0\ \ 0] + a_0[1\ \ 0\ \ 0] = [0\ \ 0\ \ 0]$$

$$[0\ \ \lambda_2^3\ \ 0] + a_2[1\ \ \lambda_2^2\ \ 0] + a_1[0\ \ \lambda_2\ \ 0] + a_0[0\ \ 1\ \ 0] = [0\ \ 0\ \ 0]$$

$$[0\ \ 0\ \ \lambda_3^3] + a_2[0\ \ 0\ \ \lambda_3^2] + a_1[0\ \ 0\ \ \lambda_3] + a_0[0\ \ 0\ \ 1] = [0\ \ 0\ \ 0]$$

These three carefully tailored row matrices can be combined by partitioning into a set of three matrix relations, i.e.,

$$[D]^3 + a_2[D]^2 + a_1[D] + a_0[I] = [0] \qquad (D.34)$$

where $[D]$ is diagonal with respect to the eigenvalues. The expression was developed earlier in Sec. D.5 as Eq. (D.28).

Last, the similarity transformation as posed by Eq. (D.30) can be performed on Eq. (D.34) by pre- and postmultiplication using the modal matrices in the following manner:

$$[P][D][P]^{-1} = [A]$$

$$[P][D]^2[P]^{-1} = [P][P]^{-1}[P][D][P]^{-1}[P][P]^{-1} = [A]^2 \ldots$$

which yields, finally, that

$$[A]^n + a_{n-1}[A]^{n-1} + a_{n-2}[A]^{n-2} + \cdots + a_1[A] + a_0[I] = [0] \qquad \text{QED}$$

An interesting feature of the Cayley–Hamilton theorem [Eq. (D.33)] is its application for finding the inverse of a matrix. Premultiplication of Eq. (D.33) by $[A]^{-1}$ allows one to obtain for a (3×3) example that

$$[A]^{-1} = (-1/a_0)[[A]^2 + a_2[A] + a_1[I]] \qquad (D.35)$$

The obvious advantage of this technique is that the inverse of the matrix may be found by a reduced number of multiply operations, a feature considered of import for inverting large matrices during the early introduction of digital computers (cf. Pipes[3]).

The use of the Cayley–Hamilton theorem to develop the inverse of a matrix, as shown by Eq. (D.35), sets the stage for performing a similar operation to obtain a higher power of the matrix $[A]$ as an algebraic expression. Recognition of this capability will prove invaluable later in establishing state-space fundamentals. For an example, assume that the matrix $[A]$ is of order $n = 3$. By the Cayley–Hamilton theorem, one can state that

$$[A]^3 + a_2[A]^2 + a_1[A] + a_0[I] = [0]$$

This relation can also be expressed as

$$[A]^3 = -a_2[A]^2 - a_1[A] - a_0[I]$$

The fourth power of the matrix $[A]$ will be recognized as

$$[A]^4 = [A][A]^3 = -a_2[A]^3 - a_1[A]^2 - a_0[A]$$
$$= (a_2^2 - a_1)[A]^2 + (a_2 a_1 - a_0)[A] + a_2 a_0[I] \qquad (D.36)$$

The interesting implication from Eq. (D.36) is that a higher power of a matrix can be expressed as a linear combination of powers of the matrix $[A]$ up to powers of $(n - 1)$ for an nth-order square matrix. A logical extension would be to obtain the fifth, sixth, seventh, and higher powers of the matrix [A], all neatly stated in a manner similar to Eq. (D.36). In general, for [A] being an nth-order matrix, $[A]^m$ (where $m \geq n$) can be stated as

$$[A]^m = c_{n-1}[A]^{n-1} + c_{n-2}[A]^{n-2} + \cdots + c_1[A] + c_0[I] \qquad (D.37)$$

A practical usage for the statement of Eq. (D.37) is to compact an infinite power series summation of matrix terms into an equivalent one having a finite number of matrix powers. As an example, consider the transition matrix as defined by an infinite power series, i.e.,

$$\phi(t) = e^{At} = \sum_{k=0}^{\infty} \frac{(At)^k}{k!} \qquad (D.38)$$

Application of the Cayley–Hamilton theorem for an nth-order plant matrix [A], as shown by Eq. (D.37), alters $\phi(t)$ to a finite power series having the form

$$\phi(t) = e^{At} = \sum_{k=0}^{n-1} f_k(t) A^k \qquad (D.39)$$

This principle, as embodied by Eq. (D.39), was used in Sec. 7.7.2 during the development of the controllability concept. Techniques for determining the $f_k(t)$ coefficients are shown by example in numerous books describing state-space methods for finding the transition matrix (e.g., Ogata[1]).

D.9 Linear Independence and Rank

A set of vectors (or columns of a matrix) are said to be linearly independent if a linear sum of the vectors can equal zero only if the constant coefficients are all zero, i.e.,

$$a_1\{x_1\} + a_2\{x_2\} + \cdots + a_n\{x_n\} = \{0\} \qquad (D.40)$$

where each of the constants $a_1, a_2, \ldots, a_n = 0$.

In case Eq. (D.40) can be satisfied when some of the constants are nonzero, then the system of vectors is said to be linearly dependent.

Necessary and sufficient conditions for linear independence by a set of vectors $\{x_i\}$ for $i = 1, 2, \ldots, n$ with each having m components include the following: $n \leq m$ and when a matrix [B] is formed by nesting the n vectors, i.e.,

$$[B] = [\{x_1\} \quad \{x_2\} \quad \cdots \quad \{x_n\}]$$

then the determinant of [B] must satisfy

$$|B| \neq 0$$

The latter requirement for linear independence of vectors has a counterpart when describing the rank of a matrix. Consider a matrix [B], not necessarily square, having n columns. The rank of [B] is the largest number of linearly independent columns of $[B]$. It should be apparent that the rank cannot be greater than n, but it certainly can be smaller than n. If matrix [B] were of order $m \times n$ (i.e., m rows and n columns), then the rank of $[B] \leq \min(m, n)$.

The rank of a matrix $[B]$ is the dimension of the largest nonzero determinant formed by deleting rows and columns from $[B]$. The determination of the rank of a matrix corresponds to finding the largest nonsingular determinant embedded in the matrix. An implication from these statements is that an $n \times n$ matrix will be of rank n when the columns internal to the matrix form a linearly independent set that satisfies Eq. (D.40). An $n \times n$ matrix $[B]$ with rank $(n - 1)$ will be singular and, consequently, an inverse of that matrix would not exist. Therefore, when employing a matrix inversion in a linear algebra expansion, a frequent statement made to ensure existence is that $[B]^{-1}$ = inverse of a nonsingular matrix $[B]$.

The numerical determination of a matrix rank can be difficult when done manually. Available numerical software such as the *rank* function in MATLAB makes the rank determination of a matrix a relatively easy task.

References

[1]Ogata, K., *Modern Control Engineering*, 2nd ed., Prentice–Hall, Englewood Cliffs, NJ, 1990, pp. 688–693, Appendix A.

[2]Kane, T. R., Likins, P. W., and Levinson, D. A., *Spacecraft Dynamics*, McGraw–Hill, New York, 1983, pp. 47–49.

[3]Pipes, L. A., *Applied Mathematics for Engineers and Physicists*, 2nd ed., McGraw–Hill, New York, 1958, pp. 91–93.

Appendix E
Usage of MATLAB Programs

E.1 Introductory Remarks

The subject of aircraft flight dynamics depends on 1) the formulation of the governing equations of motion, 2) finding solutions to a set of coupled differential equations, and 3) interpreting the results. The focus in this text relates to the use of linear algebra concepts as a way of making and solving problem statements in accord with the modern control theory approach.

Solution techniques involve the use of MATLAB software in conjunction with personal desktop computers that are typically available to both engineers and students. The author views MATLAB software during the learning process of flight dynamics as being more direct as compared to using other, earlier languages such as Fortran, Basic, or C. However, a Basic program has been provided in Appendix C for solving elementary wing span load problems as support to aerodynamic concepts expressed in Chapter 2. Its implementation does not require the user to learn another software language; instead, the program just requires the ability to execute a program. Additionally, it is assumed that most disk operating systems provide access to the Basic language thereby creating little difficulty in implementing the SPANLD.BAS program.

E.2 Elementary Use of MATLAB

The first listing of MATLAB statements in support of the text material appears in Sec. 6.2 during the introduction to aircraft longitudinal dynamics (i.e., Example 6.1). MATLAB commands and output results are shown by bold-faced type with instructions usually expressed by lower-case mnemonic statements. Entry of data input, such as the matrix [A], must occur prior to execution of any MATLAB commands. For a (2 × 2) matrix, the following entry will provide a sample data input of two (1 × 2) row matrices separated by the ; (semicolon), i.e.,

```
A=[ 1.0 2.0; 3.0 4.0 ];  <CR>
```

where ⟨CR⟩ implies use of the keyboard enter command. The ; at the statement end is optional for suppressing the otherwise automatic print out on the computer screen of the data entry and/or result. To obtain an echo-check of the entry, the user has the option of the display command.

```
disp(A)  <CR>
   1.0   2.0
   3.0   4.0
```

[Ainv] is the inverse of the nonsingular, square matrix [A], which may be either real or complex natured. The inverse of the square matrix is found using the *inv* command. Note that any title may be used to describe the variable of interest (e.g., A in this case). Although program style varies among users, the author finds it good practice to keep the first letter of a variable in upper case. Nonexecutable comments in the form of string character messages can be entered providing that they are preceded by the % (percent) character,

```
Ainv = inv(A); <CR>
% Echo Check of the inverse of matrix [A] <CR>
disp(Ainv) <CR>
     -2.0000    1.0000
      1.5000   -0.5000
```

Other commands that are useful for solving the eigenvalue problem include the following.

The *poly* command returns a $[1 \times (n + 1)]$ row matrix representing the coefficients in decreasing order for the characteristic polynomial $|\lambda I - A|$ of the nth order plant $[A]$,

```
P = poly(A)
```

The roots of the characteristic polynomial, which correspond to the eigenvalues for the plant $[A]$, are returned by this command as an $(n \times 1)$ column matrix for the nth order plant,

```
R = roots(P)
```

The undamped natural frequencies Wn and the dimensionless damping ratios Z of the roots R may be found by the use of the *damp* command. They are returned as $(n \times 1)$ column matrices for an nth order plant,

```
[Wn,Z] = damp(R)
```

The eigenvalue problem for the plant described by the square matrix $[A]$ may be solved by the use of the *eig* command. Two square matrices are returned by execution of this command, i.e., the $[V]$ eigenvector matrix and the corresponding diagonal $[D]$ eigenvalue matrix,

```
[V,D] = eig(A)
```

The absolute value and phase angle of the first column of the eigenvector matrix $[V]$ are returned as column vectors by the use of the *abs* and *angle* commands. These commands are especially useful when the eigenvector is complex natured,

```
MAG = abs(V(:,1))
PHASE = angle(V(:,1)); % in radian measure
```

The solution for the aircraft dynamics problem may be found either in the time or the frequency domain. This feature is a valuable property provided by modern control theory approaches as applied to the statements describing the coupled differential equations, i.e.,

$$\{\dot{x}\} = [A]\{x\} + [B]\{u\} \quad \text{(E.1)}$$

and the output given by

$$\{y\} = [C]\{x\} + [D]\{u\} \quad \text{(E.2)}$$

A block diagram corresponding to these statements is shown in Fig. 5.15.

Useful commands for solving the time response problem include the following.

The *expm* command, as shown next as an example, returns the transition or exponential matrix for the plant matrix $[A]$ at $t = 2.0$ s,

```
E2A = expm(2.*A)
```

The next expression creates a regularly spaced vector t extending from $t = 0.0$ to $t = 10.0$ by increments of $\Delta t = 0.10$. The example t row vector will be of order (1×101).

```
t = 0.0:0.10:10.0
```

Vector arrays representing the time-history responses for the system defined by Eqs. (E.1) and (E.2) due to 1) initial condition, 2) impulse response, and/or 3) step response are typically provided by the following statements after the definition of the plant and input/output relations has been made.

The state vector response X and output vector response Y are provided by execution of the *initial* command due to the initial condition, $X0$, for the regularly spaced time values indicated by the t row vector,

```
[Y,X,t] = initial(A,B,C,D,X0,t)
```

The impulse/step response to application of the IUth control for time values indicated by the t row vector are provided by the *impulse* and *step* commands,

```
[Y,X,t] = impulse(A,B,C,D,IU,t) or
[Y,X,t] = step(A,B,C,D,IU,t)
```

The response to a predefined control input U that is consistent with the regularly spaced time vector t is provided by the *lsim* command. The control input matrix U must have as many rows as the control effectiveness matrix B has columns. This command is particularly useful in providing the simulated response of a linear system due to pulses, doublets, and other types of control waveforms.

```
[Y,X,t] = lsim(A,B,C,D,U,t)
```

It is convenient to create an .m file for use with the *lsim* command when investigating the linear system response to a pulse. Implementation of an input vector U representing a unit pulse can be obtained by

```
% Enter the pulse command
pulse <CR>
How many elements in the time vector? 101 <CR>
No. of time steps in the unit pulse? 10 <CR>
% Above entries created a U input pulse vector
```

where the pulse.m file is described by

```
% File is: PULSE.M
m=input('How many elements in the time vector? ');
n=input('No. of time steps in the unit pulse? ');
for i=1:n,
U(i)=1.0;
end
for i=n+1:m,
U(i)=0.0;
end
```

A similar file can be written for a doublet, and if desired, implementation of the pulse/doublet with a time delay can be included when editing the .m file.

The transfer function for the output response of Eq. (E.2) has been shown in Chapter 5 to be

$$G(s) = C[sI - A]^{-1}B + D = \frac{\text{Num}(s)}{\text{Den}(s)} \qquad (E.3)$$

where in Eq. (E.3), the denominator corresponds to the system's characteristic polynomial. Useful commands for obtaining the transfer function and the corresponding steady-state frequency response (when $s = i\omega$) are as follows.

The command *ss2tf* is the state space to (2) transfer function command, which provides information as to the numerator and denominator polynomials for the plant and output as defined by Eqs. (E.1) and (E.2) and in response to the IUth control input,

```
[Num,Den] = ss2tf(A,B,C,D,IU)
```

The *bode* command returns the response magnitude and phase angle (degrees) for the transfer function defined by the Num(s) and Den(s) polynomials of Eq. (E.3) at frequency values (radians per second) defined by the frequency row vector w. The alternate statement directly involves the system statements of the plant and output as defined by Eqs. (E.1) and (E.2) for the IUth control input.

```
[Mag,Phase,w] = bode(Num,Den,w) or
[Mag,Phase,w] = bode(A,B,C,D,IU,w)
```

Pole placement principles were developed in Sec. 7.7 following an introduction to controllability concepts. Ackerman's formula, Eq. (7.47), was shown as a convenient pole placement method for a SISO system. MATLAB provides the *acker* command to execute Ackerman's equation, i.e.,

K = acker(A,B,R)

This command determines the [K] row matrix feedback law for a SISO system that moves the eigenvalues of an unaugmented plant [A] having a control vector {B} to closed-loop poles {R} in accord with the augmented plant as described by

$$[Aaug] = [A] - \{B\}[K]$$

The *lyap* function solves Lyapunov's equation [Eq. (5.75)] for applications to 1) the verification of a system's asymptotic stability (cf. Sec. 5.7) and 2) the study of the random response behavior of an aircraft system when excited by turbulence (cf. Sec. 9.5). Application of the *lyap* function to the latter situation is important for determining the response covariance matrix, i.e.,

Q = lyap(A,B*R*B')

The Q output is the covariance matrix due to white noise excitation, $w(t)$, with mean-square amplitude R applied to a system defined by [cf. Eqs. (9.53) and (9.54)]

$$\dot{x} = [A]x + \{B\}w(t)$$

and

$$E[w(t)w'(\tau)] = R\delta(t - \tau)$$

The MATLAB plot instruction provides a convenient screen display of either the time history or frequency response information. Further details on the use of the *plot* command, along with more specific information on the usage of other commands, may be found in the MATLAB software user manuals, e.g., Refs. 1 and 2.

E.3 Use of MATLAB Function Files

Several nonlinear equations of motion were described in Chapter 8 with specific aircraft examples being wing rock (cf. Sec. 8.3) and stall dynamics (cf. Sec. 8.4). These examples can be described by a system of nonlinear ordinary differential equations, i.e.,

$$\dot{x} = f(x, t) \tag{E.4}$$

where x is the state vector, t is time, and $f(x, t)$ is the functional relationship. A Runge–Kutta integration scheme could be written to solve Eq. (E.4) using techniques such as described by Gerald and Wheatley.[3] However, MATLAB provides several equivalent instructions for solving nonlinear ODEs. The routine selected for use in Chapter 8 was *ode23* to obtain time-history solutions.

E.3.1 Wing-Rock Limit Cycle

The specific MATLAB statement to illustrate a wing-rock limit cycle (cf. Example 8.3) was

```
[t,x] = ode23('wrock',t0,tf,xin)
```

where

wrock = function file describing Eq. (E.4)
$t0$ = initial time
tf = final time for the solution, s
xin = Initial condition of the state vector x

The function file wrock.m was used to establish the derivative dx/dt for specified values of t and x in the *ode23* equation solver routine. A listing is provided for reference purposes.

```
function xdot = wrock(t,x)
% Applicable to Example 8.3
% ALD = Lat-Dir. plant matrix
% ALG = Long. plant information
% MU = Inertial ratios
ALD=[-0.083 -0.001 0.145 -1.004;
-3.735 -0.433 0.009 0.015; 0.0 1.0 0.0 0.0;
1.902 0.036 -0.021 -0.029];
% ALD plant matrix set for Zeta-DR = -0.03
ALG=[-0.307 1.0 0.144; -7.91 -0.48 -0.144 ];
% M-alpha set for Freq. Ratio = 2.0
MU=[0.3587 0.3648 -0.6406];
Xtemp= ALD(1,1)*x(1)+ALG(1,3)*cos(x(7))*sin(x(3))-x(4) ;
xdot(1)= Xtemp + x(2)*x(5) ;
Xtemp2= ALD(2,1)*x(1) + ALD(2,2)*x(2) + ALD(2,4)*x(4) ;
xdot(2)= Xtemp2  - MU(1)*x(4)*x(6) ;
xdot(3)=x(2)+(x(6)*sin(x(3))+x(4)*cos(x(3)))*tan(x(7));
Xtemp4= ALD(4,1)*x(1) + ALD(4,2)*x(2) + ALD(4,4)*x(4);
xdot(4)= Xtemp4 - MU(2)*x(2)*x(6) ;
Ytemp= ALG(1,3)*cos(x(7))*cos(x(3)) - 1.) ;
xdot(5)= ALG(1,1)*x(5) + x(6) + Ytemp - x(2)*x(1) ;
Dtemp= ( ALG(2,1) + ALG(2,3)*ALG(1,1) )*x(5) ;
Dtemp1= Dtemp + ( ALG(2,2) + ALG(2,3) )*x(6) ;
Dtemp2= Dtemp1 + ALG(2,3)*Ytemp ;
xdot(6)= Dtemp2 - ( MU(3)*x(4) + ALG(2,3)*x(1) )*x(2) ;
xdot(7)= x(6)*cos(x(3)) - x(4)*sin(x(3)) ;
```

where, in accord with Sec. 8.3, the state vector components are

$x(1)$ = sideslip angle, β, rad
$x(2)$ = roll rate, p, rad/s

x(3) = bank angle, Φ, rad
x(4) = yaw rate, r, rad/s
x(5) = angle of attack perturbation, α, rad
x(6) = pitch rate, q, rad/s
x(7) = pitch attitude, Θ, rad

E.3.2 Demonstration of Stall Dynamics

An illustration of stall dynamics during the determination of minimum aircraft speed was provided by Example 8.6 in Sec. 8.4 using the nonlinear state equations, Eq. (8.17).

The numerical simulation also included nonlinear models for the elevator angle, the lift and drag force coefficients, and the pitching moment coefficient. The nonlinear set of equations were solved using the following *ode23* statement:

```
[t,x] = ode23('stall',t0,tf,xin,1.E-5)
```

where new problem specific terms include

stall = function file describing Eq. (E.4)
tol   = optional tolerance for problem accuracy, 1.E-5

The function file stall.m representing Eq. (8.17) for finding t and x in the *ode23* equation solver routine is shown next using values consistent with Example 8.6.

```
function xdot = stall(t,x)
DE = elev(t,x);
CL = lift(x(1),DE);
CD = drag(x(1));
CM = moment(x(1),DE);
V = sqrt( x(4)^2 + x(5)^2 );
Q = 0.00087775*(V^2); % Alt. = 10,000 ft.
% Consts.: S/m = 0.64348, Sc/Iy = 0.085613, c/2 = 3.4245
QSM = 0.64348*Q ; QSIY = 0.085613*Q ; C2V = 3.4245/V ;
G = 32.174 ; % Gravitational const.
TM = 2.60767 ; % Thrust/mass = T/m = Const.
TEMP1=(-TM*sin(x(1))-QSM*CL+G*cos(x(1)-x(3)))/V + x(2) ;
xdot(1)= TEMP1 ;
xdot(2)=( CM+C2V*(-8.0*x(2) -4.0*TEMP1) )*QSIY ;
xdot(3)=x(2) ;
TEMP2= ( CL*sin(x(1))-CD*cos(x(1)) )
xdot(4)=TM + QSM*TEMP2 -G*sin(x(3)) - x(5)*x(2);
TEMP3= ( CL*cos(x(1))+CD*sin(x(1) ))
xdot(5)=-QSM*TEMP3 + G*cos(x(3)) + x(4)*x(2);
% For Stall Porpoising Studies:
% Set DE=elev1(t), V=constant, Disable xdot(4) & xdot(5)
```

where, in accord with Sec. 8.4, the aircraft state vector components during the stall analysis are

x(1) = angle of attack, α, rad
x(2) = pitch rate, q, rad/s
x(3) = pitch attitude, θ, rad
x(4) = longitudinal velocity component, u, ft/s
x(5) = vertical velocity component, w, ft/s

Four other functions are used in the *stall.m* function to describe the nonlinear aerodynamics and longitudinal control. These functions, tailored specifically for the aircraft modeled in Sec. 8.4, are used as a subfunction in *stall.m* and are shown in order to complete the text's example.

The *lift.m* function provides a value of C_L as a function of the angle of attack and longitudinal control setting.

```
function y = lift(x,DE)
% Lift determines CL
% x = alpha (rad)
% y = lift coefficient
% DE = Elev. angle (rad)
% DL = Delta CL jump at stall
DL = 0.0;
if x<0.135 ; % Upper limit of linear CL curve
y = 0.380 + 4.60*x + 0.42*DE ;
elseif x>0.135
y = 1.40 - 13.227*((x-0.3089)^2) + 0.42*DE ;
end
if x>0.3089 ; % x = 17.70 deg at stall
y = 1.40 - DL -80.0*((x-0.3089)^2) + 0.42*DE ;
end
if x>=0.37 ; % x = 21.20 deg
y = 1.101 -DL -0.66*(x - 0.37) + 0.42*DE ;
end
```

The *drag.m* function provides a value of C_D as a function of the angle of attack.

```
function y = drag(x)
% DRAG determines CD
% x = alpha (rad)
% y = Drag coefficient, CD
if x<0.250 ; % Upper limit on quadratic CD variation
y = 0.0188 + 0.0774*((0.38 + 4.60*x)^2) ;
elseif x>0.250
```

```
y = 0.20 + 1.88*(x - 0.250) ;
end
```

The *moment.m* function yields a value of C_m as a function of angle of attack and longitudinal control setting.

```
function y = moment(x,DE)
% MOMENT determines Cm
% x = alpha (rad)
% y = moment coefficient
% DE = elev. angle (rad)
if x<0.3089; % Alpha for Cm curve break
y = 0.084 - 0.48*x - 0.90*DE ;
elseif x>=0.3089 ;
y = -0.06427 -1.00*(x-0.3089) - 0.90*DE ;
end
```

Two elevator control input functions are listed. The first, *elev.m*, simulates pilot control motion from an initial trim setting to represent the aircraft stall demonstration. The second, *elev1.m*, represents a control actuation from a trim setting at stall in order to initiate a stall porpoising limit cycle.

The *elev.m* input control function includes a feedback term representing the pilot's suppression of the phugoid mode during the stall approach.

```
function y = elev(t,x)
% Alter this routine for various elevator rates, etc.
% y0 = DE0 = elev. angle for trim near to 1.3 Vs
% ydot = DEdot = rate for elev. angle change
% A0 = alpha initial condition (rad)
% T0 = theta for level flight i.c. (rad)
% Q0 = pitch rate initial condition (rad/sec)
A0 = 0.0; Q0 = 0.0; T0 = A0;
y0 = (+2.534)*0.01745 ;
ydot = (-0.170)*0.01745 ;
% Suppress stall demonstration for 10 seconds
if t<=10. ;
y = y0 ;
end
if t>=10. ;
y = y0 + ydot*(t - 10.0) ;
end
% Add state feedback to suppress Phugoid mode,
% Phugoid Zeta = 0.707
y=y-0.1458*(x(1)-A0)+0.0446*(x(2)-Q0)+0.1814*(x(3)-T0) ;
```

The *elev1.m* function provides a single sinusoidal control doublet with a 2.0-s period following an initial 1.0 s of trim at the stall conditon.

```
function y = elev1(t,x)
% y0 = elev. angle for trim at max. CL
% amp = amplitude of elevator control doublet
amp = (-0.250)*0.01745 ;
if t<1. ; y = y0 ; end
if t>=1. ; y = y0 + amp*sin(pi*(t-1.)); end;
if t>=3. ; y = y0 ; end
```

In addition to demonstrating the stall maneuver by a simulated pilot application of longitudinal control, Example 8.6 also determined the time history of true aircraft velocity and load factor. Once the *stall.m* function had been completed, the size of the time vector was identified by using the MATLAB *whos* function. This information combined with the stall data was used as input to the *velocity.m* function as shown next.

```
function [V,nz,DE] = velocity(n,t,x)
% n = No. of rows in the x matrix
% t(i) = value of the time vector (sec)
% x(i,1) = alpha (rad)
% x(i,3) = theta (rad)
% x(i,4) = u velocity component (ft/sec)
% x(i,5) = w velocity component (ft/sec)
for i=1:n,
V(i) = sqrt( x(i,4)^2 + x(i,5)^2 );
DE(i) = elev( t(i),x(i,1:3) );
CL = lift( x(i,1),DE(i) );
Q = 0.00087775*( V(i)^2 ); % Alt. = 10,000 ft
QSM = 0.64348*Q ;
TM = 2.60767 ; % Thrust/mass ratio
G = 32.174 ;
an = -TM*sin(x(i,1))-QSM*CL+G*cos(x(i,3)-xi,1)) ;
nz(i) = 1. - (an/G) ;
end
```

References

[1] Anon., *386-MATLAB User's Guide,* The MathWorks, Inc., Natick, MA, 1991.
[2] Anon., *Control System Toolbox,* The MathWorks, Inc., Natick, MA, 1990.
[3] Gerald, C. F., and Wheatley, P. O., *Applied Numerical Analysis,* 4th ed., Addison-Wesley, Reading, MA, 1989, Chap. 5.

Index

Acceleration
  lateral, 229–232
  normal, 24, 73, 81, 198–204
Ackerman's formula, 248, 250–252, 270
Aerodynamic influence coefficients, 53–55, 355–358
Aileron control, 43–46, 60–61, 211
Airfoil theory, 46–49
  aerodynamic center, 47
  ground effect, 64
  panel model, 48, 64
  rearward aerodynamic center, 48
  thin airfoil, 46
Airspeed
  equivalent, 4, 82, 292, 328
  true, 3–4
Angle of attack, α, 6,7
  α perturbation, 17, 20, 294
  $\dot{\alpha}$ damping, 16, 26–29
  derivatives, 19–23, 26
Asymptotic stability, 158
Atmospheric gusts, 291
Atmospheric properties, 11–12, 331–332
Autocorrelation, 309–311

Basic, 56, 358
Biot–Savart law, 47, 50, 356
Block diagram, 154, 238, 244, 255
Bode plot, 194, 197, 203, 233, 235, 242, 247

Cayley–Hamilton theorem, 249, 251, 253, 381–383
Center of gravity, 22, 71, 80, 173, 277
Characteristic
  determinant, 152, 374
  equation, 189, 200, 215, 216, 220, 250, 261
  polynomial, 156, 166, 179
Closed-loop system, 238
Coefficients
  dimensionless, 5
Constraints, 178, 181, 219, 221
Controllability, 186–187, 205, 248–252, 257
Controls, 4
  response, 187–198, 209–214, 223–229
Convolution integral, 126, 129, 299
Coordinate
  system, 2
  transformations, 7–8, 12, 169, 170, 187

Correlation length, 311
Covariance matrix, 320–322, 326

Damping ratio, 134, 136, 167, 217, 222, 237
Decibels, 195
Dihedral, effect
  sweepback, 34
  vertical tail, 35
  wing, 32
  wing–body interference, 33
Dirac delta function, 120, 130, 223
Discrete gust response, 293–302
  one-minus-cosine, 296–302
  sharp edged, 293–296
Downwash angle, 21, 23
Drag counts, 173
Dryden gust model, 312, 319, 323–325
Duhamel superposition integral, 297
Dutch-roll mode, 207, 216, 222
  modal approximation, 219–223, 261
Dynamic pressure, 4, 17–19

Eigenvalue, 146, 150, 167, 170, 174, 215, 262
  problem, 373–376, 379–380
Eigenvector, 147, 167, 170, 217–219
  estimate, 179, 183, 216
Equations of motion, 107
  linearized, 107–110
Ergodic hypothesis, 307
Error function, 306
Euler angle, 269
  body axis, 93
  inertial axis, 93
  transformation, 94–97
Euler's formula, 122
Expectation, 304, 306
Exponential matrix, 150–151, 163

Feedback, 237, 245, 250, 255, 281
First-order system, 128–132, 209
Flight radius, 24
Flying qualities, 175–177, 236–237, 275
Fourier transform, 308, 310, 312
Frequency
  corner, 125, 213, 234
  damped natural, 135
  undamped natural, 134, 167, 217

INDEX

Gain function, 131, 140–141, 156, 193, 213
Gaussian probability distribution, 305, 314
Gilbert's method, 186, 206
Gust envelope, 82, 292–293
   design guidelines, 328–329

Harmonic response, 130–133, 139–141, 155–157, 162, 194, 202, 212–214, 232–236
Helmholtz theorem, 49
Homogeneous solution, 129, 135, 137, 150, 154
Hopf bifurcation, 267, 273, 287

Impulse response, 130, 138, 223
Inertial cross coupling, 208, 257–265
Input function, 146, 188, 209

Kinetic energy, 100–102
Kussner function, 294, 299
Kutta–Joukowsky law, 47, 50

Laplace transform, 119–128, 148
   attenuation rule, 124
   exponential function, 121
   final value theorem, 125, 190, 200, 226, 229
   impulse function, 120
   initial value theorem, 126, 200
   inverse, 126–128, 152
   shift rule, 124
   step function, 121
Lateral control, 4, 43–46, 83, 208–214
   derivatives, 43, 209
   span load, 44
Lateral-directional control, 83–90
   sideslip, 84–86
   thrust asymmetry, 86–90, 92
Lateral-directional dynamics, 114–118, 214–219
Limit cycle, 157, 265, 272, 285–287
Linear algebra, 367–369
Linear independence, 367, 383
Linearization, 18, 108–110
Load factor, 76, 78, 81, 283, 292, 295, 298
Log-decrement procedure, 136
Long period (phugoid) mode, 170–172, 183
   modal approximation, 181–183
Longitudinal control, 4, 68–81
   maneuvering, 73–81
   velocity change, 68–73
Longitudinal dynamics, 111–114, 165–167
Lyapunov
   equation, 159, 293, 319–322, 326
   stability principles, 157–161

Mach Number, 11
   critical, 18
Maneuver
   envelope, 81–82, 292
   point, 77, 174

Mass moment of inertia, 101, 105, 258
   principal axes, 105–107
MATLAB, xii, 160, 167, 179, 191, 195, 224, 226, 227, 252, 266, 385–394
Matrix
   diagonal, 150, 367, 378
   orthogonal, 8, 20, 96, 373, 376
   partitioning, 245, 260, 326
   rank, 250, 367, 383
   skew symmetric, 100, 104–105, 371
   symmetric, 102, 375, 379
   transformation, 8, 12, 97, 373
   unsymmetric, 379
Mean aerodynamic chord, 9–10,12–13
Minimum control speed, 88, 90
Modal matrix, 150, 169, 377, 380
Modern control theory, 144
Momentum conservation, 93, 102–105, 132, 209, 259

Neutral point
   stick fixed, 22, 173
   stick free, 22, 90
Nonlinear oscillator, 266–267
Nonminimum phase, 200, 226, 232

Open-loop system, 153, 259
Orthogonality, 8, 373
Output equation, 153, 199, 230, 326

Parseval's theorem, 308
Particular solution, 129–132
Period, 136, 170, 182, 217
Phase angle, 131, 140–141, 156
Phase plane, 144, 266–267, 272–273
Phasor diagram, 148, 168, 171, 218
Pitch angle, θ, 2
   θ perturbation, 179
   $\dot{\theta}$ damping, 15, 17, 23–26, 59
   derivatives, 16, 23
Pole placement, 250–252, 270, 281
Pole-zero cancellation, 227–229
Positive definite, 158–160
Power spectrum, 307–310
Prandtl–Glauert transformation, 55, 56, 358
Probability function, 303

Quasi-steady aerodynamics, 20, 293, 299

Random process, 302–313
   stationary, 306
Resolvent matrix, 149, 152, 188
Roll angle, ϕ, 2, 212
   ϕ perturbation, 214
   damping, 15, 38–42, 60, 209
   derivatives, 39, 209
   roll helix angle, 39
   span load, 41
Roll mode, 207, 216
   modal approximation, 208–214, 221–222

Root locus, 175, 185, 261
Routh's method, 143
Rudder control, 5

Scale of turbulence, 311
Second-order system, 132–141
Short period mode, 168–170, 180
  modal approximation, 177–181, 188, 261, 301, 326
Sideslip angle, β, 6–7
  derivatives, 30–35, 85
Sidewash angle, 31
Similarity transformation, 150, 187, 377–378
Sound speed, 11
Span load, 53
  additional loading, 56–59
  antisymmetric, 54, 60–61, 207
  symmetric, 54, 56–59
Spiral mode, 207–216
Stability, 67, 83, 142–143, 157–161, 262
Stability axis, 2, 18, 20, 268
Stability derivatives, 333–354
  dimensional, 17–39, 114, 117, 165, 178, 215, 224, 227
  dimensionless, 19–39
Standard deviation, 305
Stall dynamics, 275–285, 291, 391
Stall porpoising, 285–287
State, 145
  equation, 144–146, 153, 165, 214, 269, 278
  response, 149, 153–157
  space, 144
  vector, 144, 165, 214, 270, 278
Static aeroelastic effects, 62
Static margin, 22, 80
Static pressure, 11
Static value, 130, 131, 138, 14, 190–193, 210
Statistical averages, 304
Step response, 130, 138–139, 189, 210–212
Sylvester's criterion, 159

Tail volume coefficient
  horizontal tail, 22
  vertical tail, 31
Taylor series, 15–17, 111, 114
Time constant, 130, 210, 218, 227
Transfer function, 155, 189, 193, 200, 226, 243, 313
Transition matrix, 150–153, 190, 209, 224, 249, 383
  properties, 151, 191
Tuck under, 19, 184–186
Turbulence, 291
  isotropic, 311
  response, 314–319, 325–328

Variance, 305
V–n diagram, 81–82
Vortex models
  thin airfoil, 47–49
  wing, 49–52, 64, 356

Wagner function, 293, 299
Washout filter, 241–248
Weiner–Kintchine relation, 310
White noise, 313, 320–323
Wing rock, 157, 208, 257, 265–275, 390
Wing theory, 49–56
  aerodynamic center, 22, 51
  horseshoe vortex model, 50–52
  panel model, 55, 355
  rearward aerodynamic center, 51
  strip theory, 41, 45, 52
  torsional divergence, 62
  Weissinger approach, 52–54, 355

Yaw angle, ψ, 2
  ψ perturbation, 215–218
  $\dot\psi$ damping, 15, 35–38
  derivatives, 36
Yaw damper, 208, 237–241, 252–254

Texts Published in the AIAA Education Series

Introduction to Aircraft Flight
Dynamics
Louis V. Schmidt 1998

Aerothermodynamics of Gas Turbine
and Rocket Propulsion,
Third Edition
Gordon C. Oates 1997

Advanced Dynamics
Shuh-Jing Ying 1997

Introduction to Aeronautics:
A Design Perspective
*Steven A. Brandt, Randall J. Stiles,
John J. Bertin, and Ray Whitford* 1997

Introductory Aerodynamics and
Hydrodynamics of Wings and Bodies:
A Software-Based Approach
Frederick O. Smetana 1997

An Introduction to Aircraft
Performance
Mario Asselin 1997

Orbital Mechanics, Second Edition
V. A. Chobotov, Editor 1996

Thermal Structures for Aerospace
Applications
Earl A. Thornton 1996

Structural Loads Analysis for
Commercial Transport Aircraft:
Theory and Practice
Ted L. Lomax 1996

Spacecraft Propulsion
Charles D. Brown 1996

Helicopter Flight Dynamics:
The Theory and Application
of Flying Qualities and
Simulation Modeling
Gareth Padfield 1996

Flying Qualities and Flight Testing
of the Airplane
Darrol Stinton 1996

Flight Performance of Aircraft
S. K. Ojha 1995

Operations Research Analysis in
Quality Test and Evaluation
Donald L. Giadrosich 1995

Radar and Laser Cross Section
Engineering
David C. Jenn 1995

Introduction to the Control of
Dynamic Systems
Frederick O. Smetana 1994

Tailless Aircraft in Theory
and Practice
Karl Nickel and Michael Wohlfahrt 1994

Mathematical Methods in Defense
Analyses,
Second Edition
J. S. Przemieniecki 1994

Hypersonic Aerothermodynamics
John J. Bertin 1994

Hypersonic Airbreathing
Propulsion
William H. Heiser and David T. Pratt 1994

Practical Intake Aerodynamic
Design
*E. L. Goldsmith and J. Seddon,
Editors* 1993

Acquisition of Defense Systems
J. S. Przemieniecki, Editor 1993

Dynamics of Atmospheric
Re-Entry
*Frank J. Regan and Satya M.
Anandakrishnan* 1993

Introduction to Dynamics and
Control of Flexible Structures
John L. Junkins and Youdan Kim 1993

Spacecraft Mission Design
Charles D. Brown 1992

Rotary Wing Structural Dynamics
and Aeroelasticity
Richard L. Bielawa 1992

Aircraft Design: A Conceptual Approach,
Second Edition
Daniel P. Raymer 1992

Optimization of Observation and
Control Processes
*Veniamin V. Malyshev, Mihkail N.
Krasilshikov, and Valeri I. Karlov* 1992

Texts Published in the AIAA Education Series (continued)

Nonlinear Analysis of Shell Structures
Anthony N. Palazotto and Scott T. Dennis — 1992

Orbital Mechanics
Vladimir A. Chobotov, Editor — 1991

Critical Technologies for National Defense
Air Force Institute of Technology — 1991

Defense Analyses Software
J. S. Przemieniecki — 1991

Inlets for Supersonic Missiles
John J. Mahoney — 1991

Space Vehicle Design
Michael D. Griffin and James R. French — 1991

Introduction to Mathematical Methods in Defense Analyses
J. S. Przemieniecki — 1990

Basic Helicopter Aerodynamics
J. Seddon — 1990

Aircraft Propulsion Systems Technology and Design
Gordon C. Oates, Editor — 1989

Boundary Layers
A. D. Young — 1989

Aircraft Design: A Conceptual Approach
Daniel P. Raymer — 1989

Gust Loads on Aircraft: Concepts and Applications
Frederic M. Hoblit — 1988

Aircraft Landing Gear Design: Principles and Practices
Norman S. Currey — 1988

Mechanical Reliability: Theory, Models and Applications
B. S. Dhillon — 1988

Re-Entry Aerodynamics
Wilbur L. Hankey — 1988

Aerothermodynamics of Gas Turbine and Rocket Propulsion, Revised and Enlarged
Gordon C. Oates — 1988

Advanced Classical Thermodynamics
George Emanuel — 1988

Radar Electronic Warfare
August Golden Jr. — 1988

An Introduction to the Mathematics and Methods of Astrodynamics
Richard H. Battin — 1987

Aircraft Engine Design
Jack D. Mattingly, William H. Heiser, and Daniel H. Daley — 1987

Gasdynamics: Theory and Applications
George Emanuel — 1986

Composite Materials for Aircraft Structures
Brian C. Hoskins and Alan A. Baker, Editors — 1986

Intake Aerodynamics
J. Seddon and E. L. Goldsmith — 1985

Fundamentals of Aircraft Combat Survivability Analysis and Design
Robert E. Ball — 1985

Aerothermodynamics of Aircraft Engine Components
Gordon C. Oates, Editor — 1985

Aerothermodynamics of Gas Turbine and Rocket Propulsion
Gordon C. Oates — 1984

Re-Entry Vehicle Dynamics
Frank J. Regan — 1984

**Published by
American Institute of Aeronautics
and Astronautics, Inc.
Reston, Virginia**

Printed in the United States
142149LV00004B/3/A